Health Related Effects of Phyllosilicates

NATO ASI Series

Advanced Science Institutes Series

A series presenting the results of activities sponsored by the NATO Science Committee, which aims at the dissemination of advanced scientific and technological knowledge, with a view to strengthening links between scientific communities.

The Series is published by an international board of publishers in conjunction with the NATO Scientific Affairs Division

A Life Sciences B Physics	Plenum Publishing Corporation London and New York
C Mathematical and Physical Sciences D Behavioural and Social Sciences E Applied Sciences	Kluwer Academic Publishers Dordrecht, Boston and London
F Computer and Systems Sciences G Ecological Sciences H Cell Biology	Springer-Verlag Berlin Heidelberg New York London Paris Tokyo Hong Kong Barcelona

Series G: Ecological Sciences Vol. 21

Health Related Effects of Phyllosilicates

Edited by

Jean Bignon

INSERM Unité 139, CHU Henri Mondor
51, Avenue du Maréchal-de-Lattre-de-Tassigny
94010 Creteil Cedex, France

Springer-Verlag Berlin Heidelberg NewYork
London Paris Tokyo Hong Kong Barcelona
Published in cooperation with NATO Scientific Affairs Division

Proceedings of the NATO Advanced Workshop on Health Related Effects of Phyllosilicates held in Paris (France), March 16–18, 1989

ISBN -13 : 987-3-642-75126-4

Library of Congress Cataloging-in-Publication Data
Health related effects of phyllosilicates/edited by Jean Bignon. p. cm.—(NATO ASI series. Series G, Ecological sciences; no. 21) "Proceedings of the NATO Advanced Workshop held in Paris, France, March 16–18, 1989"—Copr. p.
ISBN -13 : 987-3-642-75126-4 e-ISBN -13 : 987-3-642-75124-0
DOI: 10.1007/987-3-642-75124-0
1. Phyllosilicates—Toxicology—Congresses. 2. Lungs—Dust diseases—Congresses. I. Bignon, Jean. II. Series. RA1231.P52H43 1990 615.9'25683—dc20 90-9934

Softcover reprint of the hardcover 1st edition 1990

2131/3140-543210 – Printed on acid-free-paper

PREFACE

Considerable progress in understanding how inhaled minerals cause disease in man has been made in the past two decades. This is mostly due to the great amount of human, animal and cell multidisciplinary studies carried out on silica, asbestos and asbestiforms all around the world.

Two previous NATO Workshops on "In Vitro Effects of Mineral Dusts on Cells", have been published in the NATO ASI Series (1985 and 1989). The present NATO-INSERM workshop has focused specifically on a group of silicates, named phyllosilicates because of their sheet structure, of which health related effects have been poorly and sporadically investigated. These silicates are presently largely used as filling materials (kaolin, talc, chlorite), insulating materials (vermiculite, micas), adsorbants (sepiolite, attapulgite) and in many other industrial applications. The estimated annual world production is presently 5.5 million tons of talc (1.8 million for Europe) and only in the United Kingdom about 3.5 million tons of kaolin.

Three of these minerals, talc, attapulgite, sepiolite have been evaluated for the carcinogenic risk by a group of experts of the International Agency for Research on Cancer in 1986 (IARC Monograph on the evaluation of the carcinogenic risk of chemicals to humans. N° 42, 1987). However, so far the phyllosilicates have not been examined as a group in order to compare the toxicity of the different species when inhaled by humans at their place of work and in the domestic or general environment.

Therefore, this first international NATO-INSERM workshop was designed to initiate such an evaluation by combining human, animal and cellular data. The best experts in the field of mineral toxicology have been gathered altogether. We hope that they will succeed in classifying phyllosilicates according to their toxicity for humans, regarding pneumoconiosis and respiratory cancer risk. These health related effects will also be discussed according to the specificity of the minerals themselves or to possible contaminants, such as asbestos or iron oxides. This will provide a scientific basis for establishing control limit values at place of work, valuable for most of the NATO industrialized countries.

Jean BIGNON

Chairman : J. BIGNON (France)

Organizing committee :

P. BROCHARD (France)

G. CHIAPPINO (Italy)

J.E. CRAIGHEAD (USA)

R.O.FELIUS (Netherlands)

M.C. JAURAND (France)

A. LANGER (USA)

A LUMBROSO (France)

J.C. Mc DONALD (UK)

W.K.C. MORGAN (Canada)

H.W. SCHLIPKOTER (FRG)

P. SEBASTIEN (France)

M. TURPIN (France)

The workshop was sponsored by the ARW Program of the NORTH ATLANTIC TREATY ORGANIZATION under the directorship of Dr Craig SINCLAIR and INSTITUT NATIONAL DE LA SANTE ET DE LA RECHERCHE MEDICALE.

A number of other organizations contributed financially to the workshop and deserve special mention. These include :

- Ministère de l'Environnement (France)
- Ministère de la Recherche et Technologie (programme Technologie, Emploi, Travail)
- Ministère du Travail, de l'Emploi et de la Formation Professionnelle.
- CERCHAR (Centre d'Etude et de Recherche des Charbonnages, France)
- Industries (Scientific Association of European Talc Industry, Rhone Poulenc), France

and under the patronage of Societé de Pneumologie de Langue Française.

The chairman and the organizing committee wish to thank Ms C. Vaslin who was in charge of the entire workshop secretariat with the help of Mrs J. Beaurain and C. Benard. Mrs Vaslin contributed to the editing work of these proceedings.

Finally, we thank Springer-Verlag for their interest in publishing these proceedings and for the excellent printing.

TABLE OF CONTENTS

I - PHYSICO-CHEMISTRY OF NON FIBROUS PHYLLOSILICATES

1. R.O. Felius : The mineralogy of phyllosilicates 3
2. A. Meunier, J.H. Thomassin, A. Decarreau : Geological occurrence of phyllosilicates. Application to kaolinite, talc, sepiolite and palygorskite deposits 15
3. G.M. Clarke : Phyllosilicates as industrial minerals 31
4. H. Pezerat : Surface properties of phyllosilicates 47
5. A.M. Langer, R.P. Nolan, F.D. Pooley : Phyllosilicates : associated fibrous minerals 59
6 K. Rödelsperger, B. Brückel, R. Patrzich, F. Pott, H.J. Woitowitz : Concentrations of fibres in phyllosilicates 75
7. U.M. Schenk, P. Golob, H. Kolmer : Talc. A morphological study on pseudomorphs 85
8 S.B. Schyma : The physical characterization of muscovite and kaolinite dusts 97
9 A. Marconi, U. Verdel : Asbestos content of talcs from italian mines and fibre concentration in various commercial talcum powders used in Italy 107
10. E. Moulin, P. De Vuyst, J.C. Yernault : Occurrence of Phyllosilicates in tobacco 117
11. J. Rosmanith, W. Hilscher, S.B. Schyma : The effect of the surface quality on the fibrogenicity of the phyllosilicates muscovite and kaolinite 123
12. D. Costa, J. Guignard, H. Pezerat : Oxidizing surface properties of divalent iron-rich phyllosilicates in relation to their toxicity by oxidative stress mechanism 129
13. G. Larsen : Solubility and surface properties of talc in a physiological fluid. Results of *in vitro* experiments at 37°C 135
14. J. Ferret, P. Moreau : Mineralogy of talc deposits 147

II - CLINICAL AND EPIDEMIOLOGICAL EVIDENCE OF HEALTH EFFECTS AFTER OCCUPATIONAL EXPOSURE.

1. P. Cullinan, J.C. Mc Donald : Respiratory disease from occupational exposure to non-fibrous phyllosilicates.. 161
2. J.E. Craighead : Pathological features of pulmonary disease due to silicate dust inhalation............................ 179
3. W.K.C. Morgan, A. Donner, I.T.H. Higgins, M.G. Pearson, W. Rawlings, Jr : Clinical aspects of kaolin pneumoconiosis.. 191
4. P. Leophonte, A. Didier : French talc pneumoconiosis.............. 203
5. G. Chiappino, G. Rivolta, A. Todaro, E. Trillo, A. Forni : Talc pneumoconiosis in Italy............................ 211
6. A.R. Gibbs : Human pathology of kaolin and mica pneumoconioses.. 217
7. G.K. Sluis-Cremer, P.A. Hessel : Palabora vermiculite............. 227

III - HEALTH RELATED EFFECTS AFTER NON OCCUPATIONAL EXPOSURE

1. C. Boutin, J.R. Viallat : Talc : its cosmetic, addictive and therapeutic uses and their effects on health................. 237
2. J.C. Pairon, M.C. Jaurand, A. Gaudichet, P. Brochard J. Bignon : Therapeutic and domestic uses of attapulgite and sepiolite... 249
3. P.L. Kalliomäki, O. Taikina-aho, P. Pääkkö, S. J. Sivonen, K. Kalliomäki : Lung retained minerals correlated with smoking, emphysema and lung cancer... 265
4. L. Paoletti, M. Falchi, D. Batisti, M.P. Carrieri, M.G. Petrelli, C. Ciallella, G. Donelli : Mineral particulate in the lung parenchyma of subjects not occupationally exposed to dust.................................... 273

IV - FIBROGENESIS - CARCINOGENESIS. ANIMAL AND CELL RESPONSES TO PHYLLOSILICATES.

1 J.C. Barrett, P.W. Lamb, R.W. Wiseman : Hypotheses on the mechanisms of carcinogenesis and cell transformation by asbestos and other mineral dusts.... 281

2. J.C. Wagner : Review on pulmonary effects of phyllosilicates after inhalation.... 309

3. F. Pott, B. Bellmann, H. Muhle, K. Rödelsperger, R.M. Rippe, M. Roller, M. Rosenbruch : Intraperitoneal injection studies for the evaluation of the carcinogenicity of fibrous phyllosilicates.... 319

4. M.C. Jaurand, J.C. Pairon : Review of short-term tests used for the toxicological evaluation of phyllosilicates.... 331

5 M. Holopainen, V. Vallyathan, M. Hedenborg, M. Klockars : Toxicity of phlogopite and muscovite *in vitro*.... 349

6. A. Ghio, T. Kennedy, M. Baser, W.R. Rawlings, N.V. Rao, J.R. Hoidal : Kaolin generates °OH and causes hemolysis by acting as a Fenton reagent.... 361

7. S. Endo-Capron, J. Fleury-Feith, M. Nebut, R. De Neef, M.C Jaurand : Some *in vivo* and *in vitro* studies carried out with talc samples.... 369

8. Q. Rahman, M.F. Khan, S.V. Singh : Slate dust toxicity.... 377

9. A.P. Sahu : Biological effects of mica dust in experimental animals.... 387

10. R. Begin, S. Massé, P. Sebastien : Assessment of the fibrogenicity of attapulgite.... 395

11. A. Wastiaux, H. Daniel : Pulmonary toxicity of kaolin in rats exposed by inhalation.... 405

V - IMPLICATIONS IN INDUSTRIAL HYGIENE AND IN PUBLIC HEALTH

1. P. Sebastien : The scientific basis for controlling exposures to phyllosilicate dust at the workplace.............. 417

2. R.W. Niemeier : Control limit values in different industrialized countries : USA.. 429

3. M. De Smedt : Limit values : the European dimension............... 435

4. N.G. West : Occupational exposure limits for phyllosillicates in the United Kingdon... 441

5. B. Herve-Bazin : Occupational exposure limits in France............ 445

CONCLUSIONS : J.C. McDonald.. 449

PART I

PHYSICO-CHEMISTRY OF NON FIBROUS PHYLLOSILICATES

IERALOGY OF PHYLLOSILICATES

ROBERT O. FELIUS, Institute of Earth Sciences,
State University at Utrecht, P.O. Box 80.021,
NL-3508 TA Utrecht, The Netherlands

JCTION

introduction to other contributions, dealing with the /, the industrial use and the health effects of phyllo- :es, an outline of the mineralogy of these minerals is in this paper.

this outline special attention is payed to the crystal ıres of the phyllosilicates, be it in a schematical way. eans that certain minerals, notably the micas, smectites lorites are treated in groups. Not every mineral of the is mentioned separately, but one or several of them are as an example. Little attention is payed to the sometimes complex variations in chemical compositions within each groups, due to substitutions in the octahedral sheets pecially in the interlayers, occasionally compensated by cutions in the tetrahedral sheets as well.

er the crystal structures are given in a simplified way, t going into details like rotation of the tetrahedra or ation of the octahedra in the corresponding sheets due to l chemical influences. Nor the different modifications of he mica and chlorite structures are given either.

so, a useful overview of the relationship between the ct phyllosilicate structures is obtained. Some further gs are suggested at the end of this paper, in order to he possibility to take knowledge of what has not been d here.

NG UNITS IN PHYLLOSILICATES

licium in more than 98% of the silicates and silicium is surrounded by four oxygens each, through the centra ch the corners of a tetrahedron can be thought (Fig. 1). licium atom is situated in the centre of the tetrahedron. iO_4]-tetrahedra are corner-linked with three other tetra- in phyllosilicates, mostly forming sheets with (pseudo-)

NATO ASI Series, Vol. G 21
Health Related Effects of Phyllosilicates
Edited by J. Bignon

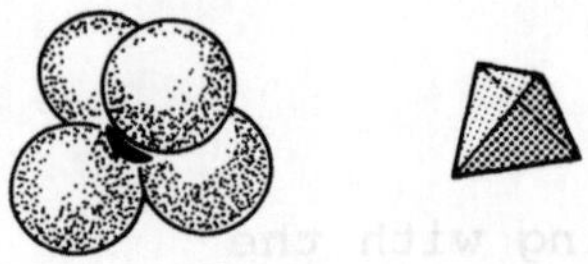

Fig. 1. $[SiO_4]$-tetrahedron

hexagonal symmetry (Fig. 2). The free corners of the tetrahedra (not depicted in Fig. 2A) are pointing towards the same direction of the sheet (Fig.'s 2B and 2C), but that may change periodically in some phyllosilicates. The structure formula of the <u>tetrahedral sheet</u> is found by counting the number of atoms per period: $[Si_4O_{10}]^{4-}$..(1)
The (OH)-group in the centre of each silica hexagon belongs rather to an adjecent octahedral sheet than to the silica sheet itself. The consideration of the tetrahedral sheet together with the (OH)-group as one complex anion however, gives the structure formula: $[Si_4O_{10}/(OH)_2]^{6-}$......................(2)
The mentioned <u>octahedral sheet</u> is formed by two layers of (OH)-groups in closest packing, with middle sized cations (mostly bi- or trivalent) in their octahedral interstices. The free oxygens of an adjecent tetrahedral sheet replace 2/3 of the

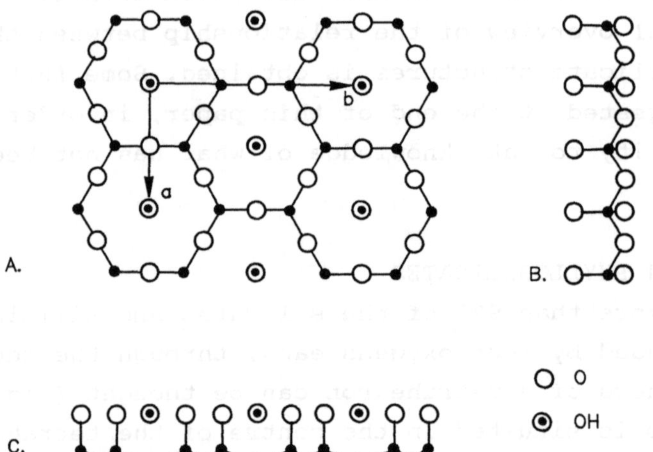

Fig. 2. Idealized tetrahedral sheet in phyllosilicates, seen perpendicular to the a-b plane (A), in the b direction (B) and in the a direction (C)

(OH)-groups either at one side (as in e.g. the minerals kaolinite and serpentine, Fig. 3) or at both sides of the octahedral sheet (as in e.g. the minerals talc and pyrophyllite, Fig. 5). The combination of one tetrahedral sheet adjecent to one octahedral sheet forms the building unit of the <u>1:1 type phyllosilicates</u> with the structure formula:

$[Si_4O_{10}/(OH)_2/(OH)_6]^{12-}$ ....(3)

The combination of two tetrahedral sheets with one octahedral sheet inbetween them forms the building unit of the <u>2:1 type phyllosilicates</u>, the structure formula of which is found by doubling (2): $[Si_4O_{10}/(OH)_2]_2^{6-}$ ....(4)

If the octahedral interstices are occupied by trivalent cations (e.g. Al^{3+} in kaolinite (Fig. 3A) or pyrophyllite (Fig. 5A)) only 2/3 of them are filled. Therefore these phyllosilicates are called <u>dioctahedral phyllosilicates</u>.

If the octahedral interstices are occupied by divalent cations (e.g. Mg^{2+} in serpentine (Fig. 3B) or talc (Fig. 5B)) they are all filled, so 3/3 of them are occupied. Therefore those phyllosilicates are called <u>trioctahedral phyllosilicates</u>.

The room in between the 1:1 or 2:1 type layers is called <u>the interlayer</u>. The interlayers of the phyllosilicates mentioned up to now are empty, but they may contain H_2O in between the 1:1 layers (halloysite). The interlayers in between the 2:1 layers may contain big cations (micas), middle sized to big cations plus H_2O (vermiculite and smectites), or even an extra octahedral sheet (chlorites).

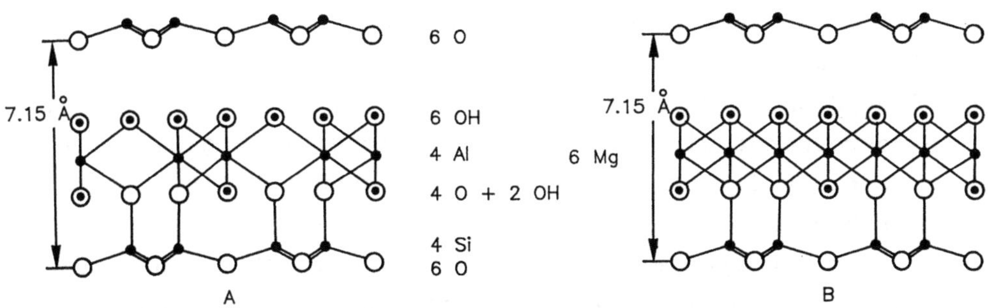

Fig. 3. Idealized crystal structure of kaolinite (A) and serpentine (B), seen in the direction of the a-axis

The $[SiO_4]$-tetrahedra are always corner-linked with three other tetrahedra in phyllosilicates, but the direction into which the free tetrahedron corners are pointing may alternate periodically (sepiolite (Fig. 9) and palygorskite (Fig.10)). Thus the octahedral sheet is split up into small strips in between two succeeding tetrahedral sheets, forming 2:1 chains, laterally coupled in a layered structure.

THE 1:1 TYPE PHYLLOSILICATES

Kaolinite (Fig. 3A) is a dioctahedral 1:1 type phyllosilicate. It has the structure formula: $Al_4[Si_4O_{10}/(OH)_2/(OH)_6]$.......(5)
The chemical composition of kaolinite may be expressed in a simplification of its structure formula: $Al_2Si_2O_5(OH)_4$.
The 1:1 layers are piled up obliquely in the kaolinite structure, which lowers the symmetry to triclinic, though the morphology remains pseudo-hexagonal.
The bonding between the layers in kaolinite, which is very weak indeed, is generally ascribed to Van der Waals bonds only, but certainly it is caused by the polarity of the electrostatic neutral layers in the first place.
Kaolinite is one of the frequent occurring clay minerals. Its application as raw material in china-ware is well known. It has no colour, as it contains no colouring cations. Therefore it is very suitable as a filler component as well.
The crystal structure of the clay mineral halloysite can be thougt of that of kaolinite with water in adition in the interlayer, in between the 1:1 layers. The morphology of halloysite tends to be elongated in the direction of the a-axis.
The 1:1 phyllosilicate with bivalent cations (Mg^{2+}) in 3/3 of its octahedral interstices is serpentine (Fig. 3B). Its structure formula containes Mg_6 in stead of Al_4 in (4):
$Mg_6[Si_4O_{10}/(OH)_2/(OH)_6]$,..................................(6)
which corresponds with the chemical composition $Mg_3Si_2O_5(OH)_4$.
The completely filled octahedral sheet does not fit well with the tetrahedral sheet, which causes the sheets to warp. The misfit is not compensated completely by a constant curvature in the antigorite modification of serpentine (Fig. 4), but grows to a maximum in the regions PP', QQ' and RR'. Here the curved

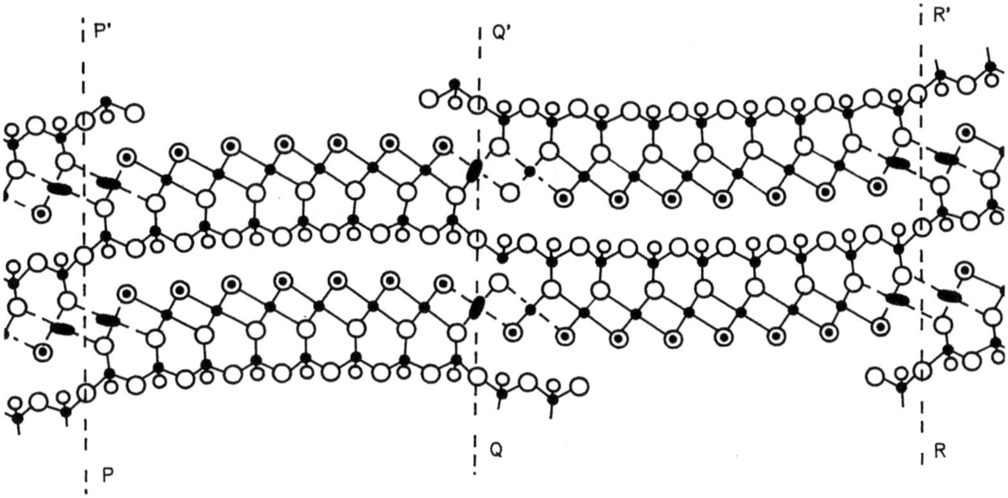

Fig. 4. Idealized crystal structure of antigorite, seen in the direction of the b-axis. The misfit between the octahedral and tetrahedral sheets grows to a maximum in the regions PP′,QQ′ and RR′, causing the curved layers to reverse polarity

layers reverse polarity in order to prevent the misfit to become too large. The result is a wavy structure.
The polarity of the curvature is not reversed in the chrysotile modification of serpentine, but the mismatch is here eventually controlled by an increase of the curvature. Chrysotile fibres thus appear to be long slender concentric tubes or rolls.
The most frequent occurring modification of serpentine is lizardite, the measurements of the tiny crystallites of which are ruled by the mismatch between its tetrahedral and octahedral sheets.
Most serpentine is the product of hydrothermal or metamorphic reactions in ultrabasic rocks, in which the Mg-rich silicates olivine and orthopyroxene react with SiO_2 and H_2O to give serpentine. Whole peridotite bodies may have reacted in this way to serpentinites, from which the chrysotile sometimes can be mined as asbestos.
Ultrabasic rocks may be relatively rich in Ni. This element may be concentrated in garnierite, a serpentine in which the Mg

partly is replaced by Ni. Garnierite deposites may be important Ni-ores.

THE 2:1 TYPE PHYLLOSILICATES

Pyrophyllite is a dioctahedral 2:1 phyllosilicate with Al in 2/3 of its octahedral interstices and an empty interlayer (Fig. 5A), so its structure formula follows from (4) by adding Al_4:

$Al_4[Si_4O_{10}/(OH)_2]_2$ ....(7)

and the chemical composition $AlSi_2O_5OH$ from (7) by simplification.

Talc is a trioctahedral 2:1 phyllosilicate with mainly Mg^{2+} in the octahedral interstices (Fig. 5B). It has an empty interlayer, like pyrophyllite, so its structure formula follows from (4) by adding Mg_6: $Mg_6[Si_4O_{10}/(OH)_2]_2$ ....(8)

The 2:1 layers are piled up obliquely in both minerals, which lowers the symmetry to triclinic. The morphology remains nevertheless pseudo-hexagonal. The only difference between these minerals is the filling in of the octahedral sheet, which means that their physical properties are very similar. Therefore it is understandable that talc and pyrophyllite are used in the same way as industrial minerals, depending on which of the two is available. The electrostatic bonds between the 2:1 layers are so weak, that they do not only have the lowest hardness in Moss' hardness scale, but even possess a greasy feel. The white appearence of these minerals in powdered form makes them suitable, like kaolinite, as fillers.

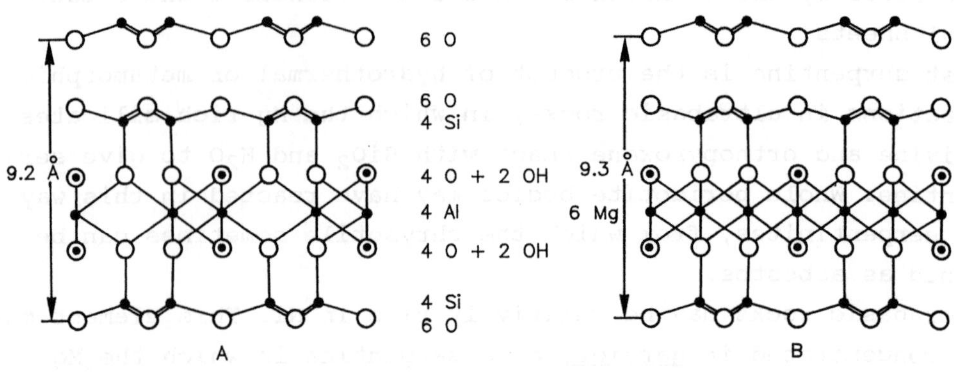

Fig. 5. Idealized crystal structure of pyrophyllite (A) and talc (B), seen in the direction of the a-axis

Pyrophyllite is a low grade metamorphic mineral, formed by the reaction kaolinite + SiO_2 -> pyrophyllite + H_2O, or a retrograde metamorphic mineral formed at the expense of other Al-silicates like andalusite or disthene, both Al_2SiO_5 polymorphs. Talc can be formed in considerable amounts by metamorphism of silicious dolomites or on a more modest scale by hydrothermal reactions in basic or ultra-basic rocks.

The micas are 2:1 phyllosilicates with large cations like K^+, Na^+ or Ca^{2+} in their interlayers. The most common micas are the K-micas muscovite and biotite (Fig. 6). Muscovite (the name is because of the moscovian use of this mineral as window glass in former times) is a dioctahedral mica with structure formula:

$K_2Al_4[Si_3AlO_{10}/(OH)_2]_2$ .... (9)

and chemical composition $KAl_3Si_3O_{10}(OH)_2$. Its structure is, as a matter of fact, that of pyrophyllite with K added to the interlayer. This causes a valence surplus, which is compensated by the replacement of 1/4 Si by Al in the tetrahedral sheet (Fig. 6A).

The trioctahedral K-mica is biotite. It has the structure formula (Fig. 6B): $K_2M_6[(Si,Al)_4O_{10}/(OH)_2]_2$ .... (10)

with a range of chemical compositions $KM_3(Si,Al)_4O_{10}(OH)_2$, where M stands for a variety of middle sized cations of which Mg, Fe (both Fe^{2+} and Fe^{3+}), Al and some Mn, V and Ti are the

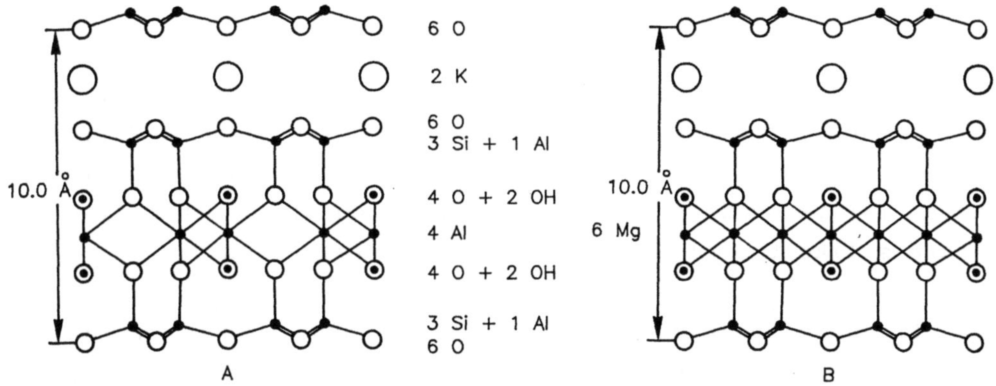

Fig. 6. Idealized crystal structure of muscovite (A) and biotite (B), seen in the direction of the a-axis

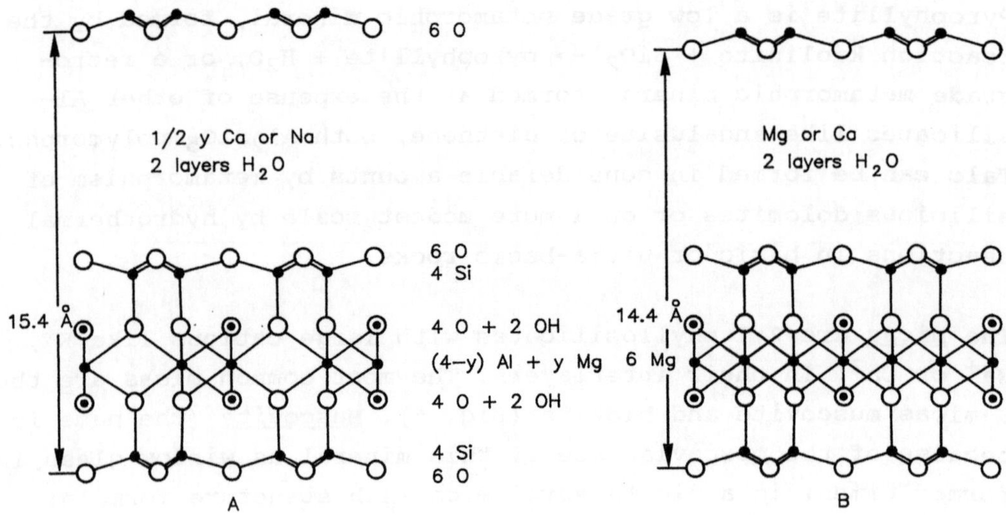

Fig. 7. Idealized crystal structure of montmorillonite (A) and vermiculite (B), seen in the direction of the a-axis

most common ones. Phlogopite is the (almost) pure Mg end member of biotite.

Because of their excellent cleavage into large, thin, plan-parallel plates, they may be used as e.g. heat resistant windows in stoves or as electric insulators inbetween condenser plates.

Micas do occur in nearly all kind of rocks, either magmatic, sedimentair or metamorphic, but really big, exploitable crystals are restricted to veins as pegmatites.

A frequently occurring 2:1 type clay mineral is illite. It has essentially the structure of muscovite, but it is usually less well crystallized and contains always a deficiency in K. Therefore a varying amount of water may be attracted into the interlayer. Illite occurrs as a weathering product of other Al-rich minerals like the Al_2SiO_5 polymorphs or feldspars.

The structure formulas of the smectite group clay minerals are rather complicated because of the variety of middle sized cations that may occur in the octahedral sheet as well as the ex-

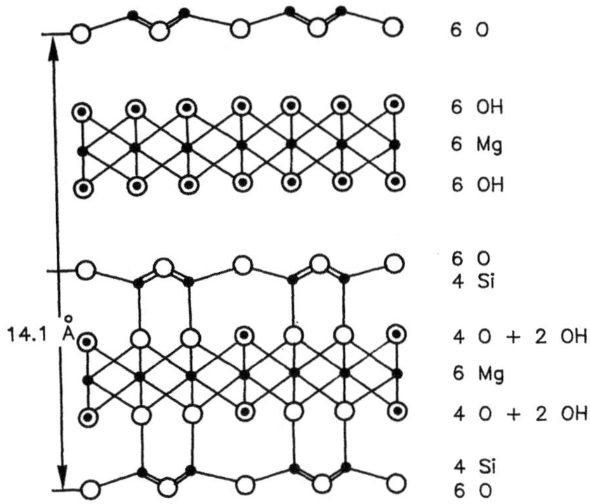

Fig. 8. Idealized crystal structure of Mg-chlorite, seen in the direction of the <u>a</u>-axis

change of big cations together with variable amounts of water, which is possible in the interlayer. The structure formula of e.g. <u>montmorillonite</u> (Fig. 7A) is:

$(\frac{1}{2}Ca,Na)(Al,Fe^{3+},Fe^{2+},Mg)_4[(Si,Al)_4O_{10}/(OH)_2]_2 \cdot nH_2O$........(11)

The cations in the interlayer of these minerals can be reversibly rereplaced by other cations. Therefore they are used as cation exchangers. They may be applied as swelling material as well, for dried smectites will swell when they absorb water again.

Smectites are, like other clay minerals, weathering products from other minerals. The thick smectite deposits are called bentonites.

The weathering product of (e.g.) biotite is <u>vermiculite</u>. It is an essentially trioctahedral 2:1 phyllosilicate with Mg and H_2O in the interlayer (Fig. 7B). The cations in the octahedral sheet however may be trivalent as well, which gives vermiculite a slight tendency towards the bioctahedral 2:1 phyllosilicate type.

In <u>chlorites</u>, we find a second octahedral sheet intercalated as interlayer (Fig. 8). Note that this sheet does not share oxygens with any tetrahedral sheet. It is an intercalation of two closed packed layers of (OH)-groups with middle sized cations in the octahedral interstices. Therefore a pure Mg-chlorite with Mg in both octahedral sheets would have the structure formula:

$Mg_6Mg_6[Si_4O_{10}/(OH)_2/(OH)_6]_2$,................................(12)

which means that the chemical composition of pure Mg-chlorite equals that of serpentine: $Mg_3Si_2O_5(OH)_4$. However real pure Mg-chlorite does not occur in nature, though the chlorites show a broad scala of chemical compositions and certainly can be Mg-rich. The most common middle sized cations in both octahedral sheets are: Mg, Fe^{2+}, Fe^{3+} and Al.

If Ni replaces the mentioned cations in a considerable way, as in the Ni-chlorite <u>nimite</u>, such chlorite may become of economical interest, like and generally together with garnierite. Usually chlorites are not of industrial or economical value, but they may decrease that of talc powder. Their own colour is greyish to dark green and they are formed during low grade metamorphism (both prograde and retrograde) or under hydrothermal circumstances. So they may be formed together with talc.

As all 2:1 type clay minerals with comparable octahedral sheets possess the same structures except for their interlayers, they easily may form parallel intergrowths by forming varying intercalations inbetween the 2:1 layers. This mechanism may result in rather complicated structures or stacking faults, especially among clay minerals.

LAYERD STRUCTURES OF 2:1 CHAINS

The octahedral sheet alternates periodically with the interlayer in between two succeeding tetrahedral sheets in the phyllosilicates <u>sepiolite</u> (Fig. 9) and <u>palygorskite</u> (also known as attapulgite, Fig. 10). Thus the octahedral sheet is split up into strips, which form 2:1 chains with the adjecent tetrahedral sheets. The interlayers form long tunnels or tubes in between the chains, as known from zeolites. The oxygens of the octahedral sheet bording those tunnels are replaced by fixed H_2O molecules. Zeolitic water is found in addition in the tunnels themselves.

Sepiolite has the structure formula:

$Mg_8[Si_6O_{15}/(OH)_2]_2 \cdot 4H_2O \cdot 8H_2O$,................................(13)

and palygorskite: $(Mg,Al)_4[Si_4O_{10}/OH]_2 \cdot 4H_2O \cdot 4H_2O$..........(14)

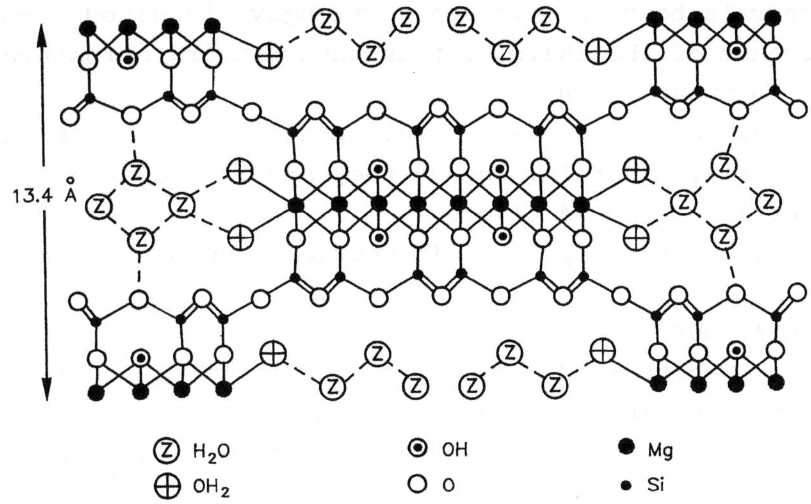

Fig. 9. Idealized crystal structure of sepiolite, seen in the direction of the a-axis

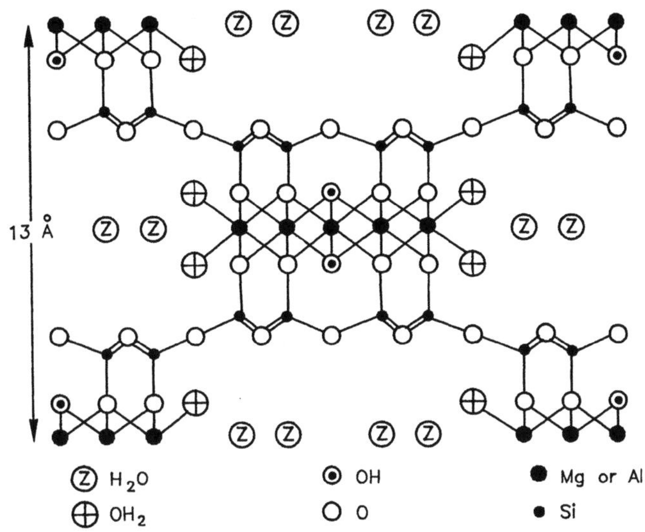

Fig. 10. Idealized crystal structure of palygorskite, seen in the direction of the a-axis

These minerals tend to develope a strongly elongated morphology, because of the chains within their layerd structures and parallel to those chains.
Both minerals are products of weathering and have to be characterised as clay minerals, though they are less common than the clay minerals discussed before.
Sepiolite is especially found in weathered serpentinites.

FURTHER READINGS

S.W. Bailey (editor): Micas. Review in Mineralogy, vol. 13. Mineral. Soc. of America, Washington D.C., 1984, 584 pp.
S.W. Bailey (editor): Hydrous Phyllosilicates (exclusive of micas). Review in Mineralogy, vol. 19. Mineral. Soc. of America, Washington D.C., 1988, 725 pp.
G.W. Brindley & G. Brown (editors): Crystal Structures of Clay Minerals and their X-Ray Identification. Min. Soc. Monograph 5, Mineral. Soc., London, 1980, 495 pp.

GEOLOGICAL OCCURRENCE OF PHYLLOSILICATES. APPLICATION TO KAOLINITE, TALC, SEPIOLITE AND PALYGORSKITE DEPOSITS

Alain MEUNIER* , Jean-Hugues THOMASSIN* and Alain DECARREAU**
* Laboratoire de Pétrologie des Altérations Hydrothermales
**Laboratoire de Pétrologie de la Surface
Université de Poitiers 40, avenue du Recteur Pineau 86022 POITIERS Cedex FRANCE

Industrial names of natural materials do not designate well defined mineralogical species even when these products are as purified as possible. A simple name like "kaolinite" or "talc" hides a wide variety of crystalline state, chemical composition and associated impurities. These characteristics mostly depend on the physicochemical conditions which prevailed during the genesis of these materials in natural environments. In other words, kaolinite or talc are known to have noticeable differences in chemical composition and crystalline state when they originate in supergene, hydrothermal or metamorphic conditions. Associated minerals are also different : the supergene environment favours the formation of expandable clay minerals, carbonates and iron oxy-hydroxides while the hydrothermal one favours the growth of less expandable clay minerals, silicates, carbonates and sulphides. This is the main reason why the commercialized natural products must be carefully studied on a petrological and mineralogical point of view in order to control their effects on human environments.

The aim of this paper, after an overview on the general conditions which prevail in the natural domains of formation of the phyllosilicates, is to study the influence of geological conditions on the crystallochemical properties of kaolinite, talc, sepiolite and palygorskite which have a broad industrial interest and subsequently ecological influences. As the available references on such a general subject are too numerous to be extensively presented here, most of the informations can be found in a few general publications : Brindley and Brown (1980) for crystal structures, Newman (1987) and Bailey (1988) for the chemical compositions and Velde (1985) for the physicochemical explanation of natural occurences.

NATO ASI Series, Vol. G 21
Health Related Effects of Phyllosilicates
Edited by J. Bignon

1 - MINERALOGICAL DEFINITIONS

Phyllosilicates constitute a wide family of minerals whose common characteristic is to contain a continuous two dimensional tetrahedral sheet of composition Si_4O_{10} (Brindley and Pédro, 1972). Each individual tetrahedron is linked to the three neigbouring tetrahedra sharing three corners and basal oxygens while the fourth oxygen forms part of the immediatly adjacent octahedron. These octahedra are organized in continuous or non continuous sheet. In the previous case, two types of structure are possible : 1/1 (7.1 to 7.3 Å thick) or 2/1 (9.1 to 9.4 Å).

Whatever be the lattice structure , the tetrahedral cation is mainly Si and substitutions are frequent with Al and possible with Fe. These replacements induce the formation of a layer charge which must be compensated by an additional interlayer sheet which is mainly composed of Ca, Na, K, Mg. Similar ionic substitutions are also frequent in the octahedral sheet : trivalent component (Al, Fe^{3+}) can be replaced by divalent ones (Fe^{2+}, Mg, Mn). This explains the very large changes of chemical compositions in the phyllosilicate minerals.

Sepiolite and palygorskite (in spite of the fact that "attapulgite clays" have been widely commercialized, the name palygorskite has priority, Bailey et al., 1971) are characterized by a discontinuous octahedral and tetrahedral sheets. These minerals can be regarded as containing ribbons of discontinuous 2:1 layer , one ribbon being linked to the next by inversion of tetrahedra along a set of Si-O-Si bonds (Jones and Galan, 1988). Such an organization of the crystalline lattice favours a fibrous habit. Sizes of fibers are quite variable, ranging from 100 Å to 3-4 µm in length, 100 to 300 Å wide and 50 to 100 Å thick.

2 - GEOLOGICAL ENVIRONMENTS OF CLAY MINERALS

According to Velde (1985), the geological environments where clay minerals growth can be basically divided into five types : weathering, sedimentation, burial, diagenetic and hydrothermal (Figure 1).

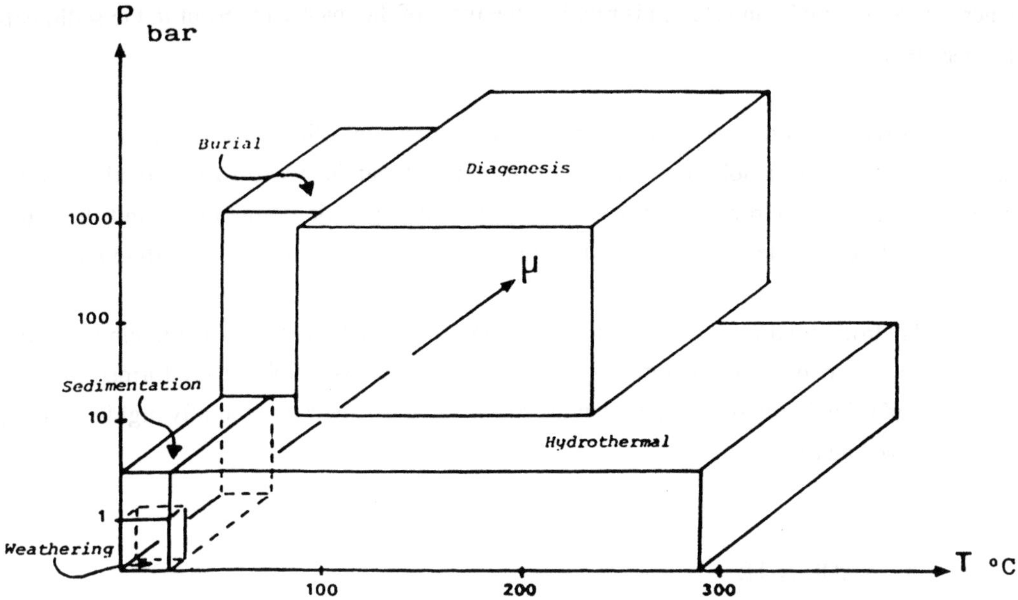

Figure 1 : Physicochemical conditions of the geological environments of clay minerals in a P-T-μ diagram (P = pressure ; T = temperature ; μ = chemical potential)

The weathering environment presents a chemical system where temperature and pressure are near constant (earth surface conditions) and many chemical components are mobile. Dissolution-precipitation processes are governed by local conditions (microsystems).

The sedimentation environments , both marine and soft water , correspond to the few centimeters thick at the sediment-water interface where dissolution-precipitation processes produce new mineral assemblages from detrital material and direct precipitation (minor process). For the most part, the composition of the aqueous solution remains constant in a given geographic zone.

After initial sedimentation, the burial of sediments is usually accomplished rapidly in the nearshore or lacustrine basin environments. The activity of ions in solution becomes dependant upon the chemistry of the silicates as the porosity decreases. The system changes from one of perfectly mobile components in the open sea to one approaching a closed system where the ionic activity in solution is ruled by the mass of the material present in the sediment-fluid systems.

Diagenetic processes involve mineral reactions controlled by temperature, pressure and activities of ions in solution. Two types of thermodynamic systems may coexist side by side in diagenetic environments : one "open" to certain mobile components and the other essentially closed to externally controlled chemical potentials of elements in the fluid phase.

Hydrothermal alteration is the most complicated and variable of the clay mineral environments due to its local nature. Essentially, the mineral assemblages are formed through the action of a fluid, usually warmer than the rocks it invades, which is initially highly charged with dissolved materials.

3 - KAOLINITE

1 - Chemical composition and crystalline structure

Kaolinite was considered to be almost strictly an aluminium-silicate whose composition is $Si_2O_5Al_2(OH)_4$. Nervertheless, when formed under conditions of intense weathering, small percentages of Fe atoms can enter into the crystalline structure. Recent experimental works showed that up to 8 % of Fe can replace the octahedral aluminium (Petit et al., 1988). The extent of the solid solution with the ferric end member ($Si_2O_5Fe_2(OH)_4$) is not known but it seems that it reduces with temperature : iron rich kaolinites are mostly of supergene origin.

In spite of the fact that chemical composition of kaolinite is in most cases very close to the aluminous end member , strong differences in the crystalline state can exist between samples of different origins. Excluding halloysite (hydrated kaolinite) whose lattice is tubular and expands after saturation with ethylene glycol, several polytypes and variable degrees of disorder are observed.

Dickite and nacrite are polytypic varieties of kaolinite with different stacking arrangements of essentially the same layers as in kaolinite itself. The conditions favouring their formation rather than kaolinite are not well known; they seem to be quite rare but this may be due to the difficulty of their identification when they are mixed with kaolinite. As the most part of industrial minerals is restricted to the triclinic variety of kaolinite, we will insist more on the order-disorder properties of this species.

A progressive disordering is mainly due to b/3 displacements parallel to the y axis of the crystalline cell and equivalent directions at ± 120° to y in less disordered form and more general displacements in more highly disordered forms. Plançon and Tchoubar (1977a and b) have developed the physical basis of this concept. This is the reason why the nomenclature of industrial material is based on a qualification like slightly disordered or highly disordered kaolinite. Recent studies (Muller, 1988) showed the relationships between the conditions of growth and the formation of crystalline defects in the kaolinite structure. The shape and the size of crystals depend on these mechanisms.

2 - Genesis and natural occurences

Kaolinite is a widespread mineral in the earth crust surface environments; it is stable throughout the range of most clay mineral environments at its bulk composition . Kaolinite is formed in soils or weathering profile developped not only on alumina rich materials like argillic, pelitic or granitic rocks but also on ferromagnesian ones. A great number of studies described its formation under different conditions of climate and parent rock composition(see Velde, 1985). In all cases, kaolinite is formed with iron oxy-hydroxides as a final product after the leaching of alkalines and magnesium. Extensive residual kaolin deposits are exploited in many countries (Murray, 1988); most of them are generated by a deep weathering of granitic rocks.

As kaolinite crystals are resistant to mechanical breaking, they are able to be transported on large distances and to be deposited in fluviatile or marine environments without great dammage. So, they were frequently discovered in sedimentary rocks (Weaver, 1959; Millot, 1964). At times, they are highly concentrated, as it is the case for the cretaceous deposits of Georgia which constitute the most important reserve of kaolinite in the world (Harben and Bates, 1984). In such sedimentary deposits, kaolinite is frequently associated with other phyllosilicates (mica, smectite) and heavy minerals (rutile, magnetite, ilmenite ....).

During the burial of the sedimentary series, the kaolinite component persists through diagenetic and weak metamorphism conditions. In fact, kaolinite does not strictly remain inert :

new crystallization from pore solutions are frequently observed in sandstones. The stacking disorder and the defects of the crystalline structure decrease with increasing temperature and pressure conditions. Kaolinite desappears from the clay paragenesis by combination with Mg from destabilized dolomite (Muffler and White, 1969), or with other phases to produce illite + chlorite or with quartz to produce pyrophyllite. Its highest temperature of stability in natural environments lies in the range between 280 to 300 °C (Winkler, 1974)

Kaolinite is a frequent product of hydrothermal alteration from near surface conditions to depths of several kilometers. It is commonly associated with mica, pyrophyllite or expanding phases in the "argillic" alteration facies developed around intrusive acid rocks. Most of the economic deposits generated by hydrothermal processes are located in granitic or rhyolitic countries and are related to ore bodies. Among the numerous non phyllosilicate accompanying minerals, pyrite, alunite and quartz are frequently observed.

4 - TALC

1- Chemical composition and crystalline structure

The chemical composition of most talcs is very close to the end-member: $Si_4O_{10}Mg_3(OH)_2$. Nevertheless, iron substitution has been observed in high pressure pelitic blueschist formation (Chopin, 1981; Nisio and Lardeaux, 1987), hydrothermal deposits (Lonsdale et al., 1980; Kager and Oen, 1983) and iron deposits (Floran and Papike, 1978). The amount of substitution is variable and depend on the physicochemical conditions of the natural environment; the highest measured amount is 1.22 Fe atoms for 3 octahedral sites that is to say 41% of the Fe end member ($Si_4O_{10}Fe_3(OH)_2$ minnesotaite). Mössbauer studies have shown that hydrothermal talcs contain only Fe^{2+} while talcs from weathering profiles contain both Fe^{2+} and Fe^{3+} (Noack et al., 1986). Minor amounts of cation substitutions in octahedral sites are possible with Mn, Ni, Cr.

Experimental works have established that about 4.0% Al_2O_3 is the apparent maximum Al substitution possible (Fawcett and Yoder, 1966). Schreyer et al. (1980) reported a natural occurence of a sodian talc-like phase with tetrahedral Al substitution for Si. Such a substitution creates a negative electric charge in the 2/1 structure which is compensated by Na in the interlayer region.

An hydrated variety of talc is frequently observed in low temperature alteration of basic or ultrabasic rocks: kerolite. Brindley et al.(1977) found the composition to be approximately $Si_4O_{10}Mg_3(OH)_2.nH_2O$ with n = 0.8-1.2. The H_2O molecules appear to be surface-held, although some may be in the interlayer region. Because of the presence of water, kerolite has an original property which distinguishes it from talc: it is able to expand to about 17 Å after a long exposure to ethylene glycol whereas in airdry d(001) = 10.6 Å .

Crystals of phyllosilicates, and especially talc, are mostly bidimensional. This specific structure allows the interstratification of layers of different species giving stacking such as chlorite-talc or saponite-talc. Interstratification is possible if the crystalline cell parameters (mainly the b parameter) of the two components are similar. This is the reason why interstratification is rare between dioctahedral and trioctahedral lattices. The order or disorder in the stacking depends on the physicochemical conditions. Interstratified minerals are not so rare as their descriptions suggest it but their identification is not easy when the stacking is irregular.

2- Genesis and natural occurences

Talc is a widespread mineral which crystallizes in a large range of temperature-pressure conditions. It appears from weathering environment to high pressure zones in the earth crust. Nevertheless the economic deposits are found in more restricted domain.

Metamorphosed rock

The most abundant occurence of talc is in metamorphosed ultrabasic rocks and in siliceous dolomitic limestones. In the first case, this mineral coexists frequently with olivine and serpentine minerals. If, during the metamorphic process, Si saturated-CO2 bearing aqueous solutions invade the rock, the crystallization of talc is strongly favoured giving nearly monomineralic deposits (Sandford, 1982). The maximum amount of Fe_2O_3 is 4%. The associated minerals are tremolite, anthophyllite, chlorite, magnesite, dolomite, magnetite and quartz. In siliceous dolomitic limestones, talc is formed by retrograde replacement of tremolite and forsterite (Brown, 1973) and is purely magnesian.

Talc is a significant phase of the high pressure metamorphosed rocks where it replaces chlorite and biotite (Chopin and Monié, 1984). The Fe-content varies from 0 to 15%

of the iron end member. Iron richer talcs and minnesotaite are found in Fe-rich metasediments where they are associated with quartz, magnetite, iron carbonates, iron sulphides and chlorite.

Sea-floor hydrothermal alteration

Talc precipitates as a result of the rapid mixing of hot silica saturated hydrothermal solutions with cold Mg-rich pore fluids. It is distinctly more Fe-rich than talc in metamorphic rocks : 19 to 25% of the Fe end member (Lonsdale et al., 1980). Temperature conditions are up to 300°C and the formation of talc is still active in the vicinity of chimneys discharging hot fluids in oceanic ridges.

Soils and weathered rocks

In spite of the fact that talc is not a common mineral constituant of soils or weathered rocks, some natural occurences were described in tropical countries as well as in temperate ones (Fontanaud and Meunier, 1983; Maynard, 1983; Noack et al., 1987).

Although a complete solid solution with minnesotaite is not well established, the Fe content of talc clearly depends on composition of the parent rock and the physicochemical conditions imposed by the geological context (figure 2).

5 - SEPIOLITE AND PALYGORSKITE (ATTAPULGITE)

1- Chemical composition and crystalline structure

On a general point of view, both sepiolite and palygorskite contain small but variable amounts of Ca, K, Fe^{2+}, Al and Fe^{3+} while Si and Mg are the dominant components. This explains why these minerals are generally considered to be magnesium silicates in spite of palygorskite may contain up to 15% of alumina.

Two points concerning the composition of natural sepiolite and palygorskite should be noticed : 1- The alkali content, although very low at times becomes important in palygorskite : 3.46% K_2O for a deep sea sample (Church and Velde, 1979). 2- Considering Si, R^{2+}

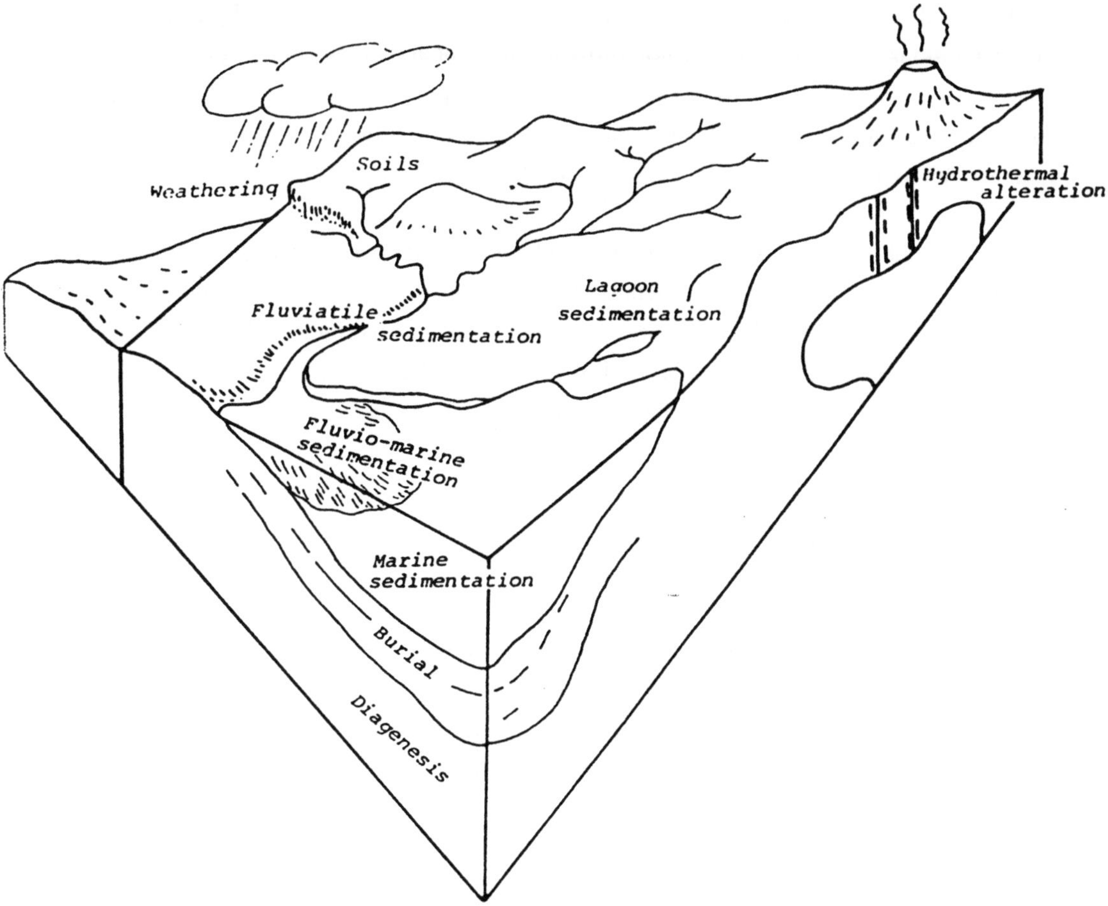

Figure 2 : Schematic representation of the geological context of clay minerals

(bivalent) and R^{3+} (trivalent) components, the compositions of sepiolite and palygorskite are quite continuous. At present, we do not know if these minerals represent a continuous series between two polymorphic forms or two distinct mineral phases. Sepiolite is considered to be trioctahedral while palygorskite is dioctahedral (Prost, 1975) although its composition field is broad enough to overlap the gap between the two domains (Paquet et al., 1987).

The ribbon organization in sepiolite and palygorskite is not always regular. Ribbon-like units of various width may fit together producing a partially ordered-partially disordered arrangement (Martin Vivaldi and Linares Gonzales, 1962). Nevertheless the fibrous habit is always conserved whatever the particular linkage of the different ribbon component. Further research about these problems may be worthwhile.

2- Genesis and natural occurences

Natural occurences of sepiolite and palygorskite were first systematically described by Millot (1964). The most recent comprehensive reviews with genetic explanation have been given by Weaver and Beck (1977), Singer (1979) , Velde (1985) and Jones and Galan (1988). Table 1 summarizes the major characteristics of sepiolite and palygorskite deposits as they are reported in these references. In many situations, the chain-structure minerals appear to be stably associated with smectites, kaolinite, serpentine, zeolites, carbonates, sulfates or other salts and free silica (cherts). The most important natural occurences of sepiolite and palygorskite are the following : saline lakes and marginal sedimentary basins, deep-sea sediments, arid soil (caliche and calcrete profiles) and hydrothermal veins in basic or ultrabasic rocks.

Saline lakes and marginal basins show a decrease in aluminous materials and an increase in magnesian silicates, as one progresses from the margin to the center of the basin (figure 3). A noticeable sequence is observed as the quantities of detrital material decreases : pure palygorskite and sepiolite precipitate away from the beach and form most of the economic deposits which are actually exploited in the world. Chert is a common accompanying phase. From detailed chemical studies, it can be assumed that a dissolution-precipitation process on a very small scale has been experienced by the sedimentary detritus which incorporate large amounts of magnesium while releasing alkalis. However the major part of the deposit is formed by a direct precipitation from concentrated saline solutions in which pH conditions are very high(over 8).

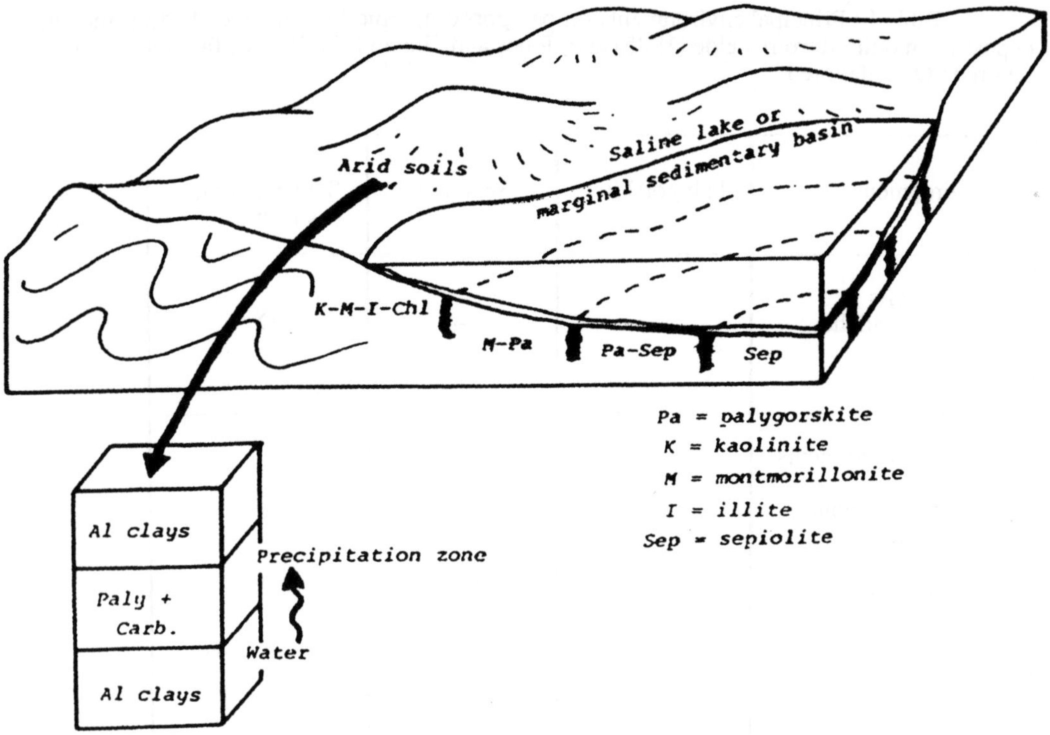

Figure 3 : Evolution of the chemical composition of clays in a saline lake environment (adapted from Velde, 1985)

Table 1 : Principal environments for palygorskite, sepiolite and trioctahedral smectite deposition modified from Velde (1985) and, Jones and Gralan (1988) . X : favored ; - : less favored ; O : unfavored

Environment	Palygorskite	Sepiolite	Trioctahedral smectite
Soils calcrete-caliche	X	---	O
Closed basin sediments			
Groundwater input dominant	O	X	---
Hypersaline	O	O	X
Marine sediments			
Lagoon or tidal	X	---	O
Deep sea	---	O	---

Soil profiles in arid climates are known to favour the precipitation of carbonate horizons produced by evaporation of the sub-surface water. Deposits of sepiolite and palygorskite were discovered in these horizons (Paquet, 1970 ; Van den Heuvel 1966). The hypothesis of the formation of such deposits leads to invoke the role of rain water. In fact, water first infiltrates rapidly through the profile deep into the bedrock and then is brought back by capillary action. The aqueous solutions become more and more concentrated until the conditions of stability of calcite and magnesian silicates are reached near but under the surface where they replace the preexisting minerals.

The deep sea sediments in the Pacific and Indian oceans are composed of detrital materials : illite, chlorite and volcanic ashes and authigenic minerals: palygorskite, dioctahedral smectite and an alkali zeolite : clinoptilolite (Couture, 1977 ; Stonecipher, 1976). At the pelitic sediment-sea water interface, the dissolved components of pore solutions react with aluminous material to form palygorskite or palygorskite-zeolite assemblages. Biogenic silica (diatoms) or volcanic glass could supply silica in sufficient amounts for this reaction proceeds.

Sepiolite and palygorskite appear to be the most often associated with carbonates in veins cutting basic or ultrabasic rocks. They are produced by the hydration reaction in hydrothermal conditions which is notoriously known as serpentinization. During this reaction, some chemical elements remain in solution (mainly Si, Ca and Mg) and precipitate in the fractures after a temperature decrease .

Velde (1985) showed that it is the silica activity which controls the formation of these minerals in the natural environments. Sepiolite and palygorskite are generated in very low temperature conditions by direct precipitation from highly Mg concentrated solutions. The strict localization of these deposits in particular places (subsurface horizon in arid soils, sediment-sea water in abyssal environment ...) is due to their absolute control by a chemical parameter : the pH of the solutions. If it is to low (under 8), aluminous clay minerals precipitate; if it is too high (above 9) saponite crystallizes.

6 - CONCLUSION

The clay minerals described here can be formed in a wide range of natural environments. The economic deposits for each of them depends on local conditions which favours their crystallization among the other stable phases. Nevertheless, whatever the degree of purity defined from industrial criteria, these materials can be very different according to their origine source.

The size, shape, crystalline properties and chemical composition of kaolinite, talc, sepiolite and palygorskite clearly depend on their conditions of formation. Besides, the set of associated minerals, potentially dangerous on the human health point of view, is also dependent of the geological environment. Consequently, the source of industrial material must be known , their processes of formation carefully studied and their crystalline and chemical compositions better studied before using.

REFERENCES

Bailey S.W. (1988) - "Hydrous Phyllosilicates(exclusive of micas)" . Reviews in Mineralogy, vol. 19, Mineralogical Society of America, 725p.

Bailey S.W. , Brindley G.W. , Johns W.D., Martin R.T. and Ross M. (1971) - Clay Mineral Soc. Report of Nomenclature Committee 1979-1980. Clays and Clay Minerals, 132-133.

Brindley G.W. and Brown G. (1980) - Crystal structures of clay minerals and their X-ray identification. Monograph n°5, Mineralogical Society , London, 480p.

Brindley G.W. and Pédro G. (1972) - Report of the A.I.P.E.A. nomenclature committee. A.I.P.E.A. Newsletter, 4, 3-4.

Brindley G.W. , Bish D.L. and Wan H.M. (1977) - The nature of kerolite, its relation to talc and stevensite. Min. Mag. , 41, 443-452.

Brown C.E. (1973) - Talc . U. S. Geol. Surv. Prof. Paper, 820, 619-626.

Chopin C. (1981) - Talc-phengite : a widespread assemblage in high grade pelitic blueschists of the Western Alps. J. Petrol., 22, 628-650.

Chopin C. and Monié P. (1984) - A unique magnesiochloritoid-bearing high pressure assemblage from the Monte Rosa, Western Alps : petrologic and ^{40}Ar-^{39}Ar radiometric study. Contrib. Mineral. Petrol. , 87, 388-398.

Church T.M. and Velde B. (1979) - Geochemistry and origin of a deep-sea Pacific palygorskite deposit. Chem. Geol., 25, 31-39.

Couture R. (1977) - Composition and origin of palygorskite-rich and montmorillonite-rich zeolite containing sediments from the Pacific Ocean. Chem. Geol. 19, 113-130.

Fawcett J.J. and Yoder H.S. (1966) - Phase relationships in the system MgO-Al_2O_3-SiO_2-H_2O. Am. Mineral. , 51, 353-380.

Floran R.J. and Papike J.J. (1978) - Mineralogy and petrology of the Gunflit Iron formation, Minnesota-Ontario : correlation of compositional and assemblage variations at low moderate grade. J. Petrol., 19, 215-288.

Fontanaud A. and Meunier A. (1983) - Mineralogical facies of a serpentinized lherzolite from Pyrénées (France). Clay Min. ,18, 77-88.

Harben P.W. and Bates R.L. (1984) - Geology of the nonmetallics. Metal Bulletin Inc. Ed., New York, 392p.

Jones B.F. and Galan E. (1988) - Sepiolite and palygorskite . In "Hydrous Phyllosilicates(exclusive of micas)" Reviews in Mineralogy, vol. 19, S.W. Bailey Ed., 631-674.

Kager P.C.A. and Oen I.S. (1983) - Iron-rich talc-opal-minnesotaite spherulites and crystallochemical relations of talc and minnesotaite. Mineral. Mag., 47, 229-231.

Lonsdale P.F., Bischoff J.L., Burns V.M., Kastner M. and Sweeney R.E.(1980)- A high temperature hydrothermal deposit on the seabed at a Gulf of California spreading center. Earth Planet. Sci. Letters , 49, 8-20.

Martin Vivaldi J. L. and Linares Gonzales J. (1982) - A random intergrowth of sepiolite and attapulgite . Clays and clay minerals, 9, 592-602.

Maynard J. B. (1983) - Geochemistry of sedimentary ore deposits. Springer Verlag, New York, 305p.

Millot G. (1964) - Géologie des argiles . Masson ed. , Paris, 510p.

Muffler L.J.P. and White D.E. (1969) - Active metamorphism of upper cenozoic sediments in the Salton Sea Geothermal field and the Salton Trough, Southern California. Bull. Geol. Soc. Amer. , 80, 157-182.
Muller J. P. (1988) - Analyse d'une formation latéritique meuble du Cameroun. Essai de traçage d'une différenciation supergène par les paragenèses minérales secondaires. Thesis University Paris VII, 188p + annexes.
Murray H.H. (1988) - Kaolin minerals : their genesis and occurrences. In "Hydrous Phyllosilicates(exclusive of micas)" . Reviews in Mineralogy, vol. 19, S.W. Bailey Ed., 67-89.
Newmann A.C.D. (1987) - Chemistry of clays and clay minerals. Monograph n°6, Mineralogical Society , London, 480p.
Nisio P. and Lardeaux J.M. (1987) - Retromorphic Fe-rich talc in low - temperature eclogites : example from Monviso (Italian Western Alps). Bull. Minéral. , 110, 427-437.
Noack Y. , Decarreau A. and Manceau A. (1986) - Spectroscopic and oxygen isotopic evidence for low and high temperature origin of talc. Bull. Minéral. , 109, 253-263.
Paquet H. (1970) - Evolution géochimique des minéraux argileux dans les altérations et les sols des climats méditerranéens et tropicaux à saisons contrastées. Bull. Serv. Carte Géol. Als. Lorraine, 30, 1-212.
Paquet H. , Duplay J. , Valleron-Blanc M.M. and Millot G. (1987) - Octahedral compositions of individual particles in smectite-palygorskite and smectite-sepiolite assemblages. Proc. Int. Clay Conf. , Denver 1985, Schultz L.G. Van Olphen H. and Mumpton F. A. ed., 73-77.
Petit S., Decarreau A., Eymery J. P. and Thomassin J. H. (1988) - Synthèse de kaolinites ferriques à 200°C. Comparaison avec les kaolinites d'altération supergène : teneur en fer, morphologie et cristallinité. C. R. Acad. Sci., Paris, t. 307, série II, 1961-1966.
Plançon A. and Tchoubar C. (1977 a) - Determination of structural defects in phyllosilicates by X-ray diffraction. I. Principle of calculation of the diffraction phenomenon. Clays and Clay Minerals, 25, 430-435.
Plançon A. and Tchoubar C. (1977 b) - Determination of structural defects in phyllosilicates by X-ray diffraction. II. Nature and proportion of defects in natural kaolinites. Clays and Clay Minerals, 25, 436-450.
Prost R. (1975) - Etude de l'hydratation des argiles : interaction eau-minéral et mécanisme Thesis , University of Paris, 100p.
Sandford R. F. (1982) - Growth of ultramafic reaction zones in greenschist to amphibolic facies metamorphism. Am. J. Sci., 282, 543-616.
Schreyer W., Abraham K. and Kulke W. (1980) - Natural sodium phlogopite coexisting with potassium phlogopite and sodium aluminium talc in metamorphic evaporate sequence from Derray, Tell Atlas, Algeria. Contrib. Mineral. Petrol. , 74, 223-233.
Singer A. (1979) - Paligorskite in sediments : detrital, diagenetic or neoformed. A critical review. Geol. Rund. , 68, 996-1008.
Stonecipher S. A. (1976) - Origin, distribution and diagenesis of phillipsite and clinoptilolite in deep-sea sediments. Chem. Geol. , 17, 307-318.
Van der Heuvel R. C. (1966) - The occurence of sepiolite and attapulgite in the calcareous zone of a soil near Las Gances, New Mexico. Clays and clay minerals, Proc. 13th Nat. Conf. , 193-208.
Velde B. (1985) - Clay Minearls . A physico-chemical explanation of their occurence. Development in sedimentology, n° 40, Elsevier Ed., 427p.
Weaver C.E. (1959) - The clay petrology of sediments. Clays and Clay Minerals, 6, 154-187.
Winkler H.G.F. (1974) - Petrogenesis of metamorphic rocks. Springer Verlag Ed., Berlin, 320p.

PHYLLOSILICATES AS INDUSTRIAL MINERALS

G.M. Clarke
Industrial Minerals Division
Metal Bulletin Plc
16 Lower Marsh
London SE1 7RJ
United Kingdom

Throughout the world the phyllosilicates represent a major group of commercially exploited minerals. They serve a wide spectrum of manufacturing and service industries that require the physical and chemical attributes derived from their individual mineralogical properties. These minerals, unlike metal ores and fuel minerals, are extracted and upgraded to enhance the value of their mineralogical performance rather than for the extraction of an elemental or chemical value.

The commercially important phyllosilicates include the clay minerals of which the dickite (kaolin and halloysite) and smectite (montmorillonite and hectorite) groups are dominant. Kaolin leads the phyllosilicate field in terms of volume with over 22m. tonnes/year produced in a variety of grades and the kaolin-containing ball and plastic clays used in refractories and ceramics enhance the mineral's commercial significance that also embraces paper, paint, plastics, rubber, and many other manufacturing industries. The mineralogical complexity of montmorillonite provides a number of commercial varieties some of which are man-made. Calcium montmorillonite, the common fullers' earth that needs treatment with soda for many of its uses, has an American cousin in the form of sodium montmorillonite, bentonite per se, which is the work-horse of the oil-well drilling mud industry amongst other fundamental applications. The ability for montmorillonite to be chemically modified without altering its crystal lattice extends applications for acid activated varieties into the food processing industry and organoclad forms into the industrial coatings business.

NATO ASI Series, Vol. G 21
Health Related Effects of Phyllosilicates
Edited by J. Bignon

Reflected in its diversity of application talc is one of the most versatile of all industrial minerals owing to its chemical inertness, platy shape, whiteness, fineness, slippery feel, and other attributes. Talc competes with kaolin in many applications and its choice depends on a combination of availability, technical, and economic factors. Some 5-6m. tonnes/year production makes talc the third most significant phyllosilicate after kaolin and montmorillonite although it is one of very few industrial minerals to be generally well known owing to its application in cosmetics. Pyrophyllite, the aluminium-containing version of talc, was commercially developed as an alternative to talc in a similar spectrum of uses. Today, pyrophyllite is important in North America and the Far East, especially Japan, and has grown away from talc's shadow owing to its rise as a major refractory lining material in steel ladles.

The micas represent a much smaller and more specialist group of industrial minerals with around 0.25m. tonnes/year production. The white translucent muscovite is the dominant mica used in plastics, drilling, special paints, etc. although the darker phlogopite-biotite is also exploited for a variety of non-colour sensitive functional filler uses. The geologically weathered variety of dark mica, vermiculite, is another phyllosilicate that almost achieves widespread consumer awareness owing to the domestic loose fill insulation application for the exfoliated variety. This, however, is dwarfed by vermiculite's other industrial applications which exploit the mineral's refractory and insulation properties although small volumes are also used in horticulture.

The serpentine mineral, chrysotile, is a major commercial phyllosilicate that belies its true sheeted crystal structure by rolling up, tube-like, into elongated fibres. As such the mineral belongs to a select sub-class known as the solenosilicates of which halloysite, a relatively rare hydrated variety of kaolinite, is another example exploited in New Zealand and Korea for world markets. Chrysotile, the prime asbestos mineral, has a highly versatile spectrum of properties that has extended its use into a myriad of industrial applications from brake linings to fire protection. Developments on health-related matters over the last two decades in particular have severely impacted chrysotile business with today's production volume at around 1.4m. tonnes/year outside the Soviet Union. However, chrysotile represents the most valuable of the phyllosilicate minerals with list prices falling within a wide range extending from £250 a tonne through £1,000 for special qualities.

Two commercially rare but important hydrated magnesium aluminium silicate minerals, palygorskite and sepiolite, are closely related to the phyllosilicates and are often included with them. However, both minerals are markedly different in terms of their structural characteristics in that they form chain-type, rather than layered, structures which give rise to their fine fibrous nature. Although mineralogists differ in their interpretation of the structural configuration of these minerals they may be regarded as akin to the amphibole minerals and transitional towards the phyllosilicates proper from the chain or ino-silicates. Attapulgite, as palygorskite is better known commercially, has annual world production in the order of 1m. tonnes and has a variety of applications in coatings, drilling muds, and markets similar to those for fullers' earth. Sepiolite, on the other hand, is much more commercially rare and world production of 300/400,000 tonnes/year is dominated by Spain for a myriad of applications.

So, these minerals constitute what is regarded as the commercially important phyllosilicate mineral group of minerals. However, within the group there is represented extremely wide diversity of industrial and consumer use although in terms of world production volume they may be split into three categories:

A	All forms of kaolin and smectite clays:	+10m. tonnes/year
B	Talc, pyrophyllite, attapulgite, chrysotile:	1-10m. tonnes/year
C	Vermiculite, mica, sepiolite, halloysite:	-1m. tonnes/year

Otherwise these minerals may conveniently be divided into the clay and non-clay minerals which provides the order for the following discussion.

CLAYS

Refined kaolin and kaolinitic clay products

Kaolin is one of the commonest minerals occurring in the uppermost 10 metres of the earth's crust. However, only exceptionally does it occur in sufficient quantity and quality to be worth exploitation. The mineral is produced as a variety of refined kaolin products or as a less pure kaolinitic clay product. In refined form it finds use as a functional mineral filler in the paper, plastics, rubber, ceramics, and other industries whereas the less pure product finds use in ceramic bodies, where the non-kaolin content is important, and in constructional applications such as bricks and cement.

Some 40 countries produce in excess of 25,000 tonnes/year of kaolin although the USA (7m. tonnes/year) and the UK (3m. tonnes/year) are the majors. Few countries produce the quality required for paper coating applications. These grades therefore command higher prices and their availability is restricted to Australia, Brazil, UK, and USA as the leaders followed by smaller volumes from France, Spain, and West Germany. Very pure kaolins, close to mineralogical purity, are produced for advanced ceramics applications such as cordierite bodies.

In recent years the kaolin market has been very buoyant owing to firm market demand especially for the lightweight coated paper and ceramics industries. This has led to increased capacities in Brazil and the USA, reactivated pits in the UK, and new operations being developed in Canada and Australia. One particular market development has been the increased use of calcined kaolin to substitute for more highly priced titanium dioxide pigment in the USA where supplies have been very tight.

With regard to kaolinitic clay products there are essentially three types to consider -- fireclays, flint clays, and ball and plastic clays all of which contain varying amounts of kaolin which contributes to the desired properties of the material as a whole. All clays with a fusion point above 1,424^{o}C. are regarded as fireclays which are used as lower duty refractory linings. Flint clays are essentially high purity very fine kaolinite which is hard, dense, weather resistant, fracture in conchoidal fashion, and do not slake in water. For application the material is often calcined, to induce mineralogical phase modifications, for use as a refractory. There are few producers primarily in Australia, China, Israel, North Korea, South Africa, and the USA. Ball and plastic clays have over 90% of their markets in the field of ceramics and refractories. They contain around 70% disordered kaolin plus illite, montmorillonite, chlorite, quartz, and carbonaceous matter. Physically they have high plasticity, high dry strength, white colour on firing, and special coating properties. Typical ball clay products are quite rare and are produced in the UK and USA (each with around 800,000 tonnes/year), Czechoslovakia, and the Soviet Union. Elsewhere plastic clays are produced in West Germany, France, East Germany, Turkey, Australia, China, etc. Calcined kaolinitic clays products, known as chamottes, are produced in France for refractory and ceramics uses.

Halloysite

This commercially very rare mineral is related to kaolin in both hydrated and dehydrated forms -- conversion to the hydrated form occurs above 50°C. Halloysite occurs in tubular form, owing to curvature of the crystal lattice, and it often occurs as a discrete component in kaolin deposits.

Halloysite rarely concentrates on its own but commercial deposits are exploited in Japan, New Zealand, South Korea, and the western USA. The spectacular deposit at Matauri Bay, New Zealand is renowned world-wide and is presently undergoing expansion from 25,000 to 33,000 tonnes/year. The crude product contains 50% halloysite and 50% silica, feldspar, etc. The final refined product for export is classified to 100% minus 2 microns whereas the coarse reject is sold for domestic filler applications. Niche markets are in high quality porcelain and bone china for the premium product which commands prices approaching US$300/tonne FOB Auckland.

Lesser qualities of halloysite are produced in Japan for refractories, paper, and ceramics where the crude material may contain a mixture of halloysite, kaolin, quartz, cristobalite, and feldspar.

Smectite

The smectite phyllosilicates represent a mineralogically and commercially complex class of minerals that are collectively known as bentonite. So far as the commercially available end products are concerned there are five essential types:

Natural refined forms:	sodium bentonite calcium bentonite
Chemically treated forms:	sodium exchanged bentonite acid activated bentonite organoclad bentonite

Bentonite is a clay consisting essentially of smectite group minerals whose physical properties are dictated by the dominant mineral regardless of origin and occurrence -- it is primarily a market definition. It should be made clear that the term fullers' earth has separate meanings in the USA, where it is used to describe an attapulgite/calcium montmorillonite range of products, and the UK, where it specifically refers to calcium montmorillonite. Mineralogically pure grades of attapulgite that are capable of forming gels are discussed separately as their market spectrum is wider than the lower priced attapulgite/montmorillonite mixtures available in the USA. Here the discussion concerns montmorillonite per se.

Montmorillonite has three essential properties that provide wide diversity of application. The extremely small flexible platy crystals provide for a very large surface area. A negative excess electrical charge leads to the presence of exchangeable cations between the alumino-silicate layers of the mineral's structure. The mineral possesses an intracrystalline swelling ability.

These attributes have allowed the development of a number of speciality modified grades of montmorillonite, so widening the application spectrum, that essentially fall into three classes. Low swelling, high calcium montmorillonites may be treated with soda to raise swelling properties -- <u>sodium exchanged bentonite</u>. High calcium montmorillonites may be treated with inorganic acids to raise specific surface area, porosity, reactivity, and purity: carbonate is removed, hydrogen replaces the exchangeable cation, and iron, magnesium, and aluminium are leached --<u>acid activated montmorillonite</u>.

Naturally occurring sodium montmorillonite (bentonite per se), or hectorite, may be treated with polar long chain molecular organic chemicals to produce hydrophobic products with thixotropic/gellant uses -- <u>organoclad clays</u>.

So far as markets for these commodity and speciality varieties of bentonite are concerned they are complex, diverse, and global. The untreated calcium variety, fullers' earth in the UK, is quarried, sized, and sometimes pelletised for uses as animal feed, pet litter, horticulture, cosmetics, pharmaceuticals, etc. An exceptionally rare white variety produced in the USA and parts of the Mediterranean is used in cosmetic creams, detergents, etc. The sodium varieties, natural and treated forms, have similar market spread -- the geologically restricted natural sodium variety of the mid western USA is extended by the ability to convert natural calcium montmorillonite to the sodium form although the natural variety generally has the edge in the market-place. Sodium bentonite is used in oilwell drilling muds, foundry mouldings, iron ore pelletisation, civil engineering, paper manufacture, and chemicals. These types of bentonite are essentially commodity workhorses for manufacturing, process, and construction industries and provide a world production volume around 10m. tonnes/year with the USA accounting for around 2.5/3m. tonnes/year.

World production of acid activated bentonite lies in the 0.5-0.6m. tonnes/year range and is dominated by just five major players that account for almost 90% of production. Markets in recent years have been buoyant and there have been capacity increases. The products are used for the refining, decolourising, and purification of vegetable oils; refining of sulphur; refining and recycling of mineral oils and organic chemicals; as catalysts in the production of insecticides and fungicides; as pigment/colour developer in carbonless copy paper; and various applications in food and beverage refining.

By contrast the organoclad clays represent a very small specialist market of 30,000-40,000 tonnes/year world-wide where a single company, NL Industries, dominates the market. These products, whether based on bentonite or hectorite, are used in the oilwell drilling industry for borehole scavenging; in paint and varnishes for thickening, stabilising, etc.; and for binding in foundries.

Attapulgite and sepiolite

The background for both these minerals has been outlined in previous sections. The USA is the major producer of attapulgite where the three leading producers of high purity gellant grades have operations in Florida and Georgia. Smaller producers of non-gellant grades produce, as already mentioned, a variety of attapulgite/montmorillonite clay mixtures. Important production also takes place in Senegal, as a by-product of phosphate mining, and Spain, Australia, and South Africa. Gellant grades are used in salt water drilling muds, suspension agents, and catalysts, etc. Other grades are used in oil absorbents, pesticide carriers, pet litter products, and animal feed.

Outside Spain the only other producing countries of note are Turkey and the western USA where output is modest be comparison with Spain's dominance. Markets are similar to those for attapulgite although performance differs. Sepiolite has large surface area of the order of 300 sq. metres/gm, large internal channel and pore surface area of around 400 sq. metres/gm, and the unprocessed form is able to absorb 2.5 times its own weight of water. These properties provide a spectrum of environmental and industrial applications that include clarifying and bleaching agents, filter aids, industrial absorbents, pesticide and catalyst carriers, thickeners, and suspension agents. Compared with attapulgite, sepiolite continues to yield on working in aqueous suspension because of unwinding of long flexible crystals whereas attapulgite breaks across the crystal structure so losing rheology. Sepiolite is also stable at higher temperatures.

NON CLAYS

Compared with the clays the commercial non clay phyllosilicates are not necessarily fine grained -- the clays are generally regarded as averaging below two microns particle size. They do not contain water within the crystal lattice and they do not become plastic when mixed with water. So, we are dealing with a more diverse group of minerals that are essentially hydrous silicates of aluminium, magnesium, iron, and lithium and potassium and may be referred to as the talc group (talc and pyrophyllite), the mica group (muscovite, biotite-phlogopite, vermiculite, and lepidolite), and chrysotile.

Talc

Although talc is a name that describes a specific hydrous magnesium silicate mineral, in the commercial world it is a word that as much describes a spectrum of diverse markets served by a range of talc-containing industrial mineral products. The non-talc component of the product may be a necessary component of the material to fulfil its function or it may be an inert material of little or no consequence to the application concerned. So, many commercial talcs contain significant proportions of carbonate minerals, quartz, mica, chlorite, tremolite, etc. The markets for talc are thus whole rock markets although some beneficiation is undertaken -- to separate talc from magnesite for use in paper making in Finland for example.

Talc competes primarily with kaolin and various types of white ground limestone in industrial markets although its use in cosmetics is virtually unrivalled. The USA is a major producer and talc from Australia travels the globe to markets in Europe although Indian material probably enjoys the highest reputation for cosmetic qualities. In Europe the spectrum of talc grades is produced for all market segments. Owing to the country's lack of sufficiently good domestic kaolin resources Finland has developed its talc industry to serve its requirement for a white filler and coating agent in the paper industry which accounts for around 360,000 tonnes/year. In Norway,

Sweden, and the UK a range of talc/magnesite products are produced for lower quality filler applications. In Norway imported Indian talc is processed for European consumption. Operations in France, Spain, Italy, and Austria are controlled by one company serving all European markets whereas another producer in Italy is renowned for cosmetic grades. Elsewhere in Germany and Turkey there are smaller deposits.

Markets for talc in the two major consuming regions of the globe, North America and Western Europe, differ considerably. Ceramics is the major market in North America whereas paper leads in Western Europe where about 450,000 tonnes/year is used for paper filler, 115,000 tonnes/year for pitch control, and 80,000 tonnes/year for paper coating. In the USA the pitch control segment has a greater proportion of the paper market. Other industrial applications include roofing felts owing to talc's hydrophobic nature that inhibits sticking, animal feed for dark talcs, fertiliser anti-caking agents, ceramics, paint and varnishes, plastics as a reinforcing agent, and pharmaceuticals and cosmetics where the purity and other quality considerations are the most stringent.

Pyrophyllite

Pyrophyllite has often been classed as the poor relation to talc owing to its history of commercial development. Nevertheless pyrophyllite is mineralogically distinct and its markets have evolved to the point where the distinction is no longer blurred with talc owing to its position today as an important refractory steel ladle lining material. The major producers are in the USA, Japan, North Korea, South Korea, and Australia. In Brazil the mineral is produced under the local name of agalmatolite. There is no production in Western Europe where the market is also minor.

Some 75% of world pyrophyllite output originates from Japan and South

Korea. In Japan, which accounts for around 1m. tonnes/year, deposits run the length of the country and are known as Roseki -- a rock mixture of sericite, kaolin, and pyrophyllite. There are many comparatively small pyrophyllite mines in the south of South Korea where the mineral is associated with quartz, kaolin, alunite, diaspore, and sericite. In the USA pyrophyllite is exploited from deposits in North Carolina and California for markets in ceramics, refractories, insecticides, and a variety of filler applications.

Mica

Commercially speaking mica embraces three minerals -- muscovite, phlogopite, and lepidolite -- although in reality muscovite is be far the most significant and lepidolite has all but disappeared from the scene. A fourth mineral, vermiculite, is here included with the micas owing to its close mineralogical association, as a surface weathering product, with biotite mica.

Two forms of muscovite mica are commercially significant -- sheet and ground mica. With the advent of solid state electronics the market for sheet mica has gradually dwindled over the past three decades. India is the clear leading producer of sheet mica with 7/8,000 tonnes/year production. The Indian Standards Institute defines some 13 grades where the finest ruby mica commands price levels in the region of £750/lb. The material, produced as blocks, thins, films, or splittings, is extracted from small labour intensive mines and fabricated using knives into the required form. Besides India this kind of mica is also produced in Brazil and Madagascar.

Of far greater economic and market significance is the ground form of mica produced largely as a by-product of kaolin or pegmatite mineral operations or as a waste by-product of sheet mica production. The USA accounts for roughly 50% of world production mainly from North Carolina lithium and feldspar operations and Georgia kaolin operations although production is also recorded in Connecticut, New Hampshire, Pennsylvania, New Mexico, and South Dakota. More than 80% of the material is dry ground material whilst the balance is wet ground. In Western Europe, France and Spain are major muscovite producers from kaolin operations and in the Far East, China is a major source for world markets.

Production of dark phlogopite mica is more restricted with Canada, Madagascar, and Finland being the important sources. This type of mica is disadvantaged compared with muscovite by virtue of the dark colour which inhibits its use in applications where pigments are to be used for colouring. In Finland phlogopite is produced as a by-product of the Siilinjarvi phosphate operation. The lithium bearing mica, lepidolite, once commanded special place as a source of lithium from the Bikita pegmatite operation in Zimbabwe. The mineral has, however, largely been superseded by other lithium containing minerals such as petalite.

As the reader may have come to expect markets for ground mica are also diverse and are largely in North America, Western Europe and the Far East. They embrace paints where dry or wet ground or micronised grades are used for external and anti-corrosion coatings, plastics where mica s reinforcing properties are exploited, fire protection boards where heat resistance is important, oilwell drilling for coarse grades where the platy shape is used to inhibit water-based mud circulation loss (purity is not important for this use), pearlescent pigments where finely divided mica acts as a substrate for coloured iron, chromium, or titanium oxides in the cosmetics and automotive industries, in textured coatings such as "Artex" used for interior decoration, and in welding rods where dry ground mica is used.

Vermiculite

Vermiculite is a surficial alteration product of biotite-phlogopite mica and is of restricted occurrence with the major producers in South Africa and the USA. Minor, but significant, production is recorded in Australia, China, Malawi, Kenya, Argentina, and Brazil. The commercial significance of vermiculite lies in its capability to expand to many times its original volume on being heated. At over 760°C. the mineral exfoliates in concertina fashion as a result of water escaping from the crystal lattice which prises the layers apart perpendicular to the mineral's cleavage. Larger flakes tend to provide better expansion properties as higher pressures are allowed to build up prior to the point of exfoliation.

Exfoliated vermiculite finds a range of applications that depend on a medium to trap air which provides for a combination of light weight, sound and heat insulation, and fire resistance. In construction vermiculite is used as a lightweight aggregate in plasters, concrete, and insulating boards as well as a protective coating substrate for construction steel in potentially hazardous environments such as oilwell platforms. Vermiculite has been the subject of considerable materials research. A chemically delaminated vermiculite has produced a 30-fold expansion using cation exchange and the resulting material used in gel form to coat glass fibres.

Chrysotile

Chrysotile distinguishes itself from the other commercially significant phyllosilicates in a number of ways. Firstly, although a true phyllosilicate the mineral disguises itself as it mostly occurs in the form of fibres which are commercially valuable. This is due to the rolling up of the sheeted crystal lattice into tubes although the sheeted form does occur in what is known as mountain leather. Secondly, chrysotile has served a multitude of applications world-wide over many decades and has been the subject of considerable study with regard to worker health which has identified the mineral as a major hazard. Thirdly, it remains the most valuable of the commercial phyllosilicates in terms of price although the producing industry as now a shadow of its former position owing to diminished markets as a result of health concern. Finally, the spectrum of markets served by chrysotile have proven not able to be served by a single material substitute and much research and development has led to a variety of minerals and materials providing substitution in specific application areas.

Nevertheless, chrysotile is still produced, albeit on much reduced scale, in Canada, USA, Brazil, South Africa, Swaziland, Zimbabwe, Italy, Greece, Cyprus, Yugoslavia, China, and the USSR. Markets have dwindled in the developed countries but the need for inexpensive infrastructure in the developing countries has maintained demand in that sector for asbestos cement construction materials in the form of sheets and pipes. As such this is the major application today although fire resistant textiles and friction products remain important. It appears that chrysotile has passed into an age of speciality status.

References

Clarke, Gerry (1989) Industrial Clays: A special review. Industrial Minerals Division, Metal Bulletin plc, London, UK.

Industrial Minerals: Many reviews on minerals discussed in this overview published over time in Industrial Minerals, Metal Bulletin plc, London, UK

SURFACE PROPERTIES OF PHYLLOSILICATES

H. Pezerat
Laboratoire de Réactivité de Surface et Structure
Université P. et M. Curie, URA 1106
4 Place Jussieu
75252 Paris Cédex 05, France

Introduction

The surface properties of phyllosilicates are related to their crystallochemical nature. A particle is generally constituted of several stacking of layers, which are more or less well oriented with respect to each other. To simplify, we can consider the particle as being constituted of only one stacking of layers. In general, the particle will be very thin but very extended in the two other dimensions.

We distinguish three types of layers :

1) Layer Te-Oc or 1/1, with one tetrahedral sheet and one octahedral sheet (fig. 1). The most common mineral in this group is the kaolinite where the octahedral sheet is occupied by Al^{3+} (two octahedra occupied over three, or dioctahedral mineral). When the octahedral sheet is occupied by divalent cations such as Mg^{2+} (serpentine) or Fe^{2+} (berthierine), all the octahedral cavities are occupied (trioctahedral minerals).

2) Layer Te-Oc-Te, or 2/1, with one octahedral sheet sandwiched between two tetrahedral sheets. The most common minerals are the micas, the vermiculites and the smectites.

In micas, which are non swelling materials, we have K^+ between the layers to compensate the deficit of charge of the layers arising from the substitution of Al^{3+} in the place of Si^{4+} in the tetrahedral sheets (fig. 2).

In vermiculites and smectites, the K^+ ions are replaced by hydrated cations (Mg^{2+}, Na^+, etc ...) between the layers (fig. 3). These minerals are swelling materials.

NATO ASI Series, Vol. G 21
Health Related Effects of Phyllosilicates
Edited by J. Bignon

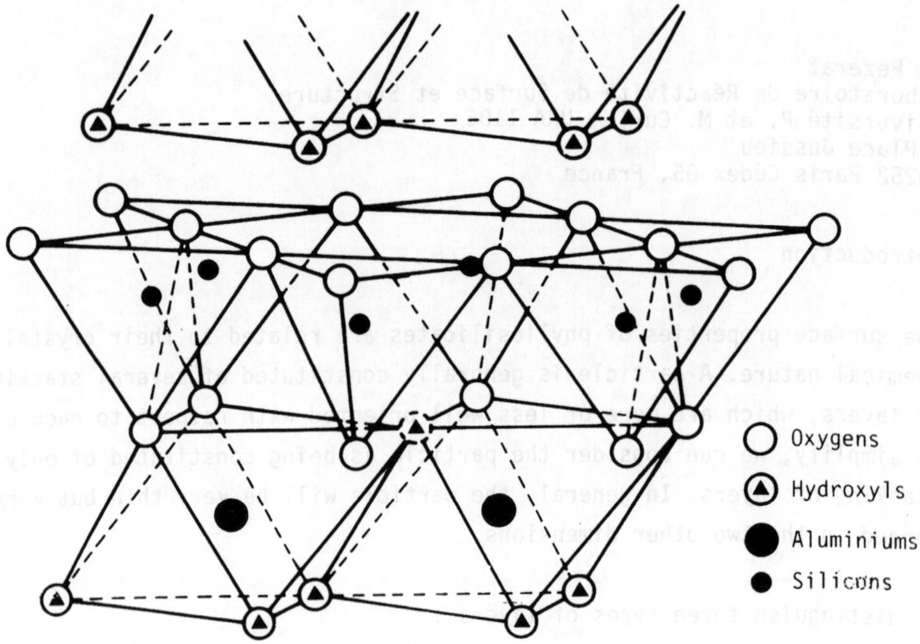

Fig. 1. Diagrammatic sketch of the structure of the kaolinite layer

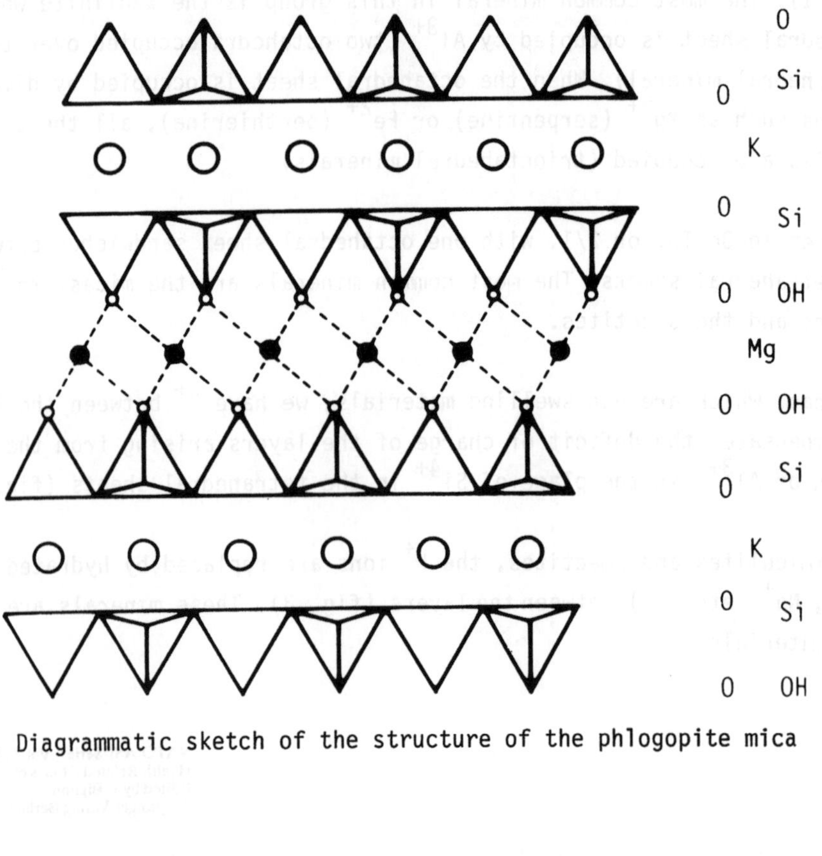

Fig. 2. Diagrammatic sketch of the structure of the phlogopite mica

As in the Te-Oc silicates, we distinguish dioctahedral minerals (such as muscovite and montmorillonite, the so called bentonite), where the octahedral sheet is principally occupied by Al^{3+}, from trioctahedral minerals (such as phlogopite, biotite, vermiculite, hectorite, the so called laponite), where the octahedral sheet is occupied principally by Mg^{2+} and Fe^{2+}.

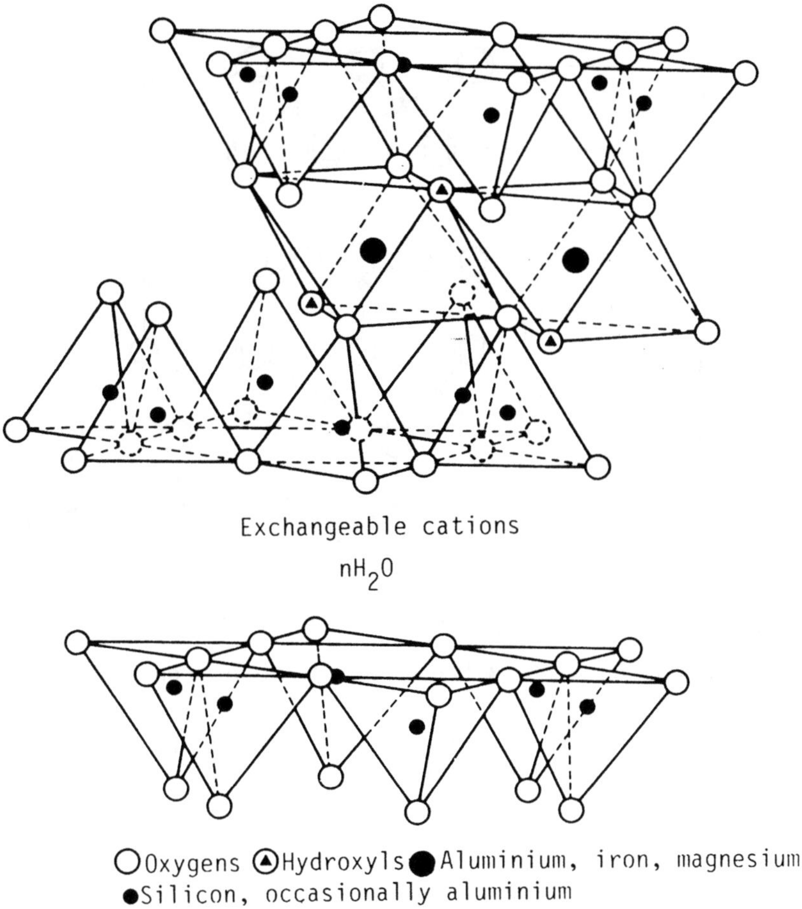

Fig. 3. Diagrammatic sketch of the structure of smectites and vermiculite

3) The third group of layers is the chlorite group (Te-Oc-Te-Oc or 2/1/1).

The schematic view is the same as in the vermiculite or smectite group, but the mineral is non-swelling because the interlayer space is not composed of mobile, exchangeable hydrated cations, but by an octahedral independent sheet, a brucite sheet of $(Mg, Fe)(OH)_2$ (fig. 4).

The chloritic minerals are generally trioctahedral.

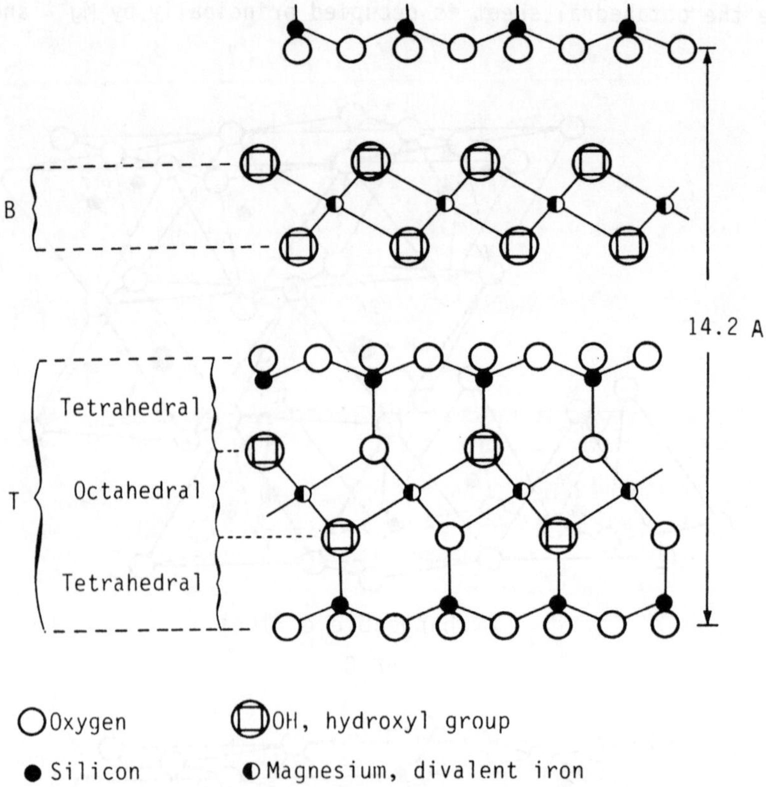

Fig. 4. Projection of the chlorite structure on a plane normal to the layer. T = talc layer, B = brucite sheet

Properties of basal surfaces

Basal surfaces - the more important in terms of area - are the external surfaces that are parallel to the layers. There are the usual cleavage faces of these silicates.

In the first group (Te-Oc phyllosilicates), for a particle containing one stacking of layers, the two basal surfaces are not similar, the one is silicic and the second is an hydroxylic surface of an octahedral sheet. In the second group both the basal surfaces are silicic, and in the third group (chlorite) the situation is variable, according to the leaching of the extra octahedral brucitic sheet.

Properties of the silicic basal surfaces

This surface is a regular network of oxygens, with a trigonal symmetry, where the oxygens are found bridging two silicons or one silicon and one aluminium atom in some particular phyllosilicates.

The properties of these basal surfaces are related to the distribution of the negative charges on the oxygens.

In the case of talc, kaolin and chlorite, without substitutions in the network, there is no deficit of positive charge and no excess of negative charge on these surfaces. So they have properties close to that of a perfect face of a crystal of silica. When a leaching of octahedral sheets occurs in biological medium, there is liberation of tetrahedral sheets, free of Al^{3+}. So, it is possible that asbestosis and talc (or kaolin) pneumoconiosis, are induced by the properties of this type of silica ... or may be, only by the properties of the silicic basal surfaces of these minerals.

In the case of montmorillonite and hectorite where there are no substitutions in the tetrahedral sheets, but only some substitutions in the octahedral sheets, (Al^{3+} by Mg^{2+} for montmorillonite and Mg^{2+} by Li^{+} in hectorite), the layer is negatively charged. This charge - compensated by exchangeable hydrated cations in the interlayer space - is very delocalized on the oxygens of the basal surface, approximately one electron for 20 oxygens. So, the silicic basal surfaces of these silicates have basic properties and they can be classified as "soft bases", by extension of the Hard and Soft Acid-Base theory of Pearson to the solid surfaces.

These properties of soft bases are responsible for the strong electrostatic bonds with the quaternary ammonium groups of some phospholipids present in the biological membranes. The trimethyl ammonium groups ($-N(CH_3)_3^+$) is indeed the prototype of the soft acids with a very delocalized positive charge. Hence, it is the delocalization of the negative charge on the basal surfaces of the montmorillonite which explains that this clay is one of the most haemolytic minerals.

In the case of micas, illites and vermiculites, approximately one SiO_4 tetrahedron over four is replaced by an AlO_4 tetrahedron in the tetrahedral sheets. Hence the properties of the basal surfaces are the ones of a silicoalumina sheet and not that of a silicic surface.

The three oxygens bridging an Al^{3+} with the three neighbouring silicon are the carriers of a charge of one electron. Consequently, we lose the property of soft base (disappearance of the very delocalized character of the negative charge), and the negatively charged basal surfaces appear as conjugate bases of relatively hard Brönsted acids, able to react with - or to adsorb - many molecules having a basic character.

<u>Properties of the hydroxylic surface of the Te-Oc silicates</u>

These properties are close to that of hydroxides and oxyhydroxides such as $Mg(OH)_2$, $Fe(OH)_2$, $AlO(OH)$ and $FeO(OH)$.

So, we have basic properties for serpentines and berthierine with strong properties of adsorption for molecules with a high dipole moment ; acidic properties for kaolinite and strong reducing properties for berthierine (see hereunder).

Properties of the lateral surface of the particles

The lateral faces of the flat particles of phyllosilicates constitute a break in the tetrahedral and in the octahedral sheets of the layers.

The lateral boundary at the level of the tetrahedral sheet presents acidic properties, weak for silanol groups (≡ SiOH) and stronger for aluminic group (≡ AlOH).

The properties of the lateral surface at the level of the octahedral sheets are very important to understand some interactions in the biological medium :

- There exists strong interactions with phosphate groups at this interface
- It is at this level that interactions with the proteins of the ferru-

ginous bodies occur. But we do not know what exactly is the nature of these interactions.

- When some Fe^{3+} ions are present in the octahedral sheet at the interface with the surrounding medium, they allow the oxidation of phenols and hydroquinone to radical species, these phenomenon being known to play a role in the synthesis of humic acids in soils (Wang and Huang, 1989). The mechanism is not well known, but probably it is the same that explains the oxidation of hydrocortisone (Cornejo and al., 1983) and the disproportionation of H_2O_2 (Zalma, 1988), after contact with palygorskite (attalpulgite). It is possible that the weak oxidizing properties of the Fe^{3+} are improved by its bonds with one or two groups (SiO_4).

- When Fe^{2+} is present at the interface of the octahedral sheets and of the medium (biotite, chlorite, berthierine), very important surface reducing properties appear on freshly generated particles, the same properties that appear on the hydroxylic basal surface of the berthierine, or on basal surface of iron-rich chlorite freshly cleaved. These reducing properties allow the reduction of molecular oxygen to activated oxygen species ($OH^{\cdot}$ or other species with equivalent oxidizing properties). Hence the freshly generated particles can induce, after inhalation, an oxidative stress, which probably is responsible for the excess of lung cancers in some metal mines, in granite quarries, in some tunneling operations, etc... The paper presented in this volume by D. Costa et al. illustrates the very oxidizing properties of these materials, by a radical pathway after interaction with molecular oxygen in an aqueous medium (see also D. Costa et al., 1989).

Clay-organic complexes and interactions

This subject concerns all the interactions taking place in the interlayer space of the swelling phyllosilicates; smectites and vermiculites, between cations, water, basal surface of layers and organic molecules.

It is a very vast field and for details it is recommended to consult some specialized reviews such as the one published by Mortland (1970), where several references are provided.

The process linked to the properties of the interlayer space in the swelling phyllosilicates probably plays an important, but as yet poorly known, role in biological medium.

There exists several processes to trap organic molecules in the interlayer space and it would be necessary to outline the bonding mechanisms in clay-organic complexes :

cationic, anionic, ion dipole and coordination, Van der Waals forces, pi bonding, entropy effects and covalent bonding.

Briefly, we discuss only some data concerning the first three mechanisms.

Cationic mechanisms

Two processess allow formation of clay complexes by this mechanism : (i) ion exchange with organic cations and (ii) protonation of organic molecules at clay surfaces.

Ion exchange : The organic cations are capable of replacing inorganic hydrated cations such as Mg^{2+}, Na^{+}, Ca^{2+}... in the interlayer space, but this mechanism implies taking into account the fact that the ionic properties of organic molecules are pH dependent, with a possibility of protonation on the amine groupe for alkylamines and aminoacids, and on the carbonyl group for urea and amides.

Within a group of primary, secondary and tertiary amines, the affinity of the alkylammonium ions for the swelling clay decreases in the series $R_3NH > R_2NH_2 > R_1NH_3$.

When introduced in vermiculite, certain quaternary ammonium cations cause gross one - dimensional swelling of the mineral in distilled water, with hundreds of angstroms separating the individual layers. With montmorillonite (bentonite) these organic cations allow the formation of industrial materials with lipophilic properties (bentones).

Protonation mechanism : This mechanism implies the acidity of the water in the interlayer space, where the dissociation degree of the water is much more important than in the normal liquid state.

The more electronegative the metal cation in the interlayer space, the more acidic will be the complex with water, and the more protonation of organic molecules will be easier.

The hydrolysis of such hydrated metal cations can be described in the following equation :

$$|M(H_2O)_x^{+n}| \rightleftharpoons |MOH(H_2O)_{x-1}|^{+n-1} + H^+$$

acid conjugate base

Urea and various amides have been observed to get protonated at montmorillonite surfaces where the exchangeable cations was H^+, Al^{3+} or Fe^{3+} but not in Na^+ or Ca^{2+} saturated clays.

As the water content of the interlayer space decreases - by drying of the clay - these polarization effects become more concentrated on the fewer remaining water molecules, causing an increase in hydrolysis and thus in their proton-donating capabilities.

Ion dipole and coordination mechanisms

The organic molecules take place of water to coordinate the inorganic cations in the interlayer space.

The greater affinity the exchangeable cations have for electrons, the greater will be the energy of interactions with polar groups of organic molecules capable of donating electrons (alcohols, ethylene-glycol, glycerol, ketones).

Ion-dipole or coordination types of interaction on clays have been noted for a wide group of polar molecules, i.e. : alcohols and particularly ethylene glycol and glycerol, NH_3, ketones, urea and amides, pyridine, nitrobenzene, aminoacids, amines, etc ...

Hydrogen bonding mechanism

This mechanism is an extremely important bonding process in many clay organic complexes, and very significant with large molecules and polymers.

We can distinguish several cases :

- water bridging for example with ketones

- organic-organic hydrogen bonding when organic cations are present in the interlayer space
- hydrogen bonding with layer surface oxygens but only when some Al^{3+} substitute Si^{4+} in the tetrahedral sheets.

Nature of some clay-organic complexes : adorption of organic acids, amino acids, proteins, purines, pyrimidines and nucleosides

The adsorption of all these compounds by clays has received a particularly great deal of attention in the past and new studies on these subjects are being regularly published in specialized journals. All the above mechanisms are implied to explain the processes, but in general little attention is conceded to the health related effects of these phenomena.

These properties are used in antidiarrhoeic drugs which are essentially composed of clay minerals such as montmorillonite and palygorskite (attapulgite). The pharmaceutical aspects of clay - organic interactions have been studied, particularly by White and Hem (1983), who presented various desirable, as well as undesirable effects of the ingestion of clays, for example concerning the latter, relating to the iron content of the clay and to the interaction with tetracycline which is inactivated by a shift of its pH domain of activity.

Catalysis reactions in the interlayer space of clays

In general the studies in this field have not employed conditions (temperature, hydratation level) compatible with biological medium, but it is sure that some catalytic reactions must be possible with interlayer cations such as Cu^{II} and Fe^{III}. For example, the catalytic decomposition of urea to ammonium ion on Cu^{II} montmorillonite was observed. Several authors have recently focussed their attention on the possible nitration of aromatic molecules in the interlayer space with the formation of a radical cation as the initial step, but till to day we are not sure of the possibility of this type of reactions in biological medium (see review in P. Laszlo, 1987).

If we take into consideration the diversity of organic reactions catalyzed by clay systems, it is tempting to consider clays as possible microreactors in prebiotic synthesis. The working hypothesis is that montmorillonite clays could have served as supports and as catalysts for the emergence of the first biomolecules, and since Bible and even before, many discussions have been devoted to this theme ... but the field is always open !

Two non-respiratory pathologies related to the surface properties of the phyllosilicates : elephantiasis of the lower legs and geophagia.

The endemic but non-filarial form of obstructive lymphopathy (elephantiasis) of the lower legs has been reported in several African countries (Price and al., 1981).

The studies showed that the clay minerals (Kaolinite, halloysite, illite) of the environmental soils are absorbed through the skin of the foot, and then reach the node via the afferent lymphatics from the foot. But, they also noted the differences in aluminium content between the clays in soils and in lymph nodes. In phagolysosomes of the macrophages, with a low pH, there is a leaching of the aluminic octahedral sheets of the clays and formation of silica, capable of provoking a fibrotic reaction in the vessels, at the origin of the pathology. In terms of toxicity to the cell, this suggests a low toxicity in the early stages after absorption of the material, with increasing toxicity as the alumina is dissolved away by the body fluids, leaving silica behind. It is probable that this transformation in biological medium (metabolisation) is the same as the one responsible for the kaolin and talc pneumoconiosis.

Geophagia, the ingestion of earth including various clays, is a very old practice (Halsted, 1968) which concerns important populations in all the continents. The medical implications of the phenomena can vary with the nature of the ingested clays but in all the cases, the authors note hepatosplenomegaly , very probably related to fibrotic reactions to siliça (Rousselot et al., 1939), anemia provoked by a blocking of the iron absorption related to the cation exchange capacity of the clay, (Minnich et al., 1968), and according to the potassium content of the clays, hypokalemia (Gonzalez et al., 1982) or hyperkalemia with chronic renal failure (Gelfand et al., 1975).

References

Cornejo J, Hermosin MC, White JL, Barnes JR, Hem SL (1983) Role of ferric iron in the oxidation of hydrocortisone by sepiolite and palygorskite. Clays and clay miner 31, 2 : 109-112.
Costa D, Guignard J, Zalma R, Pezerat H (1989) Production of free radicals arising from the surface activity of minerals and oxygen Part I : iron mine ores. Tox. and Ind. Health 5 : (under press)
Gelfand MC, Zarate A, Knepshield JM (1975) Geophagia , a cause of life-threatening hyperkalemia in patients with chronic renal failure. J Am Med Assoc 234, 7 : 738-741
Gonzalez JJ, Owens W, Ungaro PC, Werk EE, Wentz PW (1982) Clay ingestion : a rare cause of hypokalemia. Ann Intern Med 97, 1 : 65-66
Halsted JA (1968) Geophagia in man : its nature and nutritional effects. The Am J of Clin Nutrition 21, 12 ; 1384-1393
Laszlo P (1987) Chemical reactions on clays. Science 235 : 1473-1477
Minnich V, Okeuoglu A, Tarcon Y, Argasoy A, Cin S, Yorukoglu O, Renda F, Demirag B (1968) Pica in Turkey II Effects of clay upon iron absorption. The Am J of Clin Nutrition 21, 1 : 78-86
Mortland MM (1970) Clay-organic complexes and interactions. Adv in Agron, 22 : 75-117
Price EW, Mc HArdy WJ, Pooley FD (1981) Endemic elephantiasis of the lower legs as a health hazard of barefooted agriculturalists in Cameroon, West Africa. Ann Occup Hyg 24, 1 : 1-8
Rousselot LM, Thompson WP (1939) Experimental production of congestive splenomegaly. Proc Soc Exp Biol (NY), 107 : 705-708
Wang MC, Huang PM (1989) Catalytic power of nontronite, kaolinite and quartz and their reaction sites in the formation of hydroquinone - derived polymers. Appl Clay Science 4 : 43-57
White JL, Hem SL (1983) Pharmaceutical aspects of clay - organic interactions. Ind Eng Chem Prod Res Dev 22, 4 : 665-671
Zalma R (1988) Thesis."Contribution à l'étude de la réactivité de surface des fibres minérales. Relations possibles avec leurs propriétés cancérogènes". Paris VI University, France

PHYLLOSILICATES: ASSOCIATED FIBROUS MINERALS

A.M. Langer
R.P. Nolan
Environmental Sciences Laboratory
Brooklyn College
Avenue H and Bedford Avenue
Brooklyn, NY 11210

F.D. Pooley
Division of Mining and Minerals Engineering
School of Engineering
University of Wales, College of Cardiff
Cardiff
United Kingdom

The Nature of Phyllosilicates in the Marketplace

Over 45 million tons of phyllosilicates* are consumed per year in the West for a large number of industrial and consumer purposes. They may be marketed or described by generic or commercial name and sold to specific user industries on the basis of physical or chemical properties (Clarke, this conference). The important information conveyed to the consumer industry is that which pertains to application, i.e., that the phyllosilicate material possesses some specific property, e.g., particle size, color, etc. A mineral assay may accompany the data sheets, but this may be both simplified and incomplete for health hazard evaluation. The precise nature of the components of the marketed material is generally unknown and its mineral content is conveyed by impression only.

The applications of the phyllosilicates have an astonising range, from insulating material (e.g., vermiculite), to ceramic manufacture (e.g., tremolitic talc), to pet litter adsorbent (e.g., sepiolite). Single minerals commonly have multiple uses. The sheet silicates most commonly used, and their applications,

*Phyllosilicates are also known as sheet silicates (see Felius, Meunier, this conference).

NATO ASI Series, Vol. G 21
Health Related Effects of Phyllosilicates
Edited by J. Bignon

are described by Clarke (this conference). They are briefly outlined here in Table 1. Many mineralogically diverse materials enter the marketplace yearly.

It is well known in the geological community that phyllosilicate deposits are mineralogically heterogeneous (Meunier et al, this conference). Commercial products are, therefore, mixtures of minerals, some of which may not be phyllosilicates. Some component of the ore may possess biological potential. The subject of this paper is the associated fibrous minerals which may contaminate deposits of commercially exploited sheet silicates.

Mineral Complexity of Phyllosilicate Ores

Phyllosilicate deposits, whether large or small in size, are mineralogically heterogeneous. They always consist of complex admixtures of minerals. Some may be unimportant biologically while others raise concern. These admixed minerals may be phyllosilicates themselves (Fig. 1A) or naturally associated mineral fibre (Figs. 1B, 2).

The complexity of phyllosilicate ores is based on the fundamental nature of its origin, as well as other man-introduced variables. Briefly, these are:

1) The bulk chemistry of virtually all phyllosilicate deposits is such that many different minerals necessarily form, even within a narrow range of temperature and pressure. Both the concentration and activity of water seems to play an especially important role in creating different, stable, hydrous silicate minerals. The juxtaposed stability fields of these minerals are easily crossed with little energy input, making multiple minerals the rule in phyllosilicate deposits rather than the exception;
2) Oxides, unnecessary for the formation of the type mineral of the deposit, may produce quantities of biologically important mineral contaminants. Calcium oxide present in a chemical system undergoing steati-

tization (talc formation) may produce tremolite. Tremolite content of talcs may range considerably (Fig. 3A);

3) Following this line, deposit types may form from different parent precursors so that the resulting deposit may contain an even greater range of contaminant minerals. Talc may form through alteration of pre-existing minerals and a fibre may result, made up of structural mixtures (Tables 1, 2; Figs. 3B);

4) Talcboles are more common than previously recognized; the petrogenesis of many of these deposits may include

Table 2: Minerals which may commonly contaminate talc deposits

Mineral	Chemistry	Common Pre-Cursor Parent Rock Type
Tremolite*	$Ca_2Mg_5Si_8O_{22}(OH)_2$	Dolomite
Anthophyllite*	$(MgFe)_7Si_8O_{22}(OH)_2$	Ultramafic Serpentine Dolomite
Serpentine Minerals e.g., chrysotile	$Mg_3Si_2O_5(OH)_4$	Ultramafic Serpentine Dolomite
Carbonates	$CaCO_3$ $MgCO_3$ $CaMg(CO_3)_2$	Dolomite- Limestone Dolomite Dolomite
Quartz	SiO_2	----------
Mica e.g., phlogopite	$K_2(MgFe)_6(Si_6Al_2O_{20})(OH)_4$	Dolomite
Chlorite e.g., penninite	$(Mg,Al,Fe)_{12}(Si,Al)_8O_{20}(OH)_{16}$	Ultrafamic

*May occur with asbestiform, fibrous or non-fibrous mineral habit.

more than one geological process. Hence, superimposed, complex, and varied geological events tend to create similarly complex mineral assemblages. This is especially true for such deposits as talc and chrysotile. These processes may involve fluctuating temperatures and pressures over periods measured in geological time. These fluctuations are common to high crustal processes, where most phyllosilicates are formed;

5) Although not readily appreciated by biological investigation, ore deposits are mineral concentrates with finite boundaries. These physical boundaries may end at a rock type, which may be quite different mineralogically, containing amounts of minerals with known biological potential. For example, in eastern Vermont (USA), talc and serpentine are physically juxtaposed in the same deposit. The care with which these talc-barren areas, which contain fibres, are excluded from the talc mining process is dictated by the miner;
6) Single mills have been constructed to operate as regional facilities. These mills may process a compositional range of ores, and the chances of inadvertent, low-level, contamination with fibrous minerals found in one ore deposit, with relatively pure non-fibrous ore from another deposit, is greatly increased.

Phyllosilicate ore minerals are associated with other minerals brought about through both natural and man-made processes.

Focus on Fibre

Some of the associated minerals which contaminate phyllosilicate deposits are morphologically fibres (Tables 1, 3; Figs. 1-4). These include the asbestiform (asbestos) fibres and minerals which grew as fibres, yet do not possess the asbestos properties. Although these materials may be anticipated, they may occur as submicroscopic contaminants, visible only by elec-

Table 3: Examples of fibres associated with phyllosilicate deposits

Fibres Formed During Petrogenesis of Phyllosilicate Deposits

Asbestiform Fibres (Asbestos):

Fibre	Deposit Type Example and Location
Tremolite asbestos	Talc, Mojave Desert, California (USA).
Winchite asbestos	Talc, Allamore District, Texas (USA).
Richterite asbestos	Vermiculite, Libby, Montana (USA).
Anthophyllite asbestos	Talc, Finland.
Talc asbestos	Talc, mineral curiosity, California (USA).
Chrysotile asbestos	Serpentine ore bodies, some talcs, Turkey; Vermiculite, Spain.

Non-Asbestiform Fibres:

Fibre	Deposit Type Example and Location
Talc (agalite)	Talc, New York (USA).
Talcbole (structural mixture)	Talc, New York (USA).
Tremolite	Talc, Mojave Desert, California (USA).
Palygorskite (attapulgite)*	Georgia-Florida (USA).
Halloysite	Kandite deposits, Georgia (USA).
Titania (rutile)	Micas, Canada.

Fibres** Formed During Ore Processing

Acicular Cleavage Fragments:
Primarily amphiboles, including:
- tremolite, associated with talc, vermiculite.
- actinolite, associated with talc, vermiculite.
- others (e.g., tirodite, hexagonite, etc.), associated with talc.

Parting Fragments:
Primarily micas and clays, including:
- sericite, associated with clays in coal seams.
- vermiculite.
- talc.

*Palygorskite occasionally classified as an asbestiform clay.
**Fibre used in the morphological sense only (L:W$\geq$3:1). Does not imply a characteristic of growth.

tron microscopy (Figs. 1B, 2). However, most contaminants are visible by light microscopy (Fig. 3A).

Although fibres form during the petrogenesis of the ore deposit, they also form through the processes of ore manipulation (primarily crushing; Fig. 3A). Elongate objects may be produced through cleavage (breakage along crystal structure planes, which follow the crystal-chemical rules of structural failure) or parting along secondary planes (e.g., twin surfaces, etc.). The result is a mineral particle which possesses the morphology of fibre. These particles may be unanticipated.

Contaminants, Fibres, and Disease Associated with Phyllosilicate Exposure

Occasionally, reports appear in the literature suggesting that human disease is associated with exposure to a phyllosilicate dust. As an example, exposure to talc dust, in many mining and milling operations, has long been associated with the prevalence of pneumoconiosis (see review in Rohl et al, 1977). However, the nature of the pulmonary disease patterns produced in these instances have often been a matter of some interpretation. Frequently, these reports contained observations and statements which clearly implicated an agent other than talc itself. Silica (quartz) has been implicated (talco-silicosis), as well as mineral fibre, i.e., asbestos (see asbestos bodies, pleural plaques, and excess malignancies associated with "talc" exposure (Rohl et al, 1977). Even the radiographic patterns of talcosis closely resembles asbestosis, i.e., irregular interstitial scarring in the lower lobes.

With the finding of frank asbestos fibre in some talc deposits, the general impression in the scientific community was that, at least for talc, significant disease could arise from the presence of contaminant mineral fibre.

Mineralogists cautioned that many of the mineral fibres reported actually resulted from crushing and processing of the ore; that is, they were secondary artifacts produced by mineral parting and cleavage (Table 2; Fig. 3A). These, on a gross morphological level, resembled asbestos; however, were not. They do not possess the same optical properties, physical tensile strength, nor exposed surfaces (see Langer and Nolan, 1988). Recent data from the Institute of Occupational Medicine, Edinburgh, suggest these materials possess very different biological potential as well. These studies are the subject of this conference.

Discussion

Policard (1964) noted that miners and millers, as a special category of workers, were most frequently exposed to complex mixtures of mineral dust in their respective workplaces. He reasoned that small quantities of a biologically active agent, perhaps even unrecognized, admixed in a matrix of relatively "inert"* dust might produce significant pulmonary disease. He used the classic example of minor quartz contents of anthracite coals as the primary agent in the etiology of anthracosilicosis in Europe.

Experimental evidence supported this assertion and further suggested that lesions produced in animals with complex mineral mixtures were far more extensive than if the active agent had been administered at the same dose alone (e.g., Attygalle *et al*, 1956). These latter investigators amplified focus on complex mixtures, i.e., active agents and "inert" dust, as agents of pulmonary disease.

Minerals admixed in phyllosilicate ore deposits may be present in substantial quantity, as a major component, or may range down to trace. This ratio may change throughout the ore body or may be a function of the mining and/or beneficiation process. These minerals may be present as visible contaminants or as occult, submicroscopic, admixed phases. Small amounts of these latter minerals may be concentrated in the airborne dust when aerosolized, especially small-size minerals.

---

*We recognize, as our predecessors did, that virtually no dust is "inert." Hence, we use quotation marks to underscore the use of the word as a *relative* measure of inactivity.

If a phyllosilicate deposit is made of fibres to begin with, an occult exposure to fibrous dust in the workplace will be an unlikely occurrence. Even palygorskite, which is fibrous by nature, and occasionally so fine-grained that it appears to be formed of clumps when examined by optical microscopy, is unlikely to create an unrecognized fibrous dust hazard.

There are deposits, however, which contain fibres of such size that they can be resolved only by electron microscopy. Here, in these instances, exposure is occult and may go unnoticed, and therefore uncontrolled.

Every phyllosilicate deposit should be mineralogically characterized before exploitation, and after product beneficiation and processing. Milled products may contain minerals with fibrous morphology which did not exist previously in the deposit's natural state, created as the by-product of the industrial process. The use of electron beam instruments for these assays, to determine the presence of trace or submicroscopic mineral contaminants, is warranted.

Acknowledgement

The authors (AML, RPN) wish to acknowledge support from R.T. Vanderbilt and Company. RPN acknowledges support from the Stony Wold-Herbert Foundation.

Table 1: Examples of phyllosilicates used industrially, some important applications, presence of mineral fibre

Generic Phyllosilicate	Specific Mineral(s)[1] Used Industrially	Applications[2]	Natural Fibre[3]	Secondary Fibre[4]
Kandite[5] group	Kaolinite, dickite, halloysite; and "ball clay," "flint clay," "plastic clays."	Refractories, ceramics, paper, paint, rubber.	Halloysite; rutile needles.	Parting of kaolinite and dickite.
Smectite[6] group (Montmorillonite)	Montmorillonite, hectorite, bentonite,[7] Fullers earth,[8] all expandable clays, mixed-layer clays.	Filler, food processing decolorizer, pet litter absorbent, vehicle.	Volcanic glass shards; possible zeolites.	Parting of clays; fracture of disordered opal-cristobalite-tridymite leptispheres.
Vermiculite	Vermiculite.	Filler, horticulture, insulation.	Tremolite asbestos;[9] chrysotile asbestos; richterite asbestos.	Amphibole cleavage fragments; vermiculite parting.
Palygorskite group	Palygorskite (attapulgite), sepiolite.	Low-performance filler, inorganic vehicle; salt-water drilling mud; bleaching agent; pet litter adsorbent.	Palygorskite, sepiolite.	-----

(continued)

Table 1 (cont):

Generic Phyllosilicate	Specific Mineral(s)[1] Used Industrially	Applications[2]	Natural Fibre[3]	Secondary Fibre[4]
Mica group	Phlogopite, muscovite, biotite, lepidolite.	Plastics, paints, insulators, semiconductor industry; de-tackifier.	Rutile needles; sericite particles.	Parting of micas (e.g., muscovite to sericite).
Talc	Talc.	Ceramics, paper, filler, cosmetics, rubber, paint, de-tackifier.	Talc fibre[10] (agalite); talcbole;[11] talc asbestos;[12] tremolite fibre; tremolite asbestos; anthophyllite fibre; winchite asbestos;[13] chrysotile asbestos.	Talc parting; amphibole cleavage fragments (tremolite, anthophyllite, winchite, hexagonite, tirodite).
Pyrophyllite	Pyrophyllite.	Ceramics, refractory, rubber, talc substitute, de-tackifier.	May occur in fibrous form; kyanite laths; pyroxene laths.	Pyrophyllite parting.
Serpentine group	Chrysotile.	High-performance reinforcer, cements, friction pads, many others.	Chrysotile; nemalite; tremolite fibre; other fibers.	Parting of platy serpentine; amphibole cleavage.

Table 1 : Footnotes.

[1]Proper mineral names are underlined. Other names are market place, commercial, geological, or economic geological terms for same clays: ball clay = any plastic, white burning clay used as a bond in chinaware. A fine-grained kaolinic clay; flint clay = well crystallized kaolinite clay without plasticity; plastic clay = specific kaolin beds in a stratigraphic sequence in England.

[2]General, important, applications (see Clarke, this conference, for details).

[3]Fibres which may form in this type of phyllosilicate deposit. If mineral name is the same as for the specific mineral of the deposit, the mineral occurs as fibres. Therefore, halloysite, palygorskite, sepiolite, pyrophyllite, and chrysotile are natural mineral fibres.

[4]Secondary fibres (fibres used in a morphological sense) are those objects present in a phyllosilicate product which have resulted from crushing and milling of the ore. Parting and cleavage processes produce elongate particles (L:W >3:1) from a previously non-fibrous object (see discussions of fibre cleavage, asbestos in Langer et al, 1979; mineral parting in Langer, 1978.

[5]Kandite clays. Seven angstrom clays; non-expandable clays.

[6]Smectite clays. Expandable clays, generally montmorillonite clays (all varieties, based on cation substitution).

[7]Bentonite. Clay, predominantly montmorillonite, from altered volcanic ash.

[8]Fullers earth. Clay, predominantly made of montmorillonite. In Europe, is chiefly calcium-montmorillonite; in the United States, is chiefly palygorskite.

[9]Tremolite asbestos, as found in the Libby, Montana, vermiculite deposit; chrysotile asbestos, as found in the Spanish vermiculite. Richterite chemistry has been obtained on fibres at Libby.

[10]Talc fibre may be a pseudomorph after an amphibole. Especially talc forming from tremolite. May be a mixture of talc and amphibole structural domains.

[11]Talcbole is a mixture of talc and amphibole structures. Diffraction patterns obtained on <1μm diameter fibres contain elements of sheet and chain structures.

[12]Talc asbestos. Mineral curiosity. Talc fibre with properties of asbestiform mineral. Found, but rarely.

[13]Winchite asbestos described from Allamore Talc District, Texas.

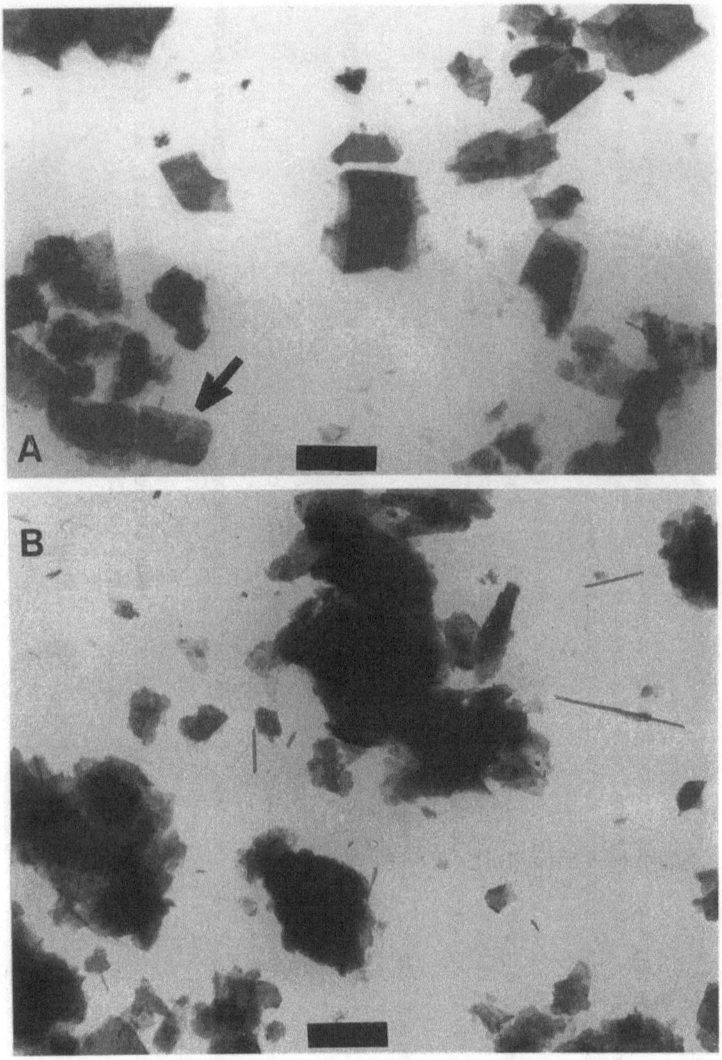

Figure 1: (A) Transmission electron micrograph of a processed talc specimen used for cosmetic purposes, from Vermont (USA). Particle population consists of ~95% talc plate, talc "fiber" (elongate object indicated by arrow, L:W>3:1), and the phyllosilicate, chlorite. No other mineral contaminants were noted by light or electron microscopy. Bar equals about 3.0 μm. (B) Transmission electron micrograph of a talc specimen from eastern Turkey. No contaminants were noted by light microscopy. Examination of specimen by transmission electron microscopy (TEM) reveals fibres/fibrils of chrysotile. Bar equals about 1.0 μm

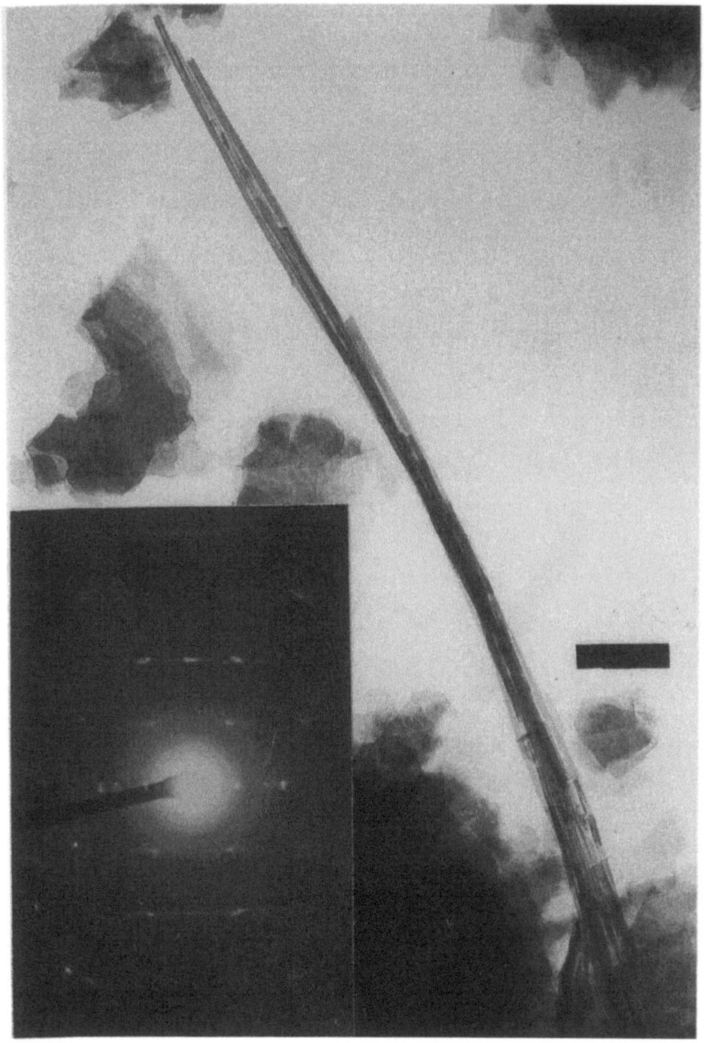

Figure 2: Transmission electron micrograph of chrysotile fibre bundle, >5.0μm in length, observed in the Turkish talc specimen. Insert, lower left, is the electron diffraction pattern obtained on the fibre. The (2kl) layer line displays "streaking" reflection, produced by chrysotile's cylindrical lattice. Bar equals about 0.3μm

Figure 3: (A) Light micrograph of processed tremolitic talc specimen, used in the making of ceramics, from New York State (USA). High-relief minerals are tremolite cleavage fragments produced by the crushing of the ore. Bar equals about 50.0µm. (B) Transmission electron micrograph of processed tremolitic talc, from New York State (USA). The high aspect ratio fibres (L:W>10:1) are talc fibres produced as pseudomorphs after pyroxenes. Bar equals about 3.0µm

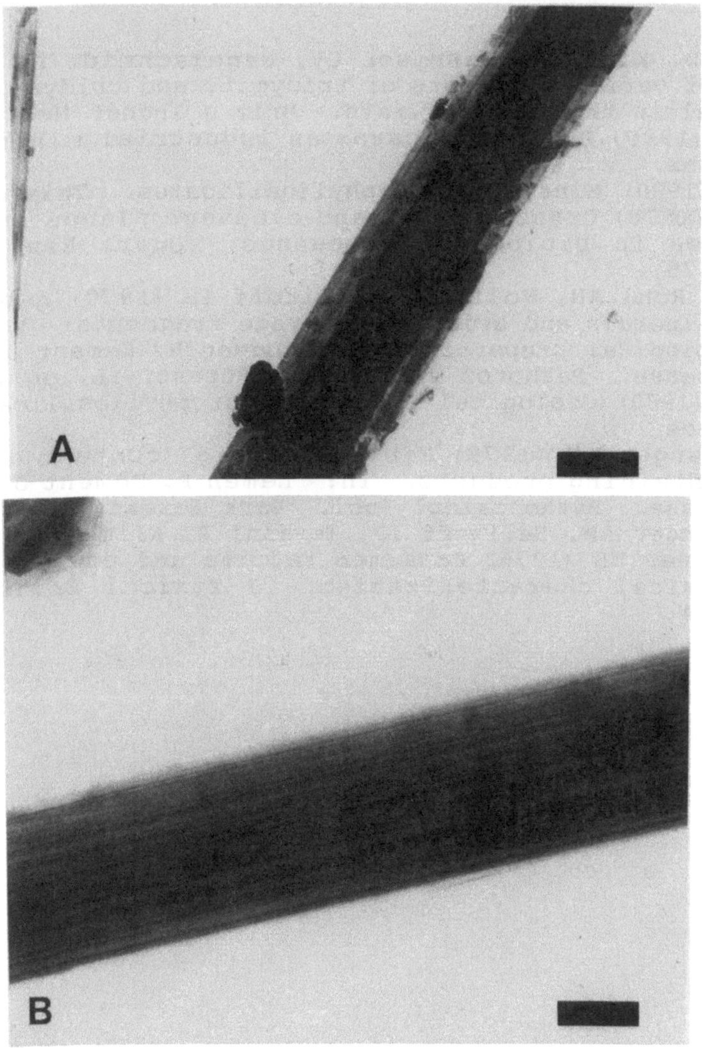

Figure 4: (A) and (B) Transmission electron micrographs of talc fibres formed from pre-existing fibrous minerals. The relative size of domains of talc to amphibole may range from those capable of resolution by light microscopy (A) to those resolved only by TEM. Bar in (A) equals about 0.8μm; bar in (B) equals about 0.25μm

References

Attygalle D, King EJ, Harrison CV, Nagelschmidt JG (1956) The action of variable amounts of tridymite and tridymite combined with coal in the lungs of rats. Brit J Indust Med 13:41-50

Clarke GM (1989) Phyllosilicates as industrial minerals. This conference

Felius RO (1989) Mineralogy of phyllosilicates. This conference

Langer AM (1978) Crystal faces and cleavage planes in quartz as templates in biological processes. Quart Rev Biophysics 11:543-575

Langer AM, Rohl AN, Wolff MS, Selikoff IJ (1979) Asbestos, fibrous minerals and acicular cleavage fragments: Nomenclature and biological properties. In: Lemen R, Dement J (eds) Dust and disease. Pathotox Publ. Park Forest, IL, pp 1-22

Meunier A (1989) Geological occurrence of phyllosilicates. This conference

Rohl AN, Langer AM (1979) Fibrous mineral content of consumer talc-containing products. In: Lemen R, Dement J (eds) Dust and disease. Pathotoxicol Publ. Park Forest, IL, pp 393-403

Rohl AN, Langer AM, Selikoff IJ, Tordini A, Klimentidis R, Bowes DR, Skinner DL (1976) Consumer talcums and powders: Mineral and chemical characterization. J Toxicol Environ Health 2:255-284

CONCENTRATIONS OF FIBRES IN PHYLLOSILICATES*)

K. Rödelsperger, B. Brückel, R. Patrzich, F. Pott+) and H.-J. Woitowitz
Institute and Outpatient Clinic for Occupational and Social Medicine of the Justus-Liebig-University University of Giessen, Aulweg 129/III, 6300 Giessen, Federal Republic of Germany

Introduction

Phyllosilicates resemble minerals of finest fibres like chrysotile, palygorskite and sepiolite; on the other hand platelet like phyllosilicates, e.g. talc, can be contaminated with tremolite fibres.

On the basis of serosa tests, Stanton (1981) and Pott (1987) suggested certain fibrous minerals as carcinogenic. Accordingly, the Commission for the Investigation of Health Hazards of Chemical Compounds in the Work Area of the Deutsche Forschungsgemeinschaft is discussing the classification of these fibrous minerals as agents with carcinogenic potency. To elucidate the consequences of such a classification, our group is studying the mineralogical composition of adsorption granulates, especially attapulgite resp. palygorskite and sepiolite, and their concentration of fibres longer than 5 um (Rödelsperger et al. 1987, 1988). Several samples of the latter materials have been tested in injection experiments (Rödelsperger et al. 1987, Pott et al. 1989).

Furthermore, the manufacturing and use of products containing asbestos is intended to be prohibited in the FRG. Products composed of

$\geq$ 0.02 weight-% chrysotile and
$\geq$ 0.005 weight-% amphibole

---

*) Supported by the German Federal Ministry of Science and Technology, Project No. 01HK076A6

+) Medical Institute for Environmental Hygiene at the Heinrich-Heine-University, 4000 Düsseldorf

NATO ASI Series, Vol. G 21
Health Related Effects of Phyllosilicates
Edited by J. Bignon

are regarded as asbestos containing (AGS, 1988). Thus, it is questionable, whether commercially available talc for cosmetic and technical purposes has to be classified as asbestos containing product (Rödelsperger et al. 1984, 1988, Henschler 1987). Moreover, the occurence of talc fibres longer than 5 um has to be considered.

Methods

Of each sample, 2 mg to 5 mg were suspended in 50 ml destilled water by sonification (about 20' to 30'). Aliquots were filtered with nuclepore filters of 0.2 um pore size. Subsequently, the specimens were sputtered with carbon, the filter material was dissolved in a Jaffee-washer and the remaining transmission sample was analysed at a magnification of 10000 times (TEM) for fibres with a length of $L \geq 5$ um. Fifty fields or 0.5 mm^2 of a 400 mesh copper grid were evaluated. To decrease the detection limit of actinolite/tremolite fibres in talc samples, which did not exhibit these latter fibres initially, the comparatively thick actinolite/tremolite fibres were counted in 100 additional fields of the grid. The attapulgite and sepiolite samples were also analysed for fibres of any length at a magnification of 29000 times (STEM). The identification of fibres in talc samples was carried out by means of qualitative and quantitative element analysis as well as x-ray diffraction. The diffraction pattern was quantitatively analysed for a portion of fibres (Miehe, 1988). Finally, the content of fibres $L \geq 5$ um with the optical qualities of actinolite/tremolite was determined in all samples of talc by the use of phase contrast microscopy.

Results

Concentration of attapulgite and sepiolite fibres

The attapulgite and sepiolite samples proved to be aggregated of fine fibres. Although the overwhelming majority of these fibres did not exceed 5 um, a smaller portion was longer than 5 um. The results of n=6 attapulgite and n=2 sepiolite samples investigated for carcinogenicity in rats as well as n=2 attapulgite and n= 3 sepiolite containing adsorption granulate samples are summerized in **Table 1.** The median of the fibre length was

L = 0.5 um to 1.3 um

for attapulgite and

L = 1.0 um to 2.9 um

Table 1: Range of the median of fibre dimensions and fibre concentrations in samples of intrapleurally and intraperitoneally injected attapulgite and sepiolite, which were investigated for carcinogenicity in rats (Rödelsperger et al. 1987, Pott et al., 1989). Comparison with fibrous adsorption granulates containing attapulgite or sepiolite. One attapulgite sample consisted of relative short fibres and one sepiolite sample of extraordinary long fibres (Rödelsperger et al. 1988).

Sample	carcinogenicity in rats	Range of results: fibre dimensions median L [um]	D [um]	L/D	fibre concentration of any length $\times 10^9$ fibres/mg	longer 5 um $\times 10^6$ fibres/mg
animal experiment						
attapulgite						
n = 4	-	0,5-0,9	0,04-0,07	7-20	0,9-72	0,06-0,61
n = 2	+	1,3-1,9	0,04-0,07	19-46	8,1-9,6	99-240
sepiolite						
n = 1	-	1,2	0,05	25	20	180
n = 1	+	2,9	0,05	64	19	5500
adsorption granulates						
attapulgite						
n = 2	not	0,7-1,3	0,03-0,05	20-29	71-110	1,8-26,4
sepiolite						
n = 3	tested	1,0-1,1	0,03-0,04	26-29	111-135	12,2-1240

for sepiolite. For both materials the median of the diameter varied between

D = 0.03 um and 0.07 um

and the ratio of length to diameter was calculated to vary between

L/D = 7 and 64.

Similar concentrations of fibres of any length were counted for both minerals. They were estimated between

0.9×10^9 fibres/mg and 135×10^9 fibres/mg;

this result as well as the fibre dimensions did not correlate with the findings on carcinogenicity obtained from the animal experiments. In contrast, the concentration of fibres longer than 5 um achieved a crucial importance if compared with results obtained from experimentally injected attapulgite and sepiolite in animals (see Pott et al., this symposium).

Analysis of mineral fibres in talc

Mineral fibres longer than 5 um were found in twelve talc samples in concentrations of 60.000 to 1.1 million fibres/mg. By the mineralogical analysis of n=374 fibres obtained from 10 talc samples, a proportion of only 8 % was detected as actinolite/tremolite and a portion of 5 % as chrysotile fibres. Beside of some chlorite (7%), the overwhelming majority of the other fibres consisted of talc fibres (56%). The remaining mineral fibres (24%) may resemble not identified talc fibres, however, other fibres as sepiolite could be also demonstrated. Asbestos fibres could be preferably demonstrated in talc samples for technical purposes, whereas talc samples for cosmetic purposes mainly contained talc- and other mineral fibres.

Concentration of asbestos fibres in talc

Figure 1 summarizes the concentration of chrysotile-, and actinolite/tremolite fibres in 1 sample of talc for pharmaceutical, 5 for cosmetic, and 6 for technical purposes. In former series at a magnification of 2500 times and a sensitivity per fibre of 500 fibres/mg (Rödelsperger et al., 1984) the concentration of actinolite/tremolite fibres was 1000 fibres/mg for the pharmaceutical sample (A') 5000 fibres/mg for one cosmetic sample (F') and 7000 fibres/mg for one technical sample (H'). In the present study no asbestos fibre was found in 4 cosmetic talc samples analysed at a magnification of 10000 times with a sensitivity of 1000 fibres/mg per one actinolite/tremolite fibre. The concentration of actinolite/tremolite as well as

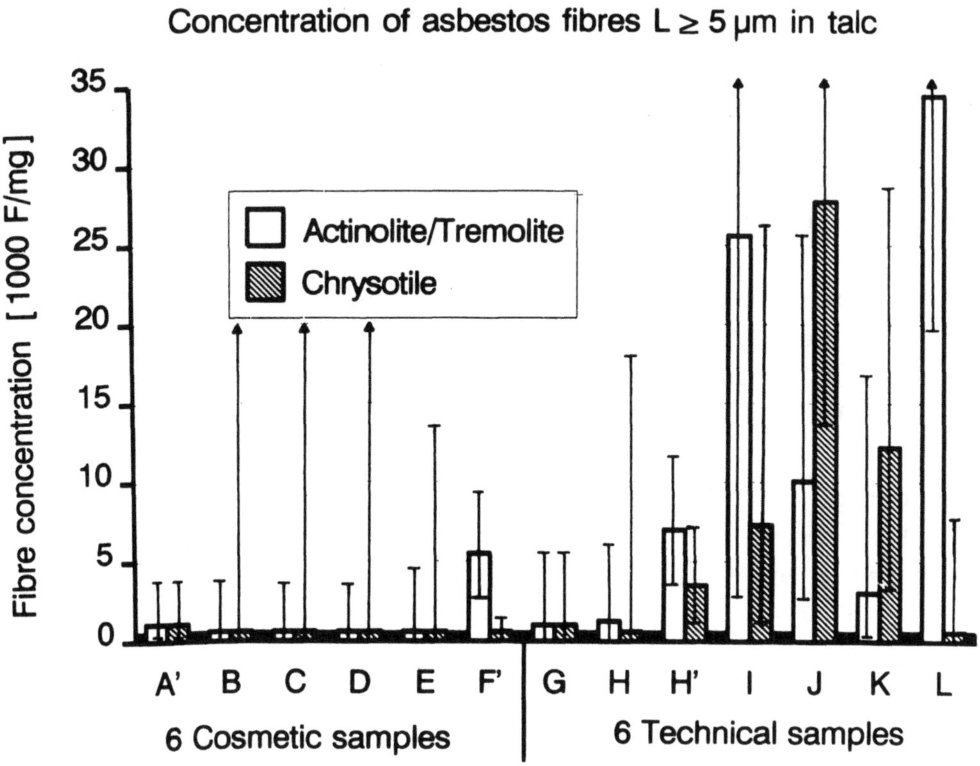

Fig. 1: Number of asbestos fibres longer than 5 um as revealed by means of TEM at a magnification of 10.000x in 6 talc samples for cosmetic and pharmaceutical as well as 6 talc samples for technical purposes. In a former investigation the samples A', F' and H' were analysed at a magnification of 2.500x. The concentrations of H and H' were measured using the same talc sample. The concentrations J and K resulted from two different samples of the same product. The error bars are estimated from the 95% confidence interval of the Poisson statistics.

chrysotile fibres in 6 of the technical talc samples was found to be up to 34000 F/mg (L) and 28.000 F/mg (J), respectively.

Discussion

Carcinogenicity of attapulgite and sepiolite fibres

Figure 2 shows the concentration of fibres longer than 5 um for samples of attapulgite and sepiolite which were tested for carcinogenicity intraperitoneally or intrapleurally in rats in relation to samples of adsorption granulates, which were not tested so far. For the precision of the results see Rödelsperger et al. (1987).

Attapulgite samples containing less then 10^6 fibres/mg longer than 5 um were found to be uneffective while concentrations of $\geq 100 \times 10^6$ fibres/mg yielded carcinogenic effects in the injection experiment.

The content of fibres longer than 5 um in a sample of Spanish sepiolite which is not shown to be carcinogenic was relatively high (180×10^6 fibres/mg); whereas a Finnish sample containing 5500×10^6 fibres/mg longer than 5 um was found to cause mesothelioma. Pott et al. (this symposium) discuss these different results in more detail.
In 4 of the 5 adsorption granulates, which were not tested so far (Figure 2, on the right) the concentrations of fibres longer than 5 um were lower than the concentrations yielding carcinogenic effects in the injection experiment. However, it was still higher, than the concentration found to be uneffective for attapulgite. It was clearly raised in the fifth adsorption granulate sample of sepiolite containing a concentration of 1240×10^6 fibres per mg. It is requested that biological effects of these materials should be tested before deposits are handled.

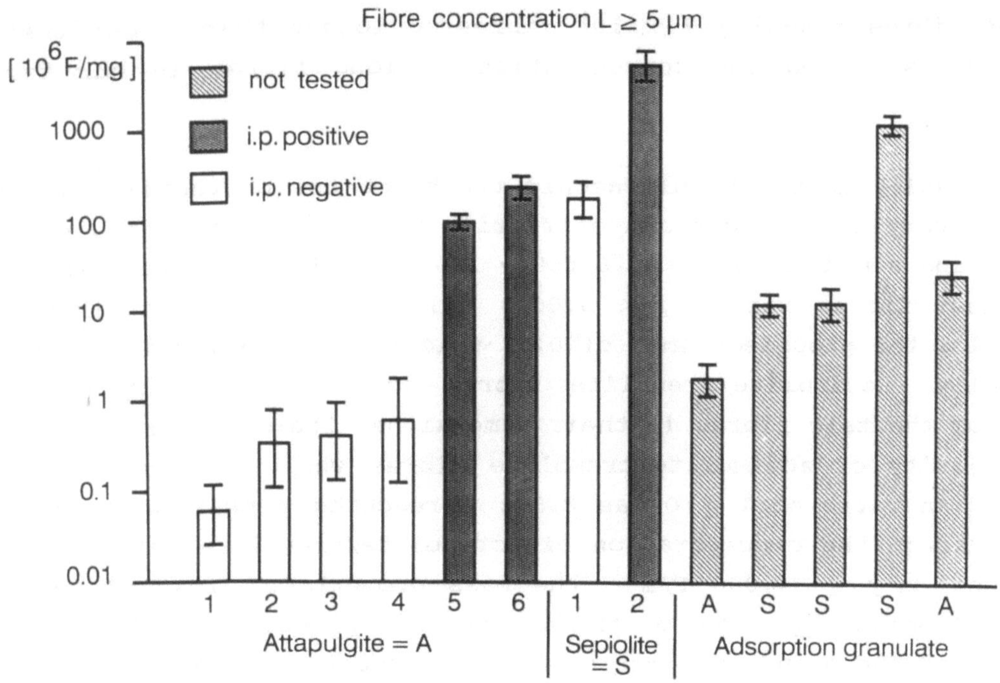

Fig. 2: Number of fibres with a length of L ≥ 5 um as revealed by TEM at a magnification of 10000x. Samples of attapulgite: Mormoiron 1, Lebrija 2, Georgia 3, 4, Torrejon 5, Cacéres 6. Samples of sepiolite: Spain 1, Finland 2. (For the intraperitoneal carcinogenicity studies of the attapulgite samples no. 1, 2, 3, 6 and the sepiolite samples no. 1, 2 see Pott et al. (this symposium), for the intrapleural carcinogenicity study of the attapulgite no. 5 see Wagner et al., 1987). Comparison with five adsorption granulates selected by means of SEM. The error bars are estimated from the 95% confidence intervals of the Poisson statistics.

Carcinogenicity of talc fibres

Although the majority of mineral fibres in talc samples consists of talc fibres, no carcinogenic effect of these fibres was suggested so far (Henschler, 1987). In rats, intrapleural administration did not indicate an increased occurence of tumours compared to control animals (Stanton et al., 1981). These authors tested seven talc samples with a concentration of fibres reaching 7 x 10^6 fibres/mg longer than 8 um; these samples exceed the concentration of long fibres in our talc samples.

Propably due to the ultrasonic treatment the chrysotile fibres are very likely composed of fibrils with a diameter of D < 0.1 um and a median ratio of L : D = 160 : 1. The estimated mass of these fibres was only < 0.0003 weight-%; this is strikingly below the proposed limit of 0.02 weight-% for chrysotile. Generally, actinolite/tremolite fibres do not remarkably differ from the talc fibres in their dimensions. The diameter D of the majority of actinolite/tremolite fibres was < 1 um and the median ratio of L : D was 8,5 : 1 reaching a peak of 67 : 1. Although the concentration of actinolite/tremolite fibres reached only ≤ 34.000 F/m^3 the mass proportion reached up to 0.18 weight-%. Two of the 6 technical products distinctively exceed the proposed limit of 0.005 weight-% for amphibole.

The identity of actinolite/tremolite fibres was investigated by both the elemental analysis and x-ray diffraction. The electron microscopical results could also be confirmed by the detection of fibres with the optical qualities of actinolite/tremolite as analysed with phase contrast microscopy. No hint was obtained, that these fibres were not related to an asbestiform structure (Breslow, 1984). Nevertheless, the identification of a small number of up to 16 actinolite/tremolite fibers in a sample of a certain commercial talc product does not generally allow the judgement of the weight content, the fibre concentration and the asbestiform habit of the actinolite/tremolite fibres of this product. Some of the actinolite/tremolite fibres appear to be asbestiform due to their raised length to diameter ratios. In contrast, the mass proportion is decisively determined by few thick fibres with a length to diameter ratio ≤ 10:1. On account of these difficulties and the expenses of this method analytical TEM-measurements cannot be propagated for routine examinations of the asbestos content in talc samples.Hence, the analytical results require a careful examination for asbestiform contaminants of commercially exploited talc deposits.

Literature

1. Ausschuß für Gefahrstoffe (AGS) (1988) Beschluß zur Umstufung von Asbest. Bundesarbeitsblatt 9: 84-86

2. Breslow L (Chairman) (1984) Asbestiform fibres. Nonoccupational Health Risks. Committee on Nonoccupational Health Risks of Asbestiform Fibres. National Academy Press, Washington D.C. 1-334

3. Henschler D ed. (1987) Talk (asbestfaserfrei). In: Gesundheitsschädliche Arbeitsstoffe. Toxikologisch-arbeitsmedizinische Begründung von MAK-Werten. Kommission zur Prüfung gesundheitsschädlicher Arbeitsstoffe der Deutschen Forschungsgemeinschaft. Verlag Chemie, Weinheim 1-18

4. Miehe G (1988) Program for Indexing Electron Diffraction Patterns (PIEP). Institut für Kristallographie und Mineralogie der Johann Wolfgang von Goethe-Universität, Frankfurt, personal communication, November 1988

5. Pott F, Ziem U, Reiffer FJ, Huth F, Ernst H, Mohr U (1987) Carcinogenicity Studies on Fibres, Metal Compounds and Some Other Dusts in Rats. Exp. Path. (Jena) 32: 129-152

6. Pott F (to be published) Intraperitoneal injection for assessing the carcinogenic potential of phyllosilicates. Paper on the first International Conference on Health Related Effects of Phyllosilicates, Paris, 16.-18.3.1989

7. Rödelsperger K, Lojewski HG, Brückel B, Woitowitz HJ (1984) Zum Fasergehalt von Pudern auf Talkumgrundlage. Staub-Reinhalt. Luft 44: 62-66

8. Rödelsperger K, Brückel B, Manke J, Woitowitz HJ, Pott F (1987) Potential Health Risks from the Use of Fibrous Mineral Adsorption Granulates. Brit.J.Ind.Med. 44: 337-343

9. Rödelsperger K, Sieben J, Patrzich R, Brückel B. Woitowitz HJ (1988) Talk, ein krebserzeugender Stoff im Sinne der Gefahrstoff-Verordnung? Verh. Dtsch. Ges. Arbeitsmedizin e.V., 28. Jahrestagung der Dtsch. Ges. f. Arbeitsmed., Innsbruck. Gentner, Stuttgart: 577-580

10. Rödelsperger K, Brückel B, Woitowitz HJ, Pott F, Strübel G (to be published) The proportion of long fibres in attapulgite and sepiolith-containing adsorption granules. Paper on the VIIth International Pneumoconiosis Conference.Pittsburgh, 23.-26.08.1988

11. Stanton MF, Layard M, Tegris A, Miller E, May M, Morgan E, Smith A (1981) Relation of Particle Dimensions to Carcinogenicity in Amphibole Asbestos and other Fibrous Minerals. J.Nat.Canc.Inst. 67: 956-975

12. Wagner JC, Griffiths DM, Munday DE (1987) Experimental Studies with Palygorskite Dusts. Brit.J.Ind.Med. 44: 749-763

TALC - A MORPHOLOGICAL STUDY ON PSEUDOMORPHS

U.M. Schenk [1], P.Golob [2], H.Kolmer [3]

ABSTRACT: Four pseudomorphic elongated talc specimens from different countries (supplied by EUROTALC) are analyzed. In the 'dimension relevant to health' (approx. < 15 microns) all samples exhibit talc feature only (platy, soft, inelastic, bendable). Properties of rare 'elongated talc particles' (ribbons, fibers, etc.) originating from cleavage fragments, disorder in the crystal lattice or pseudomorphosis are described in contrast to different features of amphibole fibers. The urgent need for a dust bank with well-defined standard samples for common research is emphasized.

In contrast to geoscience, which classifies talc as a platy phyllosilicate (e.g. STRUNZ 1978), other sources, mainly medical literature, emphasize a somehow 'fibrous habit' (STANTON et al. 1981). Although EUROTALC (The Scientific Association of European Talc Producers) knows more than 150 talc mines all over the world, socalled 'fibrous talc' in any relevant proportion is unknown apart from one exception, the Gouverneur talc district, New York State (ROSS et al. 1968, VIRTA 1985).

However, EUROTALC's efforts to find 'fibrous talc' brought up four atypically large specimens (fig. 14) of talc pseudomorphic after amphiboles. Macroscopically they all show an elongated habit and the authors were asked for analyses of these rare samples.

Besides the morphological study several questions such as the basic talc features in contrast to other minerals, especially the fibrous minerals, are discussed. Attention is drawn to the problem of insufficient availability of reference samples for common research; research, which might be of relevance to the debate on environment and health regulations.

1 Talc features

Talc, theoretical composition $Mg_3Si_4O_{10}(OH)_2$, belongs to the large group of phyllosilicates. Together with pyrophyllite, its aluminium equivalent, it is the basic form of 2:1 layer silicates. In its crystal structure assemblage a central sheet of octahedrally coordinated cations is sandwiched between two SiO_4-tetrahedra sheets whose tips are pointed inward (fig. 1). The oxygen atoms at the tetrahedra tips are shared between the octahedral and tetrahedral sheets thus combining all three sheets to a planar layered structure.

Since all hexacoordinated sites are occupied by bi-valenced magnesium ions, talc is classified as a tri-octahedral layered mineral. The structural study by RAYNER & BROWN (1973) confirmed its 1-layer triclinic structure.

Because of the absence of any interlayer material, adjacent oxygen planes can come close together and there are no geometrical restrictions for the layer stacking as with other 2:1 minerals. GIESE (1975) concluded that beside van der Waals' forces a small but in no way negligible ionic attraction also contributes to the bonding of the uncharged layers.

The low bonding forces account for the extreme softness (Mohs' hardness 1) which imparts a soapy and greasy feel commonly referred to as 'slip'. Talc is usually found in massive, foliated aggregates of thin, sometimes hexagonal plates. The single crystals range from less than one

1) Naintsch Mineralwerke, Mariahilferplatz 5, A-8020 Graz, Austria.
2) Research Institute for Electron Microscopy, Univ. of Technology, Steyrergasse 17, A-8010 Graz, Austria.
3) Inst. for Eng. Geology and Appl. Mineralogy, Univ. Technology, Rechbauerstraße 12, A-8010 Graz, Austria.

NATO ASI Series, Vol. G 21
Health Related Effects of Phyllosilicates
Edited by J. Bignon

to several hundred microns. The individual crystals easily move along the basal cleavage plane (001) into thin, flexible but inelastic plates (fig. 2, fig. 5). The colour of talc is white, grey or various shades of green with a silky lustre. Extremely rare among minerals is the hydrophobic character of talc. Further there is a remarkable affinity to organic materials even in the presence of water. Another feature is its little variation in composition and its inertness to attack from most chemical reagents. The paragenesis of large talc deposits is always in connection with a metamorphic alteration of other magnesium rich rocks. Typical precursors of talc schists are siliceous dolomitic limestones or ultramafic rocks. According to the grade and type of metamorphism (pressure of CO_2 and water, temperature) different mineral assemblages of talc together with members of the carbonate group, the amphibole group, chlorites, quartz and serpentine minerals can be formed (WINKLER 1979).

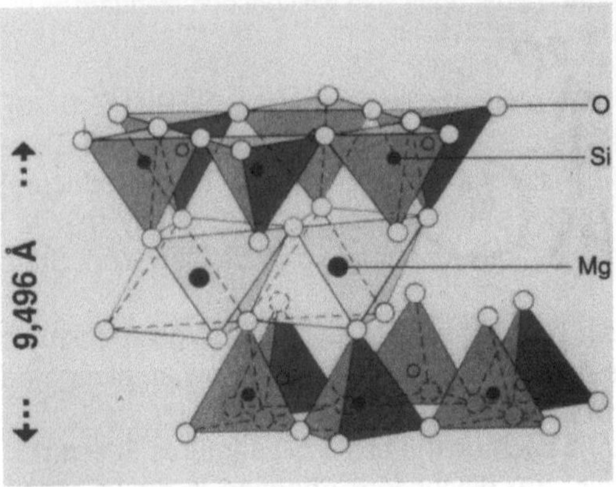

Fig.1: Talc crystal lattice

Fig.2: SEM, micro-crystalline talc, Spain

A long list of industrial applications would include paper, paints, ceramics, roofing materials, plastics, rubber, food, feedstuffs, cosmetics and pharmaceuticals (ROE & OLSON 1983). The great versatility of talc is explained by its physical and chemical properties which in turn result from the crystal structure and the composition.

2 Typical talc morphology / particle size relevant to health

Macro-crystalline talcs show their lamellar habit even on a centimeter scale (fig. 5).

Micro-crystalline talcs may be blocky or granular down to the size of 100 or 15 µm, revealing their typical platy morphology only below that dimension (fig. 2). In contrast to micas and many other phyllosilicates, the soft talc platelets are easily deformed during grinding and processing. Soft, large bends, uneven surfaces, slightly curved or rounded sides and bent or 'rolled' edges are typical features (fig. 3). The phyllosilicates chlorite and pyrophyllite have similar properties. Micas - in contrast to talc - normally have flat, even surfaces and sharp rasor-like sides. Mica particles practically do not show permanent bends. They are highly elastic and - when bent - spring back into the original flat form or may break forming two, plane particles (fig. 3, 6). Three dimensional projections avoid confusing platelets for fibers (cf. arrow in fig. 6).

Dimension relevant to health (DRTH): when evaluating the morphology of different types of talc, the morphology in the socalled macro size might differ quite substantially from the habit in the micro sizes. The term 'dimension relevant to health' of approx. < 15 µm was chosen arbitrarily merely to cover all dimensions in the respiratory system below that size. Other terms might be more adequate when evaluating for example fractions of airborne dusts.

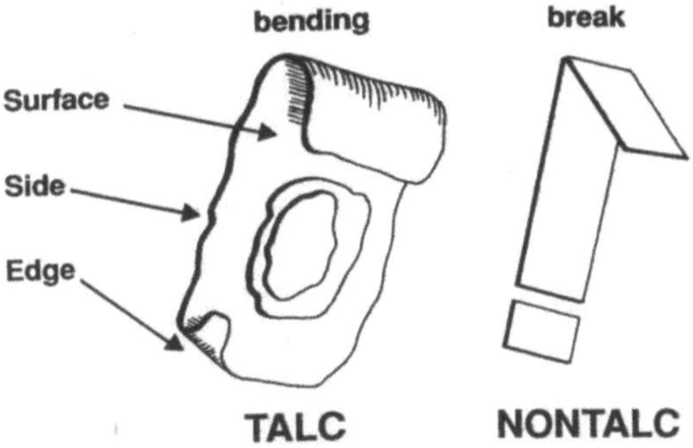

Fig.3: Talc and non-talc morphology

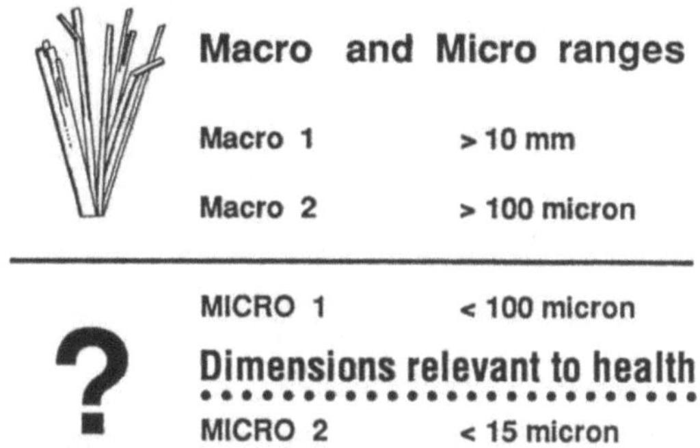

Fig.4: Dimensions relevant to health

Fig.5: SEM, macro-crystalline talc, Austria

Fig. 6: SEM, phlogopite, Canada

3 Pseudomorphism

It is not uncommon for one mineral to replace another while maintaining the external form of the first mineral. Such processes lead to a pseudomorph of one mineral after another (FRYE 1974). It is important to notice that when the first mineral is replaced by another, all chemical and physical properties are also replaced. A hard mineral might be transformed into a soft one , a brittle mineral into a ductile one, a hydrophilic into a hydrophobic one. In connection with talc there can be found quite a number of pseudomorphs such as pseudomorphic talc after dolomite, after magnesite, after quartz, etc. Regarding the discussion about 'fibrous talc', some groups of pseudomorphs might be of interest:

** Pseudomorphs after fibrous minerals in general

** Pseudomorphs where a 'fibrous habit' is not only found at the external macro form, but also after grinding of such a material.

4 Tremolite and its morphological varieties

When evaluating elongated and very elongated talc particles the morphology is compared with that of other minerals. As the mineral tremolite might occur together with talc, this example has been chosen because it is also of the highest morphological interest. Basic literature about the

Fig.7: Fibrous and blocky tremolite.

Fig.8: SEM, tremolite, NIOSH- ref. sample T-77

asbestos minerals and their non-asbestiform analogs, as well as mineralogical and morphological definitions and identification characterizations were published by AMPIAN (1976), CAMPBELL et al.(1977, 1979).

Ever since the different natural habits of tremolite were described and classified as

** *blocky*
** *prismatic*
** *acicular*
** *fibrous (non-asbestiform)* and the
** *asbestiform variety* (found in very small quantities only)

they have rarely been confused or ignored.

That there are still grey areas between these terms is of minor importance in this connection. Examples of blocky and asbestiform tremolite are shown in fig.7, whereas the particles in fig. 8 could be classified as elongated prisms and non-asbestiform.

5 Elongated talc platelets, ribbons, fibers and their physical properties

Elongated platelets in the 'dimension relevant to health' can sometimes be found with micas, with clays and with talcs. These elongated platelets can even reach the form of socalled laths, thin broad or narrow ribbons, the form of fibers and even thin fibrils might occur. These elongated talc particles in general may originate from three sources:

** *from fragments* (cleavages) from larger platelets during grinding etc.
** *from disorder in the crystal lattice* (AMELINCKX, DELAVIGNETTE 1961)
** *from pseudomorphosis* when a fibrous or asbestiform amphibole was transformed into talc maintaining, however, the fibrous habit down to the micro dimension of less than approx.15 microns. A seldom case of very narrow (approx. 1000 Å) talc fibrils - pseudomorphic after anthophyllite was described by VIRTA (1985) in particles from the Gouverneur Talc District in New York. A lath-like talc, pseudomorphic after chrysotile was described by SOUZA SANTOS and YADA (1988).

Whatever the origin of all these elongated talc particles, their chemical and physical properties are completely different from those of amphiboles. These talc particles are of the highest softness. The single crystals forming these particles are somehow mobile, producing the known gliding and lubricating effect. This property is not restricted to talc platelets only, but can be easily verified with lath-like and even with thin, narrow, ribbon-like particles (fig. 9, 10).

They have no elasticity at all, can be easily bent when thin, are deformable as a whole or at their edges and ends. The physical stability or durability in shape is very low.

Fig.9:SEM, elongated talc particle

Fig. 10: SEM, ribbon-like talc particles

The amphiboles - in contrast to talc - vary in hardness from 4 to 6 according to the Mohs' scale, i.e. amphiboles are several orders of magnitude harder than talc. Amphibole fibers (e.g. actinolite, tremolite, anthophyllite) may be flexible in terms of elasticity like springs, or brittle, but never gently bendable or ductile like talc. The former are commonly of high mechanical stability, durability, tensile strength and abrasivity.

6 Special phenomena from New York State

The unique situation around the Gouverneur talc district in New York State containing minor quantities of socalled 'fibrous talc' together with platy talc, non-fibrous tremolite and other minerals needs special consideration. About 80% of the talc literature regarding fibers in connection with talc is related to this interesting area. A great variety of assemblages of talc with several amphiboles has been found and studied (ROSS et al. 1968, VEBLEN 1980, VEBLEN and BUSECK 1981, VIRTA 1985, FELIUS 1986). Basic research was done on:

** *pseudomorphic talc after tremolite,*
** *intergrowth of talc and anthophyllite,*
** *pseudomorphic talc after anthophyllite.*

The phenomenon of such thin, long talc 'fibers' - pseudomorphic after anthophyllite - which might reach several hundred microns in length was described by VIRTA (1985). Even talc fiber bundles with splayed ends were found which could be morphologically of misleading similarity to asbestos, as long as the particles are not distorted, or bent.
It is evident that *these fibers in reality have nothing to do with asbestos* in their mineralogical, chemical and physical properties.
The carcinogenic potential effect of 72 mainly fibrous materials was studied by STANTON et al. (1981).Fibrous talc showed a 'substantially lower than expected response'.

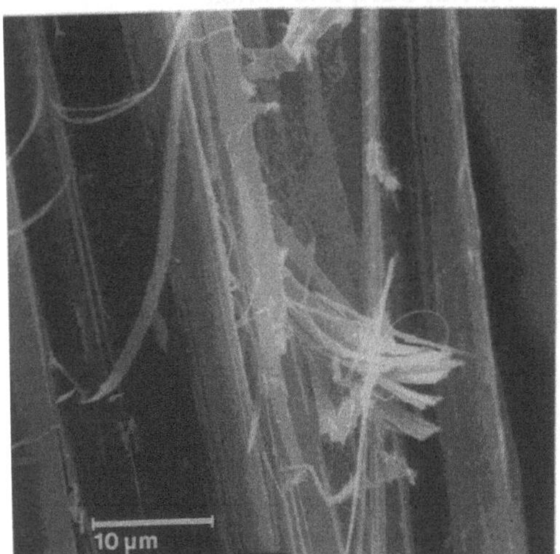

Fig.11:SEM,pseudomorphic talc after anthophyllite,NY

Further discussion about the health effect (KLEINFELD et al. 1967, LAMM et al. 1988) of materials of the Gouverneur region, containing mainly non-fibrous tremolite together with talc, continues. It is highly interesting to follow up the development of this research about morphology, and about the mineralogical, chemical, physical and especially the biological effects.

7 Pseudomorphic talc after anthophyllite from Austria

In a series of metamorphic rocks garnet-mica schists alternate with marbles. Within these ankeritic carbonate layers TUFAR (1979) observed minute radial shaped crystal aggregates of about 10 mm of anthophyllite together with talc. Anthophyllite was altered partly or completely into talc lamellae (and not into fibers) by retrograde metamorphosis. The orientation of the platelets is parallel to the length axis of the preceeding amphibole. Not grinding processes, but post-crystalline deformation caused the bending of the talc platelets (fig. 12, 13).

Fig. 12: SEM, pseudomorphic talc after anthophylite

Fig. 13: SEM, post-crystalline deformed talc

8.0 Four talc pseudomorphs from Brazil, Scandinavia, USA, and Africa

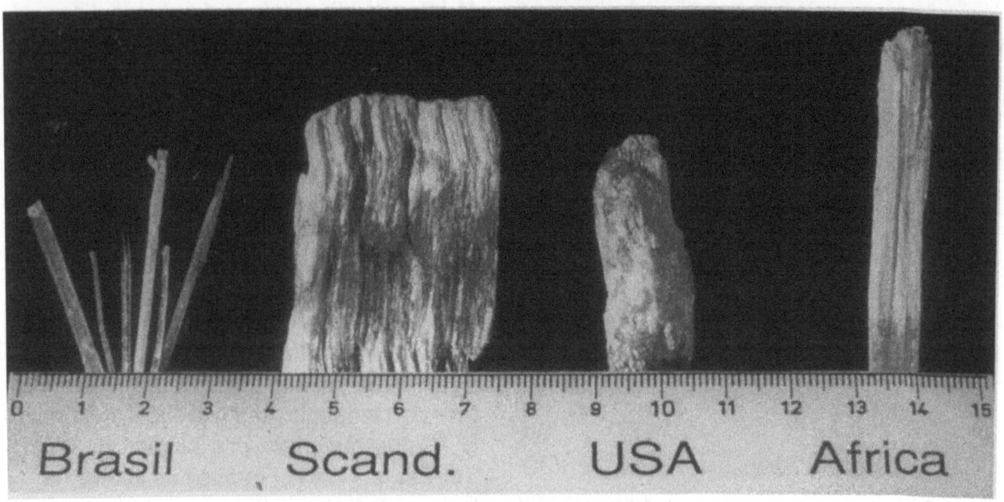

Fig. 14: The four talc pseudomorphs provided by EUROTALC

These samples, provided by EUROTALC, were carefully analyzed and a morphological study carried out stepwise from macro size down to the 'dimension relevant to health'. All 4 samples had a pronounced, elongated structure in the macro size.
Crucial question: what is the morphology in the 'dimension relevant to health' ?

8.1 The Brazilian pseudomorphic talc

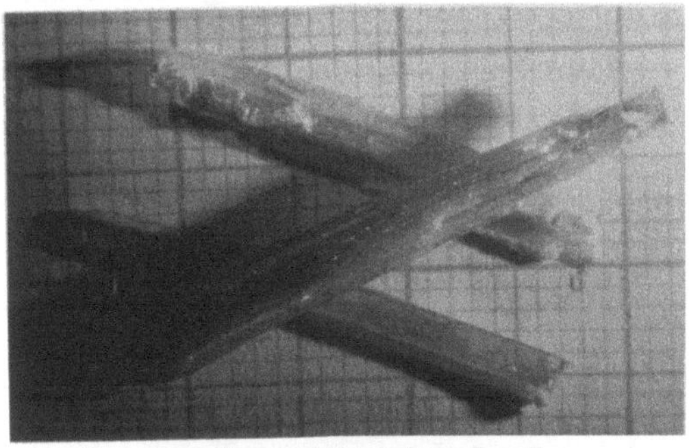

Fig. 15 a: Pseudomorphic talc from Brazil.

Fig. 15 b: SEM, pseudomorphic talc, Brazil, DRTH

The specimen (fig. 14) is a fragment from a larger, stick-like piece of about 170 millimeters in length. It is composed of strongly and loosely coherent bundles of "macro ribbons" (fig. 15a) of approx. 1 mm width. These thick macro ribbons had typical talc features, i.e.were soft, inelastic, bendable, pliable and easy to delaminate into thinner macro ribbons. In the 'dimension relevant to health' only platy talc, oriented parallel to the length axis without any ribbon-like habit, could be observed (fig. 15b).

8.2 The Scandinavian pseudomorphic talc

The specimen seems to be part of a larger sample. The preceeding 'fibrous' mineral was composed of 'fibers' pointing in three slightly different directions. On a centimeter scale plate-

Fig. 16 a: Pseudomorphic talc, Scandinavia.

Fig. 16b:Pseudomorphic talc Scandinavia, DRTH

lets or ribbons cannot be separated, but only pieces in very compact or elongated form (fig. 16a). This pronounced, very elongated habit almost disappears on a millimeter scale and only slightly elongated or almost equidimensional particles are predominant. Below the dimension of 200 microns a platy habit starts to appear. Typical talc morphology with platy habit is the charasteristic for the 'dimension relevant to health' (fig. 16b). The orientation of lamellae is parallel to the length axis.

8.3 The USA pseudomorphic talc

The USA pseudomorphic talc still shows the 'fibrous' structure of the preceeding mineral quite well (fig. 17a). Although the specimen is of very compact structure, elongated, short, thick particles of several millimeters can be separated here and there.

When grinding down to sizes of 100 or 50 microns, blocky grains, typical talc platelets as well

Fig. 17 a: Pseudomorphic talc, USA

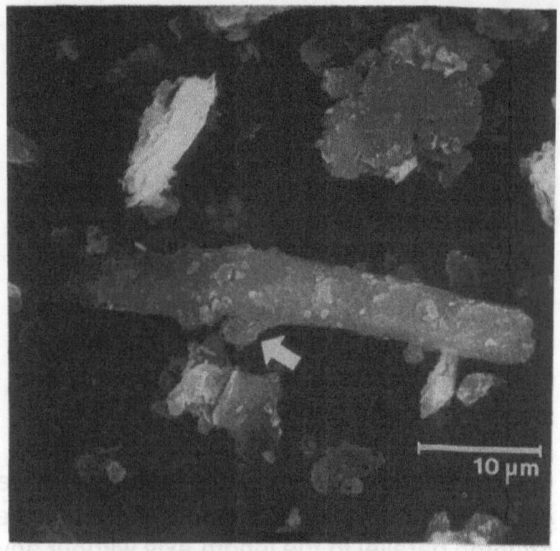

Fig.17 b :SEM , pseudomorphic talc, USA

as elongated particles appear together. One of these elongated, flat particles is shown in fig. 17b. The break (see arrow) amplified in fig. 17c, shows the platy talc crystals which

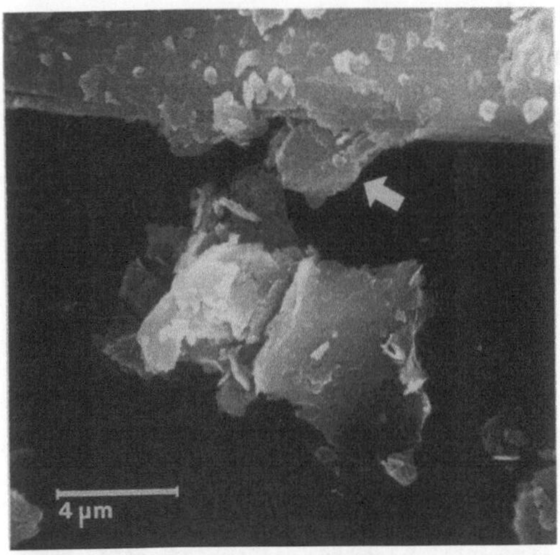

Fig. 17 c: SEM, same as 17 b, enlarged.

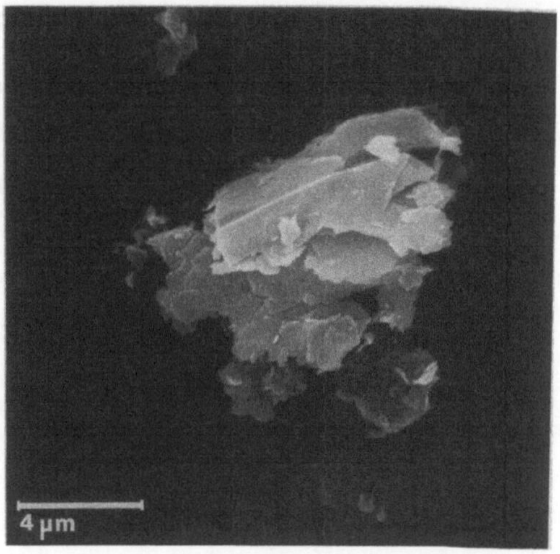

Fig. 17 d: SEM, pseudomorphic talc, USA, DRTH

build up the whole particle. Although this specimen shows some elongated habit down to the 50 microns size, there was no tendency of splayed ends or fibers in the 'dimension relevant to health', but typical talc features only (fig. 17d). Again, the platelets were oriented parallel to the length axis.

8.4 The African pseudomorphic talc

Fig. 18a: Pseudomorphic talc, Gabun

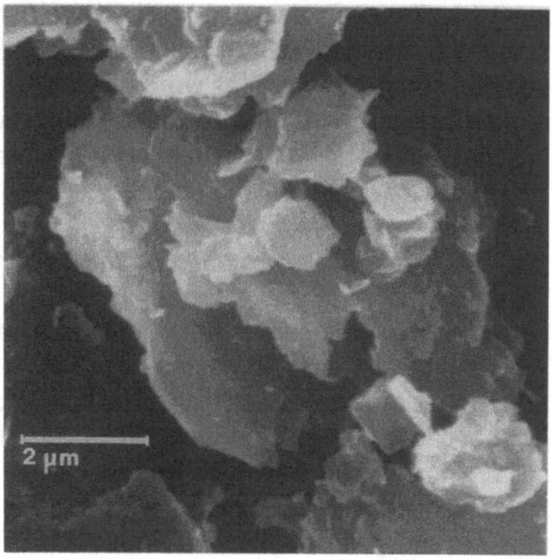

Fig. 18 b: SEM, pseudomorphic talc, Gabun, DRTH

Bundles of talc ribbons 0.5 - 2 mm wide and 65 mm long form the specimen from Gabun. The end of the specimen (fig. 18a) - very sensitive to touch - shows soft, easy to peal off pseudomorphic talc ribbons. High softness, no elasticity, bendability, pliability, etc, are the features. When grinding down, the ribbon-like morphology disappears continuously. In the 'dimension relevant to health' typical talc morphology is predominant (fig. 18b) and the platelets again are oriented parallel to the length axis of the ribbons.

9 Pseudomorphic talc - conclusions

The four talc specimens supplied by EUROTALC showed a pronounced, elongated habit in the macro dimension. When carefully ground this habit gradually disappeared, whereby in some samples ribbons of talc were observed down to a sub-millimeter scale. In the 'dimension relevant to health' (approx. < 15 microns) all samples uniformly revealed typical talc features only. The orientation of the platelets in all cases was parallel to the length axis. This corresponds with other observations on talc pseudomorphs after anthophyllite, from the Eastern Alps (TUFAR 1979) and from New York State (VIRTA 1985). Although traces of possible precursor minerals were not detected in the four specimens above, it is probable that also in these cases fibrous amphiboles (tremolite or anthophyllite) were involved. The formation of talc, pseudomorphic after amphibole minerals, took place under conditions of retrograde metamorphosis.
Under suitable geologic-petrologic conditions it can happen that the pseudomorphic talc maintains the habit of a fibrous starting mineral even down to the 'dimension relevant to health'. This would explain the trace occurrence of elongated particles in some talc products.
The fundamentally different properties of these elongated particles in general in comparison with amphibole fibers should be emphasized.
Talc particles with aspect ratios > 1 : 10 are rarely found amongst the normal particle population. Although EUROTALC members produce about one million tpy of talc, they are not in a position to provide as little as 500 grams of 'fibrous talc' for further studies.

10 Discussion / European-American dust bank

Analyzing the samples, the authors found themselves in an unpleasant and strange situation. On the one side they were confronted with an ambiguous mineralogical nomenclature, with misleading or improper definitions of fibers. On the other side necessary details about the 4 pseudomorphic talc pieces regarding geological and mineralogical circumstances were not available. Two samples were picked at talc mines just as curiosities, which is why the circumstances of the find cannot be traced back in detail. The 2 other samples from well-known, reliable, serious companies were given only on condition that name, origin and details should remain undisclosed. One of these analysed samples even had to be given back to avoid misuse, and so on. What is behind this unusual situation ?

** *Is it lack of confidence in science ?*

** *Is it fear of unfavourable results ?*

** *Is it fear of misinterpretation or improper generalization as has happened in other cases ?*

We know that the circumstances in Europe vary from country to country. Fear and mistrust amongst talc producers is not unique at all and not linked to this industry alone.

Why does such fear and mistrust exist at all ?

Although talc is the softest mineral in the world and has other quite unique, desirable properties, it might cause fibrogenic effects in overdose like other minerals, although the overdose level is still unknown. Talc, however, has never been found to be carcinogenic.

It was verified with the 4 elongated pseudomorphs from EUROTALC and with a pseudomorphic talc after anthophyllite from Austria (TUFAR 1979), that the previous, probably dangerous asbestiform amphiboles were transformed into soft talc platelets.

So there is no reason for fear.

What, if the morphology had turned out to be non-platy in the 'dimension relevant to health' ?

What if it had turned out to be of elongated habit, ribbon-like, 'fiber-like', having an aspect ratio of about 1 : 10, or even a higher one such as those rare talc particles from New York State described by VIRTA ?

Besides the fact that talc particles with an aspect ratio larger than 1 : 10 are extremely rare, there is no reason for fear because these 'fibers' have completely different characteristics in comparison with those of asbestos fibers.

There is no reason or there should be no reason for panic that these soft, inelastic, elongated particles will be classified as potentially carcinogenic on a scientific basis.

It is evident that this particular fear or mistrust makes co-operative research between science and industry very difficult and sometimes impossible.

Can't this be changed ?

There are so many open questions from industry, science and health regulatory bodies, which can only be solved in co-operation. One of these questions is:

Talc or talc together with aluminium containing chlorite is supposed to have a certain anti-synergetic effect in connection with some byminerals like QUARTZ for example - a certain buffering or masking effect. This important potential effect has never been properly verified although data and unpublished observations lead one to expect it.

To verify these basic questions, like many others about other minerals or man-made products, well-defined samples in sufficient quantities available for all scientists and interested people are necessary.

Describing the general situation, POTT (1980) refers to comments on the 1972 Lyon conference on asbestos by WAGNER & BERRY (1973), who called it disappointing that too few experiments were carried out with the UICC reference samples. In 1980 POTT adds: 'the situation has not changed since'. One of the many reasons for this state of affairs described by POTT in 1980 and 1988 is that efficient research is extremely difficult because interpreting results from ill-defined samples which are not available for all researchers is a big handicap. He therefore proposes a dust bank.

Today, 18 years after the Lyon conference, the situation in Europe is in general still the same. If scientific research has anything to do with efficiency, there are numerous reasons to spend

money on establishing a dust bank very soon. This is in the interest of the consumer, the producer and user industries, scientific researchers and so on.
Environmental and health regulatory bodies are also facing a great problem since they have to force scientists to come to conclusions based on insufficient data- a very unpleasant situation involving extremely negative side-effects for everybody.
It increases confrontation between industry, science, regulatory bodies and environmentalists, instead of giving a chance for dialogue and much better co-operation to the benefit of all parties concerned.
Those being familiar with sample preparation in research labs at universities and in industry are aware of the difficulties and the criteria which might influence both the samples and the results. There is an almost discouraging multitude of possible variables such as mineralogical circumstances, the many ways of grinding (influencing surface properties, morphology, etc.), separating, fractioning, homogenizing, plus meeting the requirements for preparing a vast range of samples.
In sample preparation, the analytical work should ideally be undertaken in parallel. It is a laborious process and should not be underestimated.
An example of excellent co-operation was given by the U.S. Bureau of Mines, the NIEHS (National Institute of Environment Health Science), industry and universitites, preparing large samples of chrysotile, amosite, crocidolite and tremolite for oral ingestion studies by NIEHS (CAMPBELL et al. 1980).
One cannot appreciate enough the efforts made to produce the UICC asbestos samples, the very large dust bank from NIOSH, which makes available free of charge small amounts of various minerals to be used as reference minerals for electron- and optical microscopy, X-Ray diffraction analyses, etc.
Such programs need the expertise and skill of all, i.e. scientists in general, universities, industry and experienced program co-ordinators.
There is an urgent need for a basic bank for dusts and other relevant substances and there is no question that working with standardized samples in the long run is not only more economic but also much more scientific as well. These programs should be elaborated on and executed with the broad co-operation of the parties concerned including norming agencies in the various countries. Considering how many people and committees need to be motivated can be quite discouraging. But can we afford to continue without a basic change ?
Can we wait another 18 years and leave the solution of one further problem to our children ? We certainly cannot build up a worldwide dust bank tomorrow.
Why not start with some more simple projects ?
Greater co-operation with our North-American colleagues, who are much more advanced in this respect, would facilitate a more rapid realization of some common pilot projects. This could be the first step on the road to a wider European - North American, and ultimately global dust bank and bank for other relevant substances.

Acknowledgements
We thank Professor W. TUFAR, University Marburg, Germany, for providing photos (fig.12, 13). We also thank Dr. F. PRÜGGER, Primarius, Sonderkrankenanstalt für interne Berufskrankheiten der AUVA, Tobelbad - Graz, Austria, for his valuable suggestions.

KEYWORDS: amphiboles / anthophyllite / asbestiform / cancer / dustbank / Eurotalc / fiber / fibrosis / health / morphology / platelet / pseudomorphy / softness / talc / tremolite

References:

Amelinckx S, Delavignette P (1961) Electron microscope observation of dislocations in talc. J Apll Phys 32: 341-351

Ampian SG (1976) Asbestos minerals and their nonasbestiform analogs. In: Asher IM, McGrath PP (eds) Proc.1st FDA office of science summer symposium on Electron microscopy of microfibers, Pennsylvania State Univ, Univ Park, PA, Aug. 23-25, 27 p

Campbell WJ, Blake RL, Brown LL, Cather EE, Sjoberg JJ (1977) Selected silicate minerals and their asbestiform varieties, mineralogical definitions and identification-characterization. Information Circular/1977 - IC8751. US Bureau of Mines, 56 p

Campbell WJ, Steel EB, Virta RL, Eisner MH (1979) Relationship of mineral habit to size characteristics for tremolite, cleavage fragments and fibers. Rep Investigations/1979-RI8367. US Bureau of Mines, 18 p

Campbell WJ, Huggins CW, Wylie AG (1980) Chemical and physical characterization of amosite, chrysotile, crocidolite, and nonfibrous tremolite for oral ingestion studies by the National institute of environmental health sciences. Rep Investigations/1980-RI8452. US Bureau of Mines, 63 p

Felius RO (1986) EM-Research into the fibrous constituents of an industrial talc powder. In: Abstr 14th General meeting IMA, Stanford Univ, Standford CA, July 13-18: p 97

Frye K (1974) Modern mineralogy, Prentis Hall, New Jersey

Giese RF jr (1975) Interlayer bonding in talc and pyrophyllite. Clays and Clay Miner 23: 165-166

Kleinfeld M, Messite J, Kooyman O, Zaki MH (1967) Mortality among talc miners and millers in New York State. Arch Environ Health 14: 663-667

Lamm SH, Levine MS, Starr JA, Tirey SL (1988) Analysis of excess lung cancer risk in short-term employees. Am J Epidemiol 127: 1202-1209

Pott F (1980) Animal experiments on biological effects of mineral fibres. In: Wagner JC (ed) Biological effects of mineral fibres. IARC Publ 30, 1: 261-272

Pott F (1988) Die krebserzeugende Wirkung anorganischer Fasern im Tierexperiment - Daten und Bewertung. Umwelthygiene 20: 97-134

Rayner JH, Brown G (1973) The crystal structure of talc. Clays and Clay Miner 21: 103-114

Roe LA, Olson RH (1983) Talc. In: Lefond SJ (ed) Industrial minerals and rocks 5th edn. SME-AIME: 1275-1301

Ross M, Smith WL, Ashton WH (1968) Triclinic talc and associated amphiboles from Gouverneur mining district New York. Am Mineral 53: 751-769

Souza Santos H de, Yada K (1988) Thermal transformation of talc as studied by electron-optical methods. Clays and Clay Miner 36: 289-297

Stanton MF, Layard M, Tegeris A, Miller E, May M, Morgan E, Smith A (1981) Relation of particle dimension to carcinogenicity in amphibole asbestoses and other fibrous minerals. J Nat Cancer Inst 67: 965 - 975

Strunz H (1978) Mineralogische Tabellen, 7th edn. Akad Verlags Ges, Leipzig

Tufar W (1979) Anthophyllit und Talk von Vorau (Oststeiermark). Joanneum Mineralog Mbl 47: 37-50

Veblen DR (1980) Anthophyllite asbestos: microstructures, intergrown sheet silicates, and mechanisms of fiber formation. Am Mineral 65: 1075-1086

Veblen DR, Buseck PR (1981) Hydrous pyriboles and sheet silicates in pyroxenes and uralites: intergrowth microstructures and reaction mechanisms. Am Mineral 66: 1107-1134

Virta RL (1985) The phase relationship of talc and amphiboles in a fibrous talc sample. Rep Investigations/1985-RI8923. US Bureau of Mines, 11 p

Wagner JC, Berry G (1973) Investigations using animals. In: Bogovski P, Gilson JC, Timbrell V, Wagner JC (eds) Biological effects of asbestos. IARC Publ 8: 85-88

Winkler HGF (1979) Petrogenesis of metamorphic rocks, 5th edn. Springer, New York

THE PHYSICAL CHARACTERIZATION OF MUSCOVITE AND KAOLINITE DUSTS

S.B. Schyma
Institut für Hygiene und Arbeitsmedizin der Medizinischen Fakultät der Rheinisch-Westfälischen Technischen Hochschule Aachen,
Neues Klinikum, D-5100 Aachen
Federal Republic Germany

An exact definition of the characteristic values of a dust is of fundamental importance in animal experimentation. Every dust sample for the intratracheal or intraperitonal animal experiment must first be classified to gain specific, possibly homogeneous fractions.

Material and methods

The very pure dust sample of muscovite without quartz contamination, derived from Lenkersmühle (Upper Palatinate, West Germany) was ground in the agate mill and divided by means of sedimentation in the Attenbergcylinder into 4 fractions.

Two fractions of one kaolinite, products of English China Clays Int., St. Austell, Cornwale, designated as an American clay (K 5,6), derived from the Institute of Occupational Medicine in Edinburgh were given to us by Bergbauforschung GmbH Essen. K 5 contains less than 1 w.-% quartz and 1 w.-% muscovite, K 6 1 w.-% quartz and 2 w.-% muscovite.

Four fractions of different size distributions (MB, MC, MD, ME) from one muscovite dust, two fraction of one kaolinite dust sample were investigated by electron scanning microscopy and characterized in a dispersed state directly by the density, the specific adsorption total surface per mass and per volume, the number and volume distribution, and indirectly by the hypothetical outer minimal surface per mass and per volume, and by the surface structure number, i.e. by the relation of the total surface to the minimal surface (Schyma 1987).

The density of dust samples with a mass smaller than 1 g was

NATO ASI Series, Vol. G 21
Health Related Effects of Phyllosilicates
Edited by J. Bignon

determined by means of a micromethod based on a quantil gas volume (Schyma 1986). The total adsorption specific surface (i.e. the outer and the inner) per mass is based on the modification of the Brunauer-Emmett-Teller-(BET) method according to the quantil differential procedure by means of a adsorption of nitrogen and krypton at a temperature of 77,4^{O}K (Schyma 1986). The total surface per volume represents the product of the density and the total surface per mass. The volume and size distribution were determined by means of a Coulter-Counter-procedure.

The dusts were applied intratrachealy or intraperitonealy in animals as suspensions in the saline solution. The physical state of applied dust suspensions in animal experiments were characterised by two different methods of quantification. The first group had separate independent values, such as surface tension, permeability coefficient and pH values, the separate independent values of the second group were the specific conductivity, the total molecule and total molecule ion numbers as well as the density (Schyma et al. 1987).

Dust that settles in the sediment is not stable, it is subject to physical changes. The dust molecules change partly in the solution, bringing about the formation of the so called density gradients and diffuse to the surface. The diffusion of the dissolved dusts from the inner part of a dust suspension on its surface generates a change in the surface tension of the solution that is dependent on the composition of the suspension.

The molecules of the dissolved substance have the property that decreases the specific surface energy of the solution fluid. The decrease of the kinetic interfacial energy of the solution molecules is parallel to the concentration of the dissolved molecules and ions as well as to the surface permeability of the solution molecules, i.e. to their ability to change from a liquid phase to a gas phase which also corresponds to the situation in the lung alveoli. The surface permeability is responsible for the energy state of the surface suspension solution and it is dependent on the partial densities of the dissolved components and on the energy state of the composition of the surface molecules.

In the new suspension state the physical values have no additional charateristics, therefore, a specific composition of the solution molecule can lead to a reduction of the total volume of the solution. The total density of the solution (i.e. that of the solution, dissolved dust and colloids) can give information about the mutual influences of the solution and dissolved molecules.

The surface tension of the dust suspensions was determined by the tensiometer LAUDA in erg/cm^2, the specific conductivity by a conductometric procedure in mS/cm (Schyma et al. 1987) and the density of the colloid suspension by a resonance procedure while the mass in a known volume of fluid was determined by a resonance frequency (Schyma et al. 1987). The total molecule number was determined by a BECKMAN method (Bonnar et al. 1958), the pH-value potentiometric. The surface permeability was determined by the coefficient of the kinetic state equation (Schyma et al. 1987).

As the dust changes from a solid to a suspension state one part of the molecule is dissolved, one part is dissociated into ions and one part remains unchanged in a solid state. With increasing number of dissolved molecules and dissociated ions the osmotic pressure of the solution increases. Therefore, the suspension state of the dust can be characterized by the estimation of the total molecule number in osmol/kg of solution as well as by the total molecule ion number in mol.cm/kg.mS and the specific conductivity in mS/cm. The total molecule ion number is given as the relation of the dissolved molecules to the dissolved ions, the specific conductivity indicates the reciprocal value of the specific electrical resistance of the solution.

Results

In the scanning electron micrographs of kaolinite were typical crystal aggregations concentrated in the coarse fraction, sometimes with parallel oriented pseudo-crystals demonstrated (Fig. 1).

Fig. 1: Scanning electron micrograph of kaolinite 5
(Magn. 12000 : 1)

Fig. 2: Scanning electron micrograph of kaolinite 6
(Magn. 12000 : 1)

In the SEM of single muscovite fraction the typical lamellar glimmer structure can be seen (Fig. 2). The platelets increase in size with increasing granule size (Fig. 3). At the base of the photo upright platelets simulate rod formed particles.

The determined values of kaolinite and muscovite dusts in a dispersed state increase with increasing finess of the single fractions. The only exception is the surface structure number of the muscovite fractions. The surface stuctur number first decreases with increasing finess of the muscovite fraction, it reaches the highest value with the finest ME-fraction (Table 1).

With increasing finess of the dust fraction the density of the solution, the specific conductivity and the total molecul number increase slightly, the permeabilty coefficient noticeably more and the muscovite fraction is more than twice as high as the kaolinite fractions. The trend of the total molecule number was the reverse. The dust fractions of the muscovite were in the neutral range and the kaolinite fractions in the acid region.

The physical characteristic of these dusts showed that the density of the solution increased with increasing total molecule number, the specific conductivity decreased with an increasing total molecule ions number.

Discussion

The determination of the density shows that the density of the coal mine fine dusts (Schyma 1986) and of the muscovite dust increases with increasing finess of the dusts whereas that of the quartz dust decreases (Schyma et. al. 1987).

The established different relationship of the intratracheal investigated dusts can be attributed partly to the different geometry of the dust particles.

The determination of the density and adsorption specific total surface indicated that the quartz consists of a solid phase that is composed of a crystal and a quasi amorphous structural state.

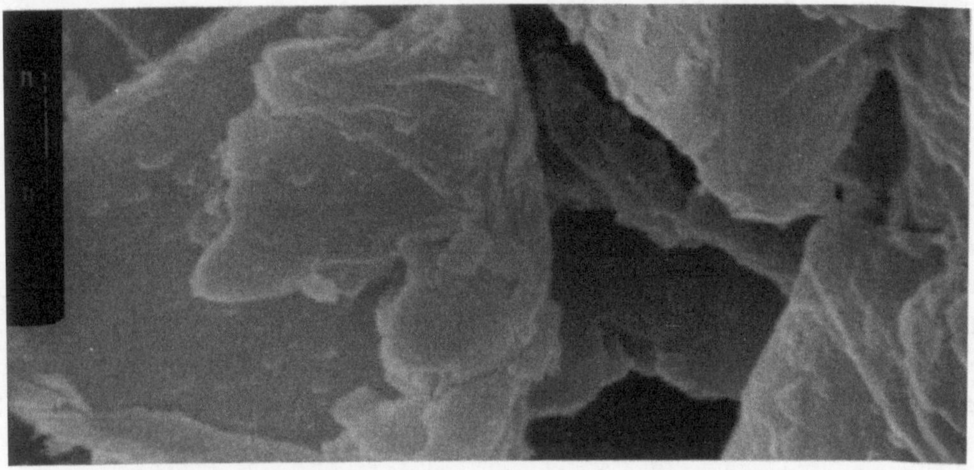

Fig. 3: Scanning electron micrograph of muscovite (MB - fraction, magn. 12000 : 1)

Fig. 4: Scanning electron micrograph of muscovite (ME-fraction, magn. 12000 : 1)

Table 1: Physical characteristic of dust samples in dispersed state

dust sample		1	2	3	4	5	6	7	8
		um	um	g/cm3	m2/g	m2/cm3	m2/g	m2/cm3	
Kaolinite	5	3,61	2,14	2,586	10,905	13,134	0,834	2,156	13,137
Kaolinite	6	2,57	2,19	2,616	16,945	14,935	1,137	2,968	14,935
Muscovite	MB	7,08	1,65	2,635	4,027	10,609	0,369	0,972	10,911
	MC	4,98	1,61	2,664	4,469	11,904	0,503	1,340	8,884
	MD	2,75	1,89	2,895	6,556	18,978	0,888	2,571	7,382
	ME	2,10	2,61	2,908	9,444	27,324	1,357	3,926	6,960

1 = mean value of volume distribution with standard deviation = 2
3 = density
4 = adsorption specific total surface per mass
5 = adsorption specific total surface per volume
6 = outer minimum specific surface per mass
7 = outer minimum specific surface per volume
8 = surface structure number

Table 2: Physical characteristic of dust samples in the suspensed state (measured in 0,9% saline + 20% glucose solution)

dust sample		1	2	3	4	5	6	7
		erg/cm2	g/cm3	joule/cm2sec .10-6	mS/cm		osmol/kg	mol.cm/kg.mS
Kaolinite	5	66,2	1,03867	5,76	5,48	4,15	0,768	0,140
	6	63,2	1,03875	6,30	5,80	4,27	0,779	0,134
Muscovite	MB	60,2	1,03606	10,63	6,10	6,14	0,749	0,122
	MC	60,2	1,03705	11,56	6,15	6,26	0,756	0,122
	MD	60,5	1,03719	12,53	6,23	6,38	0,756	0,123
	ME	57,2	1,03790	14,53	6,33	6,80	0,757	0,119

1 = surface tension of the tested dust suspension
2 = density
3 = coefficient of permeability
4 = specific conductivity
5 = pH-value
6 = total molecule number
7 = total molecule ion number

The density decreases with the fragmentation because the amorphous phase increases.

The phyllosilicate as for example kaolinite or muscovite consists of two phases, a solid and a gas phase. The solid phase belongs to a two dimensional structure group whereby the particles in the gas phase decrease and the density increases with the fragmentation.

If a relationship is established from the regression function between the change in density and the adsorption specific total surface per volume, one can determine if the density and the adsorption specific surface per volume of the single fraction represents independent or dependent dimension. In this way the proven mutual dependency can aid in a more exact characterization of the samples by the differential density. The differential density represents the mass changes per unit of the altered total surface of the particle by constant volume. This shows in comparison to those in the intratracheal tested effects that a correlation could be considered for the surface tension of the tested solution with the reaction of the total lipids and the permeability coefficient with the fibrogenic effect (Rosmanith et al 1989).

Summary

Four fractions of different size distributions (MB, MC, MD, ME) from one muskowite dust, two fraction of one kaolinite dust sample were characterized in a disperded state by the density, the specific adsorption total surface per mass and per volume, the number and volume distribution, indirectly by the hypothetical outer minimal surface per mass and per volume, and by the surface structure number. The density increased with increasing total surface per volume or with increasing surface structure number respectively.

The physical state of applied muskovite and kaolinite dust suspensions were characterised by the surface tension, permeability coefficient, pH values, the specific conductivity, the total molecule and total molecule ion numbers as well as the density. The density of the solution increased with increasing total molecule number, the specific conductivity decreased with an increasing total molecule ions number.

References

Rosmanith, J., L. Ikka, H. Breining, S.B. Schyma und W. Ehm (1989): Unterschiedliche fibrogene Reaktionen von Dörentruper Quarz und Cadmiumsulfid in Gemischen mit McIntyre-Aluminium und Polyvinylpyridin-N-oxid. Silikosebericht Nordrhein-Westfalen, Bd 17: im Druck, Essen: Steinkohlenbergbauverein

Rosmanith, J., W. Hilscher and S.B. Schyma (1989): The effect of the surface quality on the fibrogenicity of the phyllosilicates muscovite and kaolinite. First Int. Conference on Health Related Effects of Phyllosilicates, March 16 and 17, 1989, Paris, poster session.

Schyma, S.B. (1986): Mikromethode zur Bestimmung des Volumens und der Dichte fester Stoffe im dispersen Zustand. Staub-Reinh. Luft 46: 528-530

Schyma, S.B. (1987): Physikalische Charakteristik von Feinstäuben des Steinkohlenbergbaus im dispersen Zustand. In: Silikosebericht Nordrhein-Westfalen, Bd. 16: 161-168, Essen: Steinkohlenbergbauverein

Schyma, S.B. und J. Rosmanith (1987): Physikalische Charakteristik der Stäube in suspensem Zustand. In: Silikosebericht Nordrhein-Westfalen, Bd. 16: 169-178, Essen: Steinkohlenbergbauverein

Bonnar, R.U., M. Dimbat and F.H. Stross (1958): Number-average molecular weight. Interscience Publ. Inc., New York, pp. 22-29

ASBESTOS CONTENT OF TALCS FROM ITALIAN MINES AND FIBRE CONCENTRATION IN VARIOUS COMMERCIAL TALCUM POWDERS USED IN ITALY.

A.Marconi
Lab.Igiene Ambientale
Istituto Superiore di Sanità
Viale Regina Elena 299

00161 Rome
Italy

U.Verdel
Consulenza Tecnica Accertamento
Rischi Professionali
INAIL
Via Nomentana 74
00161 Rome
Italy

NATO ASI Series, Vol. G 21
Health Related Effects of Phyllosilicates
Edited by J. Bignon

Introduction

Talc naturally occurswith a variety of mineral habits: blocky, platy, polycristalline and fibrous. Due to the geological origin, metamorphic alterations of quartzite and dolomitic rocks, talc deposits have been found to contain talc mineral often in association with other minerals like serpentine and amphiboles. Health interest in the occurrence of the asbestiform varieties of these minerals is determined by the cancerogenic potential associated to the inhalation of asbestos fibres.
Several studies have indicated a close association between fibrous talc and fibrous amphibole (tremolite and anthophyllite) structures (Stemple and Brindley, 1960; Virta, 1985).
In the italian talc deposits serpentine and amphiboles, as accessory minerals, have been found in many mines, most of which are no more exploited today (Marconi et al., 1986, Verdel et al., 1985). In general italian talcs are considered very pure and not contaminated by asbestos or quartz.
The world use of talc during the last years showed a constant increase, due mainly to the increasing contribution of the oriental countries. The estimated production in Italy raised 130.000 tons in 1986, imports from other countries was about 25.000 tons and exports accounted for nearly 40.000 tons (see Figure 1).

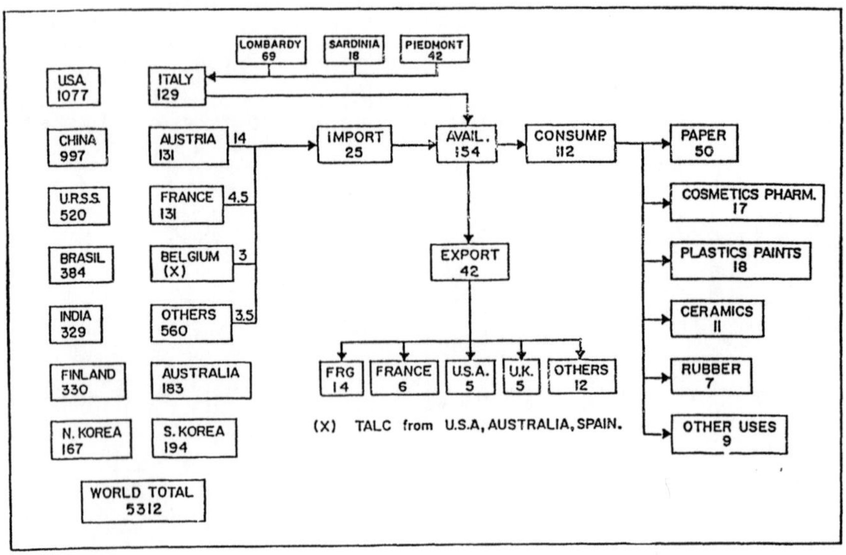

Figure 1 - Talc in the world and in Italy (data from Roskill, 1987 and Soc.Talco e Grafite Valchisone, 1989)

Italian talc active mines are localized in three main areas: Pinerolese (near Turin), Orani (east Sardinia) and Valmalenco (Lombard Alps). Mining trends have been nearly constant for the first two areas but decreasing during the last years for Orani deposits.
Several mines, no more exploited, were distributed in Piedmont (Val di Lanzo), northern Appenines (Taro Valley), Tuscany(Montenero), Calabria (Serra San Bruno) and Sardinia (S'Arenargiu, Nuoro). Domestic industrial uses are in the paper, cosmetic-pharmaceutical, paints, plastics, ceramic and rubber production.
The purpose of this study was to characterize the mineralogical composition of italian talcs from active and inactive mines, giving special attention to the occurrence of asbestiform minerals.
In addition several samples of commercial talcum powders from the market have been investigated for their mineral fibre content because more than 20% of talc used in this country comes from imports of various origin with unknown composition.

Materials and methods

Representative samples from 3 active mines and from 6 inactive mines have been analyzed for their mineral phase composition by x ray diffractometry (XRD), infrared (IR), phase contrast optical(PCDM) and scanning electron (SEM) microscopy techniques.
Analytical procedure followed were those reported in a previous paper (Verdel et al., 1985).
19 commercial, talcum powders have been analyzed by PCOM and dispersion staining methods for identification of the mineral particles using counting and sizing criteria reported in Marconi et al. (1986). In summary, asbestos fibres, identified by dispersion staining method, with diameters in the range $\leq$ 0.3 - 5 μm and aspect ratio $\geq$ 3 : 1 have been considered : chrysotile and tremolite group fibres longer than 5 μm, with diameter less than 0.3 μm have been separately recorded and aspect ratios distribution of fibres in three ranges ($\geq$ 3 : 1 - $<$ 5:1; $\geq$ 5:1 - $<$ 10:1; $\geq$ 10:1) have been reported.

Results and discussion

Tables 1 and 2 show the mineralogical composition of talcs extracted from active and inactive mines, determined by XRD and IR methods.
Samples still in production does not show the presence of serpentine or tremolite amphibole minerals, at least according

to the sensibility of analytical methods used (about 1% w/w). In samples from the inactive mines of Bagnada, tremolite and/ /or serpentine phases were found to be present, in some cases, at considerable levels (up to 15% by weight in the white type).

TABLE 1 - Mean Mineralogical Composition of Talcs from Active Italian Mines (weight %)

Active Mines	Talc	Chlorite	Calcite Dolomite Magnesite	Quartz	Production rate (100 tpa)
Piedmont					
Fontane	90÷95	3÷7	1÷2	tr.	420
Sardinia					
Sa Matta					
- white	90÷95	2÷8	–	–	40
- grey	40÷60	10÷50	2÷8	0÷2	140
Lombardy					
Valtellina	45÷55	20÷35	15÷35	–	690

Tremolite was also detected in the Val di Lanzo samples (5-10% by weight) and in Montenero and S'Arenargiu samples, but at traces levels.
Examination by microscopic techniques of the samples containing serpentine and tremolite minerals has shown, however,

Table 2 - Mean Mineralogical Composition of Talcs from Inactive Italian Deposits (weight %)

Inactive Deposits	Talc	Chlorite	Calcite Dolomite Magnesite	Serpentine	Tremolite	Mica	Quartz	Year of the end of the production
Piedmont								
Val di Lanzo	60÷ 85	5÷ 25	1÷ 5	—	5 ÷ 10	—	1	1983
Lombardy								
Bagnada								
- white	50÷ 70	—	15÷ 50	—	0 ÷ 15	—	0÷ 2	1987
- grey	45÷ 60	5÷ 20	5 ÷ 30	10÷ 20	0 ÷ 5	—	1÷,3	1987
Emilia								
Val Taro	90÷ 95	0÷ 6	2 ÷ 5	—	—	—	—	1950
Tuscary								
Montenero	30÷ 40	50÷ 60	1 ÷ 10	—	tr.	-	-	1964
Calabria								
Serra S. Bruno	20÷ 40	20÷ 50	—	—	—	20÷ 40	—	1960
Sardinia								
S. Arenargiu	40÷ 60	30÷ 50	—	—	tr.	—	—	1977

that these mineral phases were present mostly with non asbestiform morphology.
In fact they did show a platy habit (antigorite) and prismatic-acicular shape (tremolite as shown in Figure 2.

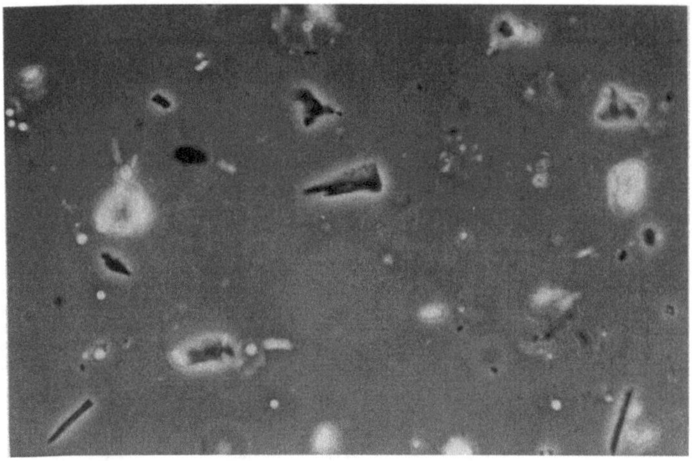

Figure 2 - Prismatic and acicular fragments of tremolite in Val Bagnada samples.

In table 3 are reported the results of the analyses obtained for commercial talcum powders.
Total mineral fibres, asbestos fibres and amphibole (tremolite) fragments (aspect ratio $<$ 3:1) concentrations, in terms of percentage by number, are included in the table.
In the majority of the samples only a small percentage of the total mineral fibres was found to be asbestos (chrysotile and tremolite) at levels ranging from $<$ 0.2 to 1.3%. In only one sample, known of chinese origin, asbestos fibre content was 3.2%. Most of the non asbestos fibres were found to be talc by SEM (see Figure 3) or cleavage particles from other

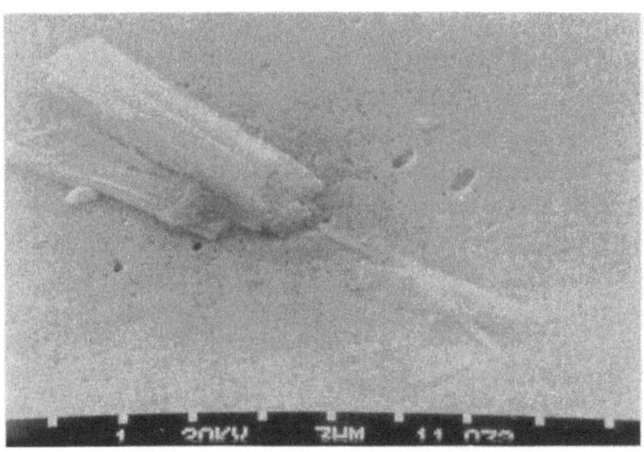

Figure 3 - Bundle of fibrous talc particles.

minerals. In all of the samples a substantial number of amphibole fragments (non fibrous) was found, probably originating from crushing and grinding operations used during the production phases (see also Pang et al., 1987). In general cosmetic powders did show asbestos concentrations more uniform and lower than the other types of powders.
From diameters distribution of amphibole fibres shown in table 4, it appears that most of the fibres have diameters in the range 0.5 - 2.0 μm, with appreciable differences among powders used for various purposes.

Table 3 - Fibre content in samples of italian commercial talcs

Sample	Total fibres (% number)	Asbestos fibres (% number)	Ratio of asbestos fibres to total fibres %	Ratio of amphibole[a] fibres to asbestos fibres %	Percentage amphibole fragments[b] in the total amphibole particles[c]	Ratio of chrysotile fibres to asbestos fibres %
Industrial						
A1	1.7	1.3	76.5	12.0	33.0	88.0
A2	5.6	0.7	12.5	n.d.	100.0	100.0
A3	1.2	n.d.[d]	n.d.	n.d.	100.0	n.d.
A4	1.2	< 0.2	16.5	100.0	90.0	n.d.
A5	2.9	< 0.2	7.0	33.0	70.0	67.0
Pharmaceutical						
B1	7.0	3.2	46.0	100.0	20.0	n.d.
B2	1.3	0.6	46.0	88.0	38.0	12.0
B3	2.4	0.9	37.5	37.0	40.0	63.0
B4	2.0	< 0.2	10.0	100.0	18.0	n.d.
B5	3.7	0.2	5.5	n.d.	100.0	100.0
B6	0.9	< 0.2	20.0	30.0	90.0	70.0
Cosmetic						
C1	2.7	< 0.2	7.5	67.0	60.0	33.0
C2	3.5	0.4	11.0	35.0	45.0	65.0
C3	1.9	0.2	10.5	95.0	45.0	5.0
C4	3.1	< 0.2	6.5	n.d.	100.0	100.0
C5	0.7	0.2	28.5	100.0	30.0	n.d.
C6	2.5	< 0.2	8.0	33.0	90.0	67.0
C7	4.2	0.5	12.0	n.d.	100.0	100.0
C8	3.3	0.5	15.0	100.0	44.0	n.d.

a - The term "amphibole" includes the minerals: tremolite, anthophyllite and actinolite.
b - Fragment means a particle with aspect ratio < 3:1.
c - Amphibole particles means fibres + fragments
d - n.d. = not detected

The differences in fibre size between chrysotile and tremolite fibres present in the samples are evident from size measurements reported in tables 5 and 6, which show the percentage of fibres longer than 5 μm and thinner than 0.3 μm and aspect ratios distribution respectively. Table 4 show that a large amount of chrysotile fibres are long and thin, while there are few amphibole fibres thinner than 0.5 μm. Similar differences exist for aspect ratios of chrysotile fibres, which mostly are ≥ 10:1, while for tremolite fibres aspect ratios fall in the range ≥ 3:1 - < 10:1 as shown in table 5.

Table 4 - Diameters (μm) distribution of amphibole fibres (Percent values)

Sample	0.3	0.31–0.5	0.51–1.0	1.1–2.0	2.1–3.0	3.0
Industrial						
A1	—	—	—	100	—	—
A2	—	—	—	—	—	—
A3	—	—	—	—	—	—
A4	—	—	—	100	—	—
A5	—	—	—	100	—	—
Pharmaceutical						
B1	—	18	35	41	—	6
B2	—	—	69	31	—	—
B3	—	15	62	12	11	—
B4	—	—	53	37	10	—
B5	—	—	—	—	—	—
B6	—	—	—	50	50	—
Cosmetic						
C1	—	—	—	100	—	—
C2	—	17	17	36	15	15
C3	16	16	26	32	10	—
C4	—	—	—	—	—	—
C5	7	21	47	18	7	—
C6	—	—	50	—	—	50
C7	—	—	—	—	—	—
C8	5	21	21	37	11	5

* Hyphen indicates the absence of fibres with the stated diameter.

Table 5 - Percentage of fibres with L $\geq$ 5 μm and D $\leq$ 0.3 μm

Sample	L $\geq$ 5 μm Chrysotile	Amphibole	D $\leq$ 0.3 μm Chrysotile	Amphibole
Industrial				
A1	90	90	29	—*
A2	> 95	n.d.	75	n.d.
A3	n.d.**	n.d.	n.d.	n.d.
A4	n.d.	$\geq$ 95	n.d.	—
A5	> 95	> 95	> 95	—
Pharmaceutical				
B1	n.d.	53	n.d.	—
B2	33	90	—	—
B3	> 95	78	33	—
B4	n.d.	74	n.d.	—
B5	> 95	n.d.	—	n.d.
B6	> 95	> 95	60	—
Cosmetic				
C1	> 95	> 95	> 95	—
C2	> 95	> 95	18	—
C3	> 95	80	> 95	16
C4	> 95	n.d.	17	n.d.
C5	n.d.	68	n.d.	7
C6	> 95	50	50	—
C7	> 95	n.d.	33	n.d.
C8	n.d.	68	n.d.	5

* Hyphen indicates the absence of fibres with the stated dimension

** n.d. means the absence of asbestos fibres of the stated type.

Conclusions

Results obtained from tha samples investigated have shown that asbestiform phases content of talcs mined today in Italy is very low. Inactive mines deposits in some case contained substantial amount of serpentine and amphibole minerals, but their predominant morphology appears to be platy and prismatic or acicular respectively.

In commercial talcum powders asbestos fibres are present at levels mostly less than 1% (by number) and likely less than 1% by weight (Marconi et al., 1986). Amphibole (tremolite) fibres occuring in these powders, however, do show sizes more consistent with prismatic, than with asbestiform shape (Campbell et al. 1979; Marconi et al., 1986).

Nevertheless, because substantial amount of talc for domestic consumption still comes from imports of various origin, China included (Verdel et al., 1985), it cannot be excluded that asbestos contamination might occur in some cases.

For these reasons it appears necessary to continue investigations to carefully control the presence of asbestiform fibres, especially in talcum powder products, which have a large use among consumers.

Table 6 - Aspect ratios distribution for asbestos fibres (percent values)

Sample	Chrysotile			Amphibole		
	≥3:1-<5:1	≥5:1-<10:1	≥10:1	≥3:1-<5:1	≥5:1-<10:1	≥10:1
Industrial						
A1	7	36	57	—*	100	—
A2	—	—	100	—	—	—
A3	—	—	—	—	—	—
A4	—	—	—	100	—	—
A5	—	<5	>95	100	—	—
Pharmaceutical						
B1	—	—	—	35	47	18
B2	—	67	33	19	69	12
B3	—	—	100	33	45	22
B4	—	—	—	37	53	10
B5	—	—	100	—	—	—
B6	—	—	100	100	—	—
Cosmetic						
C1	—	—	100	50	50	—
C2	—	—	100	17	50	33
C3	—	—	100	21	42	37
C4	—	—	100	—	—	—
C5	—	—	—	25	61	14
C6	—	—	100	100	—	—
C7	—	—	100	—	—	—
C8	—	—	—	32	58	10

* Hyphen indicates the absence of fibres with the stated aspect ratio.

References

Campbell W.J., Steel E.B., Virta R.L., Eisner M.H. (1979).Relationship of mineral habit to size characteristics for tremolite cleavage fragments and fibres. Bureau of Mines-Report of Investigations No.8367, United States Dept. of the Interior.

Marconi A., Maccione M., Rossi L. (1986). Asbesto e talco: determinazione del contenuto di particelle minerali fibrose in polveri di talco commerciali mediante tecniche associate di microscopia ottica. Med.Lav.77 (5), 496-510 (in italian).

Marconi A., Puledda S. (1987). Analisi differenziata delle varie forme di asbesto presenti in alcuni prodotti commerciali con le tecniche integrate della diffrattometria a raggi X e della microscopia ottica in contrasto di fase. G.Igienisti Ind. 12 (4), 9-19 (in italian).

Pang T.W.S., Schonfeld F.A., Nazar M.A. (1987). The determination of tremolite asbestos in talc powder samples. Ann. Occup. Hyg. 31 (2), 219-225.

Roskill Information Services Ltd. (1987). The economics of talc and pyrophyllite, 9th Ed., London.

Soc.Talco e Grafite Valchisone (1989) (Personal Communication).

Stemple I.S. Brindley G.W. (1960). A structural study of talc and talc tremolite relations. J.Am.Ceramic Soc. 43 (1),35-42.

Virta R.L. (1985). The phase relationship of talc and amphiboles in a fibrous talc sample. Bureau of Mines Report of Investigations No.8923. United States Dept. of the Interior.

Verdel U., Sperduto B., Ricchiardi P., Laurini G. (1985). Caratteristiche dei talchi italiani e possibilità di associazioni con minerali nocivi. Ind. Mineraria 2, 25-33 (in italian).

OCCURRENCE OF PHYLLOSILICATES IN TOBACCO

E. MOULIN, P. DE VUYST, JC YERNAULT
Chest Department
Erasme Hospital
Université Libre de Bruxelles
Lennik Road 808
1070 - Bruxelles
Belgium.

Introduction

Mineral particles can occur in tobacco as components added during industrial process or as soil natural contaminants accumulated on the leaves of the plant (Wynder 1967, Langer 1971). Emphasizing their possible pulmonary effects, the direct inhalation of such particles by snuffing may be associated with inflammatory reactions leading to interstitial lung diseases (as reported among North African snuffers, El Gharbi -oral communication- 1984). Considering their inhalation derived from cigarette smoke, previous works indicate a relationship between smoking habits and lung dust burden (mainly silicates) (Brody 1975, Churg 1987, Johnson 1986). Moreover, small airways lesions in smoker's lungs present morphological similarities with mineral dust airways diseases due to occupational exposure (Bignon 1987, Churg 1985, Kennedy 1985). Therefore, more information is needed concerning the nature, range and size of inorganic particles in commercial tobaccos.

Materials and methods

Samples from 8 cigarette brands (6 available in Belgium, 2 from U.S.A.) and 8 snuff brands (7 available in Belgium, 1 from Tunisia) were investigated after an incineration in a

NATO ASI Series, Vol. G 21
Health Related Effects of Phyllosilicates
Edited by J. Bignon

low temperature asher and a treatment of the inorganic residue with weak HCl (0.2M) solution to remove the natural oxalates content of tobacco leaves.

The final materials were then weighed and later suitably prepared for mineral constituents characterisation (type, range, size). Analytical electron microscopy was designed to cover all the facets of these examinations for each sample, 100 consecutive particles were analysed.

Results (Table 1, Fig 1 and 2)

The mineral residue weight ranged from 1.4 % to 2.9 % (mean 1.9 %) for cigarette tobaccos and from 2.9 to 9.1 (mean 4.4 %) for snuffs.

The inorganic residue was roughly the same in the 2 forms of tobacco.

Phyllosilicates were the major mineral components, commonly associated with cristalline silica, feldspars and Fe-Ti particles (oxides) (see Table 1).

Table 1 : Nature and range of mineral species in 100 consecutive particles from cigarette (C, n=8) and snuff (S, n=8) tobaccos.

		CRISTALLINE SILICA	ILLITE / MUSCOVITE	KAOLINITE	PHYLLOSILICATES VARIED *	FELDSPARS	Fe-Ti PARTICLES	OTHERS **
C Group n=8	R	18 to 32	27 to 45	12 to 24	1 to 14	5 to 12	6 to 13	1 to 3
	M	23	33	17	7	7	10	3
S Group n=8	R	7 to 23	15 to 61	9 to 49	1 to 27	0 to 13	8 to 19	0 to 3
	M	13	33	22	12	7	12	1
TOTAL group	M	18	33	20	9	7	11	2

* Chlorite, Talc, Ca-clay, Biotite, Bentonite...

** Corindum, Monazite, Zircon, Diatomaceous earth...

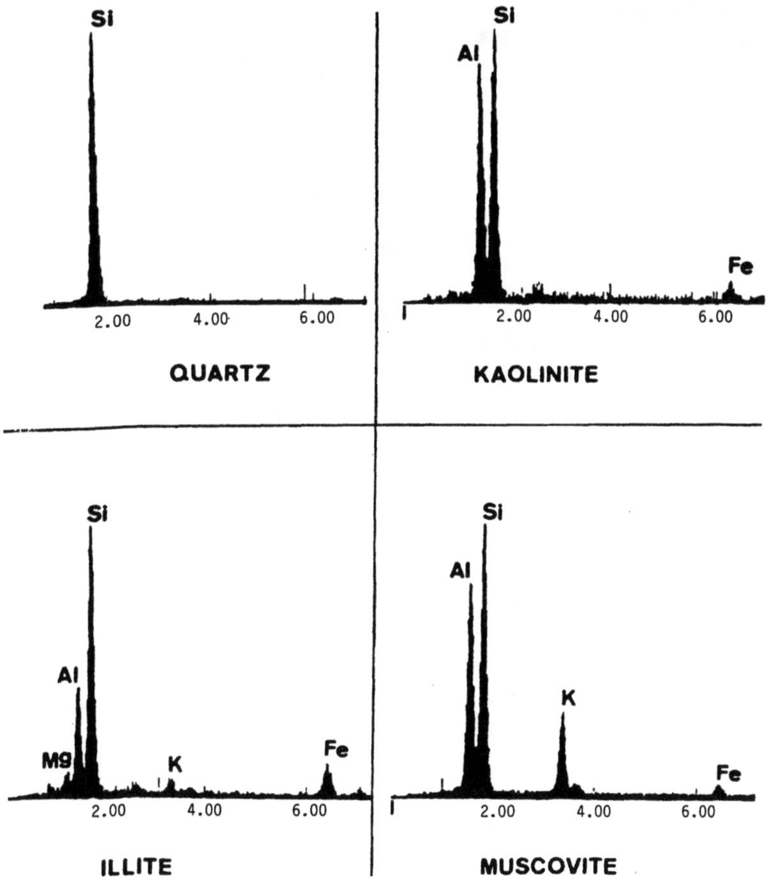

Fig 1 : Energy dispersive spectrometer spectra (in KeV) for various particles in tobaccos.

An illite muscovite group and a kaolin group were the major components among phyllosillicates (see Fig 1). Granulometric study in electron microscopy pointed out the presence of respirable particles (< 5 µm) coarser in cigarettes (mean size : 2.13 µm) than in snuff (mean size : 1.30 µm) (see Fig 2).

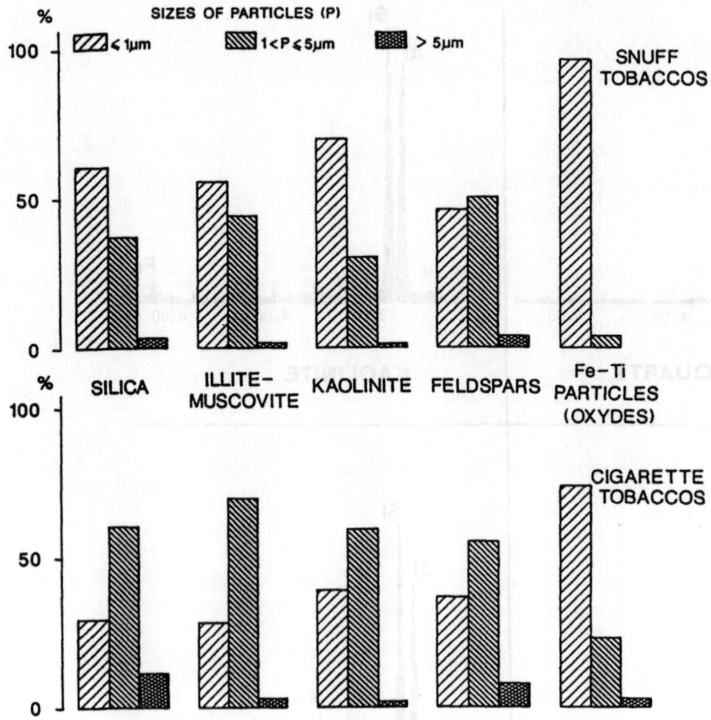

Fig 2 : Relative frequency in sizes of major mineral species from snuff and cigarette tobaccos

Conclusion

Our findings are consistent with the hypothesis that snuffing tobaccos leads to a significant inhalation of a cristalline silica and phyllosilicates mixture. Among the phyllosilicates, the more numerous particles were kaolinite, illite and muscovite.

Cigarette tobaccos contain the same type and species of mineral particles which theoretically can be inhaled. However, the cransfer of these particles in the mainstream smoke of cigarettes has to be confirmed by further studies.

References.

Bignon J, Jaubert F, Laurent P (1987) Les bronchiolites obstructives. Rev Mal Resp 4: 199-215.

Brody AR, Craighead JC (1975) Cytoplasmic inclusions in pulmonary macrophages of cigarette smokers. Lab. Invest 32: 125-132.

Churg A, Wright JL, Wiggs B, Paré PD, Lazar N (1985) Small airways disease and mineral dust exposure. Am Rev Respir Dis 131: 139-143.

Churg A, Wiggs B, (1987) Types, number, size and distribution of mineral particles in the lungs of urban male cigarette smokers. Envir Res 42: 121-129.

Johnson NF, Haslam PL, Newman-Taylor AJ, Turner-Warwick M (1986). Identification of inorganic dust particles in bronchoalveolar lavage macrophages by energy dispersive x-ray microanalysis. Arch Envir health 41: 133-144.

Kennedy SM, Wright JL, Mullen JB, Paré PD, Hogg JC (1985) Pulmonary function and peripheral airway disease in patients with mineral dust or fume exposure. Am Rev Respir Dis 132: 1294-1299.

Langer AM, Mackler AD, Rubin I, Hammond EC, Selikoff IJ (1971) Inorganic particles in cigars and cigar smoke. Science 174: 585-587.

Wynder EL, Hoffmann D (1967) Tobacco and tobacco smoke, Academic Press, New York.

THE EFFECT OF THE SURFACE QUALITY ON THE FIBROGENICITY OF THE PHYLLOSILICATES MUSCOVITE AND KAOLINITE

J. Rosmanith, W. Hilscher* and S.B. Schyma
Institut für Hygiene und Arbeitsmedizin der Medizinischen Fakultät der Rheinisch-Westfälischen Technischen Hochschule Aachen,
Neues Klinikum
D-5100 Aachen
Federal Republic Germany

The very pure dust sample of muscovite without quartz contamination, derived from Lenkersmühle (Upper Palatinate, West Germany) was ground in the agate mill and devided by means of sedimentation in the Attenbergcylinder into 4 fractions (MB, MC, MD, ME).

The investigated two kaolinites, products of English China Clays Int., St. Austell, Cornwale, designated as china clay (K 1,2) and American clay (K 5,6) respectively, derived from the Institut of Occupational Medicine in Edinburgh were given to us by Bergbauforschung GmbH Essen. K 1 contain 2 w.-% muscovite, K 2 1 w.-% quartz and 9 w.-% muscovite, K 5 less than 1 w.-% quartz and 1 w.-% muscovite, K 6 1 w.-% quartz and 2 w.-% muscovite.

Before intratracheal installation the dust fractions were physically characterized in their dispersed and suspensed states (Schyma et al. 1988, Schyma 1989).

Methods

160 female SPF-Wistar rats were divided in 8 groups: the rats of each group were instilled once, separately, intratracheally and bloodlessly with 50 mg of each dust fraction (four muskovite and 4 kaolinite fractions) as suspensions in 0.5 ml physiological saline solution. The rats were sacrified after 7 months. The lungs and the lymph nodes of the sacrified animals were removed

*Medizinisches Institut für Umwelthygiene an der Universität Düsseldorf, Gurlittstr. 53, D-4000 Düsseldorf 1,
Federal Republic Germany

NATO ASI Series, Vol. G 21
Health Related Effects of Phyllosilicates
Edited by J. Bignon

and weighed and the total lipids (Folch et al. 1957), total hydroxiprolin content (Stegemann 1958) and the dust masses retained in the lungs and lymph nodes were determined (Prajsnar et al. 1988).

Results

The wet weights of the lymph nodes and the dust masses of muscovite retained in the lungs and the lymph nodes respectively (Table 1) show a significant dependence (Table 2) on the minimum outer surface per volume (Schyma 1989), i.e. on the dust particle size: the finer the dust, the smaller the dust mass retained in the lung ($r^2 = 0,79$, $r_s = -0,89$) and the greater the wet weigts of the lymph nodes ($r^2 = 0,92$, $r_s = 0,94$) and the dust mass ($r^2 = 0,88$, $r_s = 0,94$) in the lymph nodes.

The total hydroxiproline and the total lipids after instillation of kaolinites increased only slightly in contrast to those after instillation of muscovite. The total lung hydroxiproline content increased not only after instillation of kaolinite but also after instillation of muscovite with the increasing coefficient of permeability ($r^2 = 0,58$, $r_s = 0,94$ - Table 2). In the dust suspension the coefficient of permeability shows the activity of the surface between the suspensed and the gas phase. The coefficient of permeability is not dependent on the material structure of dust (Schyma 1989).

The total lipids indicated a relationship with the surface tension of the dust suspensions (Schyma 1989). The more decreased the surface tension of the dust suspension the more increased the lipids in the lungs ($r^2 = 0,83$, $r_s = -0,99$). No correlation exists between hydroxiproline and lipid content.

Table 1: Means (x) and standard deviations (s) of of total lipids and hydroxiproline, of the wet weight of lymph nodes (wwly) and of dust masses in the lung and in the lymph nodes 7 months after intracheal instillation. For K1, K2, K5, K6, MB, MC, MD and ME see text.

dust sample	total lipids mg		hydroxiproline mg		wwly mg	dust mass lymph nodes mg	dust mass lymph nodes %
	x	s	x	s	x	x	x
K 1	59,38	7,19	6,35	1,17	140	2,34	4,7
K 2	66,86	13,52	5,50	1,46	90	0,38	0,8
K 5	58,25	11,83	5,24	0,67	70	0,64	1,3
K 6	62,26	18,43	6,22	1,52	80	0,51	1,0
M B	136,01	42,47	5,81	0,72	40	0,03	0,1
M C	130,77	42,88	6,30	1,19	50	0,25	0,5
M D	102,21	46,23	7,18	0,66	60	1,73	3,5
M E	179,54	65,20	8,34	1,46	140	5,54	11,1
contr.	45,05	3,15	3,32	0,63	35		

Table 2: Physical characteristic of dust samples in the dispersed (1,2) and suspensed (3,4) state (measured in 0,9% saline + 20% glucose solution). For K1, K2, K5, K6, MB, MC, MD and ME see text.

dust sample		1	2	3	4
		um	um	$\frac{erg}{cm^2}$	$\frac{joule}{cm^2 sec} \cdot 10^{-6}$
Kaolinite	5	3,61	2,14	66,2	5,76
	6	2,57	2,19	63,2	6,30
Muscovite	MB	7,08	1,65	60,2	10,63
	MC	4,98	1,61	60,2	11,56
	MD	2,75	1,89	60,5	12,53
	ME	2,10	2,61	57,2	14,53

1 = mean value of the volume distribution
2 = standard deviation
3 = surface tension of the tested dust suspension
4 = coefficient of permeability

All dust samples instilled intratracheally showed a fibrogenic response, the K 2-, K 6- as the MB-dust sample gave rise to gigant cell like granulomas in the lungs (Fig. 1). All tested dusts induced a fibrogenic reaction in lungs areas where dust was deposited, interweaved with hyalinic collagen fibres. In these areas, dust particles were located within macrophages. Af-

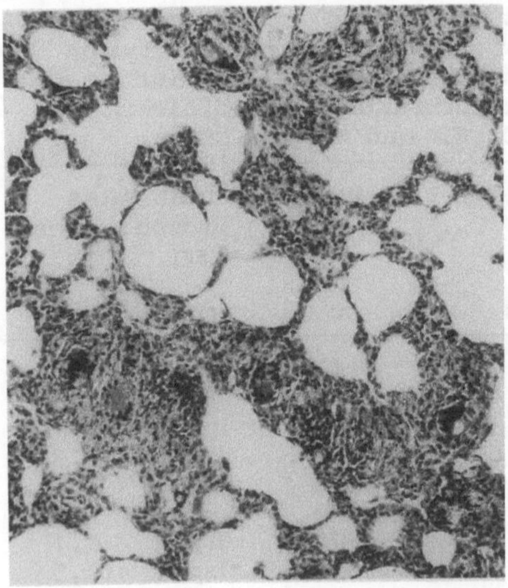
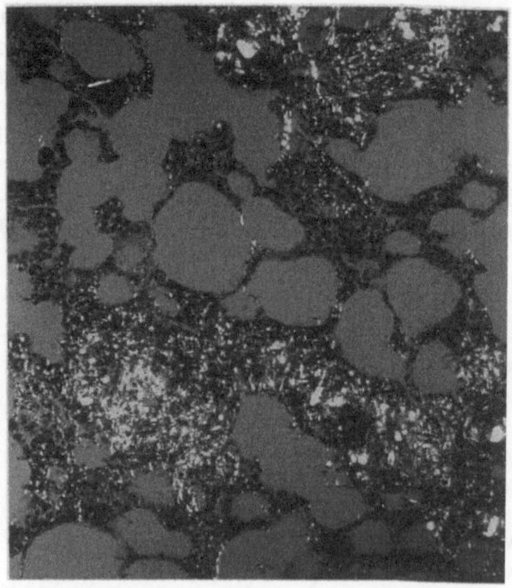

Fig. 1: Dust containing granulomas with gigant cells 7 months after intratracheal instillation of K 6-dust fraction. On the right hematoxiline-eosine-staining, on the left polarisation

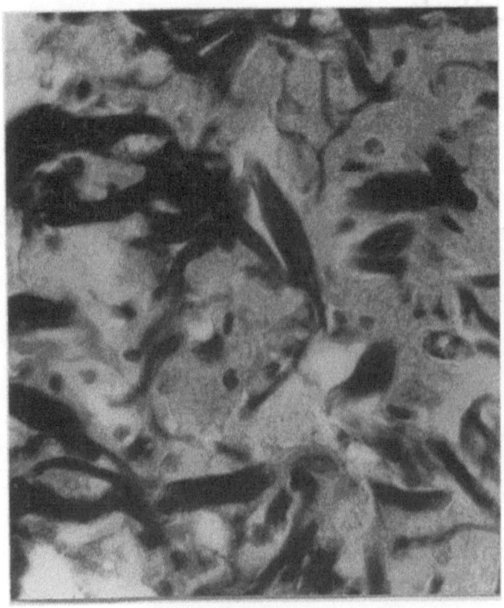
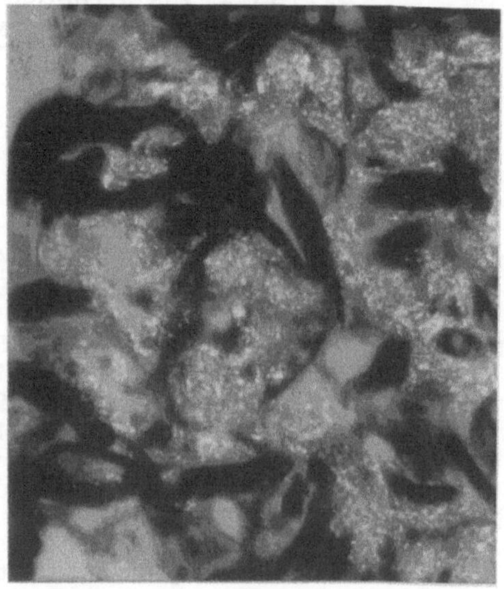

Fig. 2: Fibrogenic areas with dust containing macrophages and thick collagen fibres 7 months after intratracheal instillation of ME-fraction. On the right hematoxiline-eosine-staining, on the left polarisation

ter instillation of both of kaolinite and muscovite the amount of collagen fibres increased with the fineness of the dusts.

The finest muscovite fraction ME caused the greatest fibrogenic reaction with the thickest collagen fibres (Fig. 2). After instillation of muscovite a very great number of mast cells without alveolar lipoproteinosis was observed.

Discussion

The results showed that not only the clearence of phyllosilicate dust mass from the lung of rats but also its penetration into the lymph nodes just as that of the coal mine fine dusts or of Dörentruper quartz (Rosmanith et al. 1988) correlated with the minimal outer surface per volume. The hydroxiproline content in the lungs increased of course after intratracheal instillation of phyllosilicates or quartz with the fineness of the dust particles but it does not do it after application of coal mine fine dusts (Rosmanith et al. 1988). The minimum outer surface per volume or per mass represents only the hypothetical dimensions that only gives expression to the present degree of dispersity and therefore is the minimal outer surface dependent on the material structure and does not reflect in general the fibrogenity of dusts. A strict correlation between the coefficient of permeability and the content of hydroxiproline or the fibrogenic reaction in the lungs of rats respectively was found also at many others dusts with different structure (Rosmanith et al. 1989). The permeability was determined by the coefficient of permeability, i.e. by the determination of the specific heat of the solution and of the part of the mass, that passes over in 1 sec from 1 cm^2 surface of the solution in the gas state (Schyma 1989).

The reaction of lipids in the lungs of rats did not reflect the fibrogenic effect of the tested dusts. No relationship was found between the content of lipids and of hydroxiproline in the lungs. The lipid content decreased with increasing surface tension of the tested dust suspensions. Analog decreases the surface tension of the surfactant system of the lung. The more the dust decreases this surfactant system the more the lipids increase as a defensive reaction to the inhaled dust.

Summary

Dust samples of different particle size of muscovite and kaolinite were investigated by means of intratracheal instillation in rats. All tested samples caused fibrogenic reactions. The greatest reaction was observed after instillation of the finest dust fraction of muscovite. The dust masses retained in the lungs or penetrated into the lymph nodes respectively were dependent on the minimum outer surface per volume. After instillation of muscovite the hydroxiproline content in the lungs increased with the increasing coefficient of permeability, the total lipids content in the lungs correlated with the surface tension of the tested dust suspensions.

References

Folch, J.M., M. Less and S. Sloane (1957): A simple method for the isolation and purification of total lipids from animal tissue. J. Biol. Chem. 226: 497-509

Prajsnar, D., H. Breining, J. Rosmanith, S.B. Schyma und Th. Volf (1988): Mikromethode zur Staubrückgewinnung aus Lungen und Lymphknoten der Ratte. Einfluß des Ameisensäureaufschlusses auf die Oberflächenbeschaffenheit und die fibrogene Wirkung von Quarz. Wissenschaft und Umwelt 1: 68-72

Rosmanith, J., S.B. Schyma, W. Hilscher und W. Ehm (1988): Abhängigkeit der biologischen Wirkung von physikalischen Kenngrößen getesteter Stäube (Intratrachealversuche an Ratten). Wissenschaft und Umwelt 1: 33-38

Rosmanith, J., L. Ikka, H. Breining, S.B. Schyma und W. Ehm (1989): Unterschiedliche fibrogene Reaktionen von Dörentruper Quarz und Cadmiumsulfid in Gemischen mit McIntyre-Aluminium und Polyvinylpyridin-oxid. Silikosebericht Nordrhein-Westfalen Bd. 17, Essen: Steinkohlenbergbauverein, in press

Schyma S.B. und J. Rosmanith (1988): Physikalische Charakteristik der in Tierversuchen intratracheal instillierten Stäube in dispersem und suspensem Zustand. Wissenschaft und Umwelt 1: 39-57

Schyma, S.B. (1989): Physical characteristic of the phyllosilicates in the dispersed and suspensed states. Firts Int. Conference on Health Related Effects of Phyllosilicates, March 16 and 17, 1989, Paris

Stegemann, H. (1958): Mikrobestimmung von Hydroxiprolin mit Chloramin T und p-Dimethylaminobenzaldehyd. Hoppe Seylers Z. Physiol. Chem. 312: 41-45

OXIDIZING SURFACE PROPERTIES OF DIVALENT IRON-RICH PHYLLOSILICATES IN RELATION TO THEIR TOXICITY BY OXIDATIVE STRESS MECHANISM

D. Costa, J. Guignard and H. Pezerat
Laboratoire de Réactivité de Surface et Structure
Université Pierre et Marie Curie
CNRS, URA 1106
4 Place Jussieu
75252 Paris Cédex 05, France

The excess of lung cancers among workers in some metal mines, particularly iron mines, has been highlighted in the recent years with epidemiological studies such as those on the miners of Cumberland, Great Britain (Boyd et al. 1970), or of some Swedish iron mines (Edling 1982) and of the Lorraine region, France (Pham et al. 1983). The major etiology taken into account in these studies was the exposure to radon. However, this factor needs to be revised if we consider, for example, the case of the Lorraine mines where this exposure is very low, but the SMR for lung cancers is 3.5. So the role of the inhaled mineral dusts must to be considered in this context.

The studies on asbestos and M M M F (mineral man made fibres) (Zalma et al. 1987 ; Zalma 1988 ; Pezerat 1989) show that their Fe^{2+} content in many samples is sufficient to allow, after dissolution of the oxidized surface coating, the reduction of oxygen to $OH^{\bullet}$. The studies of Leanderson et al. (1988) on the same type of fibres show that these $OH^{\bullet}$ are able to attack DNA in the C-8 of guanosine. These effects are not dependent on the size of the fibres but only on the molecular oxygen accessibility, in aqueous medium, on the Fe^{2+} surface sites of the samples. So, this possible mechanism of carcinogenicity of some fibres, unrelated to their size, can be proposed for dusts containing some iron compounds, and particularly for dusts in some iron and other metal mines. D. Costa et al (1989) in two publications have proceeded with a large screening of the iron minerals present in the ores of many metal mines to point out the ones capable of showing a strong oxidizing activity after reaction with oxygen in an aqueous buffer medium. Among the very active mineral to emerge there are some phyllosilicates presents not only in the ores of some metal mines but also in other rocks such as granite, which may have a relation with the excess of lung and stomach cancers among the granite workers (Koskela et al. 1987).

NATO ASI Series, Vol. G 21
Health Related Effects of Phyllosilicates
Edited by J. Bignon

Herein we present the results concerning some of these iron-rich phyllosilicates.

Materials

Three mineralogical species are concerned with the present study :

- berthierine, with a layer type equivalent to the one of the serpentine, but with an octahedral sheet containing essentially Fe^{2+} and not Mg^{2+} (cf fig 1 in paper of H. Pezerat in this volume). This mineral is very sensitive to air oxidation. Hence, the samples from mineralogical collections are not representative of the "fresh" minerals present in the iron ores and inhaled in mines. Consequently, we have used a berthierine rich-iron ore from Lorraine (France).
- biotite, a very common trioctahedral mica, containing a variable amount of Mg^{2+} (cf fig. 2 in paper of H. Pezerat in this volume). The biotite is present in many granites and in other rocks present, for example, in some gold mines.
- chlorite (cf fig. 4 in paper of H. Pezerat in this volume) can be fully magnesian, for example in some talc mines, or it can contain an important level of Fe^{2+} in its octahedral sheets. In the iron mines the chlorite present in the ores is rich in Fe^{2+}. The iron-rich chlorites may be present in many rocks.

Methods

When Fe^{2+} is present at the surface between the particles and the aqueous aerated medium, the reduction of molecular oxygen to activated species of oxygen can occur. With a consumption of three electrons (three reducing surface sites) we obtain the radical hydroxyl ($OH^{\bullet}$) or an oxo compound - for example $Fe^{IV}=O$ - with an equivalent ozidizing capacity (Sutton and Winterbourn 1989). This oxidizing species (A*) is characterized by a radical reaction with a target molecule (here formate anion HCO_2^-), and formation of the radical $CO_2^{\dot{-}}$ which is trapped by a spin trapping agent (DMPO) present in the aqueous medium. Then we measure the quantity of the radical adduct in the medium by ESR spectroscopy. The reaction shown below schematically illustrates the process involved.

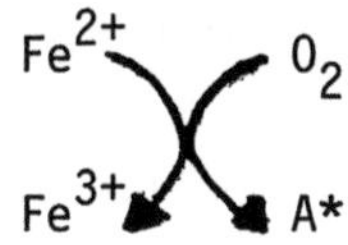

$$A^* + HCO_2^- \longrightarrow AH + CO_2^{\dot{-}}$$
$$CO_2^{\dot{-}} + DMPO \longrightarrow (DMPO,CO_2^-)^{\cdot}$$

All the details concerning the methods have been described (Zalma et al. 1988, Costa et al. 1989). The test is performed with 45 mg of the material freshly ground just before the experiment in an agate mortar, 2 ml of an aqueous potassium phosphate buffer medium (0.5 M pH = 7.4) containing the DMPO and the formate, in a reactor thermostated (37 °C) and protected from exposure to light. Aliquots of suspension are withdrawn at 25 and 60 minutes, filtered and examined by ESR spectroscopy (Varian CSE field set 3380 G, scan range 100 G, microwave power 10 mW, time constant 1 sec.). We provide a mean value of the intensity of the radical adduct $(DMPO,CO_2^-)^{\cdot}$ signal in arbitrary units, knowing that 1000 units are equivalent to the formation of 2×10^{14} radicals by 1 mg of dust.

Results

In table 1 we give the results obtained with some phyllosilicates and a quartz

Table 1.

Phyllosilicates	Fe^{II} % FeO	Intensity ESR signal	
		$(DMPO, OH)^{\cdot}$	$(DMPO, CO_2^-)^{\cdot}$
Chlorite (Pyrénées)	34.0	(-)*	1500
Chlorite (Anjou)	30.0	25	1400
Biotite (Razès)	18.9	20	1100
Iron-Ore (Lorraine) with 2/5 Berthierine	12.5	50	1700
Granite with Biotite	3.0	(-)	600
Granite with Muscovite	0.6	(-)	80
Kaolinite (St Austelle)	(-)	(-)	30
Montmorillonite (Maroc)	(-)	(-)	24
Quartz (DQ 12)	(-)	(-)	20
Blank		10-20	10-20

(-)* test not effected

We can observe that the more active compounds in our test are those rich in Fe^{2+} such as berthierine from Lorraine mine, biotite or chlorite. The phyllosilicates, very poor in Fe^{2+} (montmorillonite, kaolinite, muscovite) and quartz are weakly or non active. So, it appears that Fe^{2+} content is a very important factor of the activity in the test for the detection of oxidizing species. Results with very low values obtained in $OH^{\cdot}$ detection and very high values in $CO_2^{\dot{-}}$ detection can been explained if the A* species are not $OH^{\cdot}$ in majority but active oxygen species linked to iron on the surface.

It is not only the global content in Fe^{2+} of the mineral which is important but also the accessibility of molecular oxygen to Fe^{2+} at the interface solid-liquid. Hence, X-ray speciation and the chemical analysis are not sufficient to predict the biological activity of the samples.

Effect of air-ageing of the dusts

Tests have been performed at various times of air-ageing after several types of grinding. Characteristic results are given below table 2 for biotite, but we have observed the same effects for the chlorites.

Table 2.

Intensity of the ESR signal (DMPO, CO_2^-)$^{\cdot}$		
Hand-ground in air for 1 minute and immediately tested	Hand-ground in air for 1 minute and tested after 15 days of air-ageing	Mechanical grinding for 6 hours under nitrogen and stocked under argon
1100	80	1500

Mechanical grinding in air during a long period (2 hours or more) leads to the same passivation effect as the one observed by air-ageing.

So it appears that only the freshly generated particles are very active in this redox process. The passivation effect observed by air-oxidation is due to the fact that the Fe^{3+} formed on the surface of the particles is not completely extracted from the solid. A part of this Fe^{3+} takes the place of the Fe^{2+} at the periphery of the octahedral sheets with the occupation of two octahedra over three, and prevents accessibility of oxygen to the inner Fe^{2+}.

Effect of leaching of the oxidized coating the particles

Leaching of the particles surface allows to dissolve the Fe^{3+} species which hinder the accessibility of oxygen to the Fe^{2+}. So the composition of the aqueous medium plays an important role in the redox surface activity of the dusts.

We give in table 3 the effect of phosphate concentration for two pH values, where it appears that the increasing concentrations of the phosphate species promotes the leaching of the Fe^{3+} on the surface and consequently, increases the redox activity of the dusts.

Table 3.

Materials	Intensity of ESR signal (DMPO, $CO_2^-)^{\cdot}$			
	buffer pH = 7.4		buffer pH = 5.8	
	1 mol l^{-1}	0.2 mol l^{-1}	1 mol l^{-1}	0.4 mol l^{-1}
Chlorite (Pyrénées)	1650	650	1000	450
Biotite (Razès)	1000	450	550	320

Many other chemical complexing species other than phosphate exist in the biological fluids and these can cooperate in this process of activation of the surface redox activity.

CONCLUSION

The relation betwen the induced oxidative stress and the carcinogenicity of the dusts are now under study in several laboratories, but probably the genotoxic activity of the dusts is related to their ability - after endocytosis - to produce in a short time an important flux of oxidizing species in the cell. To test this type of dusts in -vitro or in-vivo, it will be necessary to take into account various factors such as air-ageing, aqueous-ageing , duration of grinding and ultrasonication, factors which can deplete the surface activity of the dusts.

Our results of redox activity obtained with freshly ground Fe^{2+} rich phyllosilicates are approximately the same that the one obtained with a freshly ground crocidolite or with the UICC Canadian chrysotile (Zalma et al. 1988), and the oxidative stress induced will be comparable, but with a difference in the process of quasi irreversible passivation of phyllosilicates by air-ageing, linked to the differences of structure of the materials. So, the hazards for the exposed workers would be chiefly related to the workplaces (mines, quarry, tunneling operations, etc ...) where the phyllosilicates dusts are freshly generated.

References

Boyd JT, Doll R, Faulds JR, Leiper J (1970) Cancer of the lung in iron ores (haematite) miners. Brit J. Ind Med 27 : 97-105

Costa D, Guignard J, Zalma R, Pezerat H (1989) Production of free radicals arising from the surface activity of minerals and oxygen. Part I : Iron mines ores Toxicol and Ind Health 5 : under press

Costa D, Guignard J, Zalma R, Pezerat H (1989) Production of free radicals arising from the surface activity of minerals and oxygen. Part II : Arsenides, sulphides and sulfoarsenides of iron, nickel and copper. Toxicol and Ind Health 5 : under press

Edling C (1982) Lung cancer and smoking in a group of iron ore miners. Am J of Ind Med 3 : 191-199

Koskela RS, Klockars M, Jarvinen E, Kolari PJ, Rossi A (1987) Cancer mortality of granite workers. Scand J Work Env Health 13 : 26-31

Leanderson P, Söderkvist P, Tagesson C, Axelson O (1988) Formation of 8-hydroxydeoxyguanosine by asbestos and manmade mineral fibres. Brit J of Industr Med 45 : 309-311

Pezerat H, Zalma R, Guignard J, Jaurand MC (1989) Production of oxygen radicals by the reduction of oxygen arising from the surface activity of mineral fibres In : Non occupational exposure to mineral fibres Bignon J, Peto J, Saracci R (Ed) IARC Scientific publication n° 90 100-111

Pham QT, Gaertner M, Mur JM et al. (1983) Incidence of lung cancer among iron miners. Eur J Respir Dis 64 : 534-540

Sutton HC, Winterbourn CC (1989) On the participation of higher oxidation states of iron and copper in Fenton reactions. Free radicals biology and med 6 : 53-60

Zalma R (1988) Thesis"Contribution à l'étude de la réactivité de surface des fibres minérales. Relations possibles avec leurs propriétés cancérogènes". Paris VI University, France

SOLUBILITY AND SURFACE PROPERTIES OF TALC IN A PHYSIOLOGICAL FLUID - RESULTS OF IN VITRO EXPERIMENTS AT 37 °C

Gerburg Larsen
Institut für Mineralogie und Lagerstättenlehre
Rheinisch-Westfälische Technische Hochschule Aachen
Wüllnerstraße 2
5100 Aachen
Federal Republic of Germany

NATO ASI Series, Vol. G 21
Health Related Effects of Phyllosilicates
Edited by J. Bignon

SUMMARY

Solubility and surface properties of granulated talcs and talc powders were investigated at 37° C in Gamble's solution in a closed system. All talc samples release only small amounts of Si into the physiological fluid. Ca and Mg are incorporated into precipitations and gel-like coatings covering the samples at least after 30 days of treatment. Electron diffraction analyses of treated talc powders exhibit remarkable distortions of the diffraction pattern. Comparing all the results with corresponding data for asbestos fibers a different surface reactivity of talc can be stated.

INTRODUCTION

The current discussion of possible health-related effects of those minerals, which may be incorporated into the human body, has been extended to phyllosilicates.
Talc ($Mg_3[(OH)_2/Si_4O_{10}]$) - a most versatile phyllosilicate - is used in widespread application ranges, e.g. for cosmetic purposes, as coating material on polished white rice, or as filler for tablets. Consequently talc particles may be incorporated into different parts of the human body. The retention of these particles in the tissue is controlled by several factors; solubility and surface properties are of great importance.

TALC SAMPLES

Different types of industrial talcs have been subjected to in vitro solution tests at 37 ^{o}C in a closed system. The talc

samples - 3 granulated talcs (TL 1, TL 3, TL 4; particle size approx. 3 mm x 3 mm) and 2 talc powders (TLP 1, TLP 3; grain size approx. 5-30 µm) - represent raw material provided for industrial use, e.g. refractory purposes and cosmetics.
The untreated talc samples are identical to standard talcs concerning their chemical composition (with high LOI-values) and the platy to flaky surface (DEER et al.,1971; WELTON, 1984).
The chemical composition is given below (Table 1):

Table 1: Chemical composition of the talc samples

wt. %	TL 1	TL 3	TL 4	TLP 1	TLP 3
		grains		powders	
SiO_2	65.86	65.72	65.54	64.68	64.91
Al_2O_3	0.27	0.30	0.11	0.52	0.46
MgO	32.17	32.18	31.51	30.55	30.71
CaO	---	0.01	---	0.13	0.09
K_2O	0.03	0.03	0.02	0.02	0.03
Na_2O	---	---	---	---	---
Fe_{tot}	0.14	0.59	1.84	1.26	1.37
MnO	---	0.01	---	---	0.01
TiO_2	0.01	0.01	0.01	0.05	0.04
P_2O_5	---	---	---	0.04	0.02
total	98.48	98.85	99.03	97.25	97.64

PHYSIOLOGICAL FLUID

The physiological fluid - called Gamble's solution - is composed of different salts. The cation and anion concentrations are similar to those of lung fluid (pH value approx. 8). Organic components have not been added to avoid the formation

of micro-organisms. The chemical composition of Gamble's solution is given below (Table 2).
This fluid already proved its usefulness for similar investigations with asbestos fibers, glass fibers, and sepiolite (FECK, 1984; FöRSTER, 1984; LARSEN, 1989...).

Table 2: Chemical composition of Gamble's solution

component	contents [g/l]
magnesium chloride ($MgCl_2$ $6H_2O$)	0.212
sodium chloride (NaCl)	6.171
potassium chloride (KCl)	0.311
sodium hydrogenphosphate (Na_2HPO_4)	0.148
sodium sulphate (Na_2SO_4)	0.079
calcium chloride ($CaCl_2$ $2H_2O$)	0.255
sodium acetate ($NaCH_3COO$ $3H_2O$)	1.065
sodium hydrogencarbonate ($NaHCO_3$)	2.517

METHODS

The solubility experiments were carried out following a simple but very reliable method yielding reproducible results (LARSEN, 1989).
All talc samples were used without pre-treatment. The tests were realized under closed system conditions at 37 ^{o}C in complete darkness.
The fluid/sample ratio was 100 to 1. The duration of the experiments ranged from 1 hour to 60 days.
The filtered fluids were analyzed by AAS to obtain solubility characteristics. The surfaces of the granulated talc samples were examined by SEM/EDX and XPS. TEM was applied for the talc powders.

RESULTS

Granulated talc samples

Chemical analyses (AAS) of the filtered fluids correspond to SEM/EDX results of the treated talc samples.
Figures 1a and 1b show the typical surface morphology of the untreated talc samples TL 1 and TL 4.

(a) 5 µm (b) 10 µm

Fig. 1: Typical surface morphology of untreated granulated talc samples TL 1 (a) and TL 4 (b)

Exposure in the physiological fluid first leads to a rapid solution of impurities and interlayer cations, especially small amounts of Na.
The Ca and Mg concentrations of the filtrates decrease with the duration of the experiments. Precipitations of Ca- and Mg-carbonates and/or -phosphates were detected by SEM/EDX analyses (Fig. 2a). In addition Ca and Mg are enriched in gel-like coatings developed on all talc samples at least after 30

days of treatment (Fig. 2b). This enrichment is related to the formation of gel-like Ca- and Mg-hydroxysilicates forming the coatings as well as talc material from the expanded surface layer. Similar coatings have been reported from sepiolites investigated under resembling conditions (FECK, 1984; LARSEN, 1989).

Fig. 2: (a) Ca- and Mg-rich precipitations on talc sample TL 1 after 30 days of treatment
(b) Talc sample TL 4 exhibiting gel-like coatings after 60 days of treatment

The formation of the gel-like coatings is favoured by higher temperatures. Figure 3a shows the same material TL 4 after a 50 °C experiment. Thick coatings covering the whole micrograph area can be observed. Muscovite grains - included in the 50 °C experiment series for comparison purposes - exhibit remarkable gel-like coatings, which develop on the grain surface according to crystallographic orientations (Fig. 3b).
The K-concentration of the filtered fluids varies slightly with small amounts of K being also incorporated into the coatings.

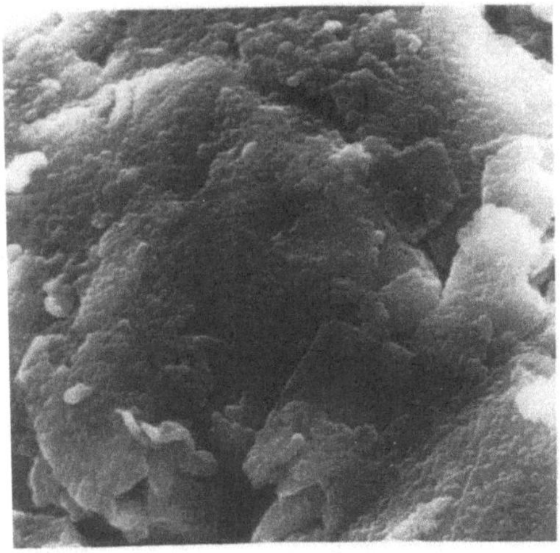

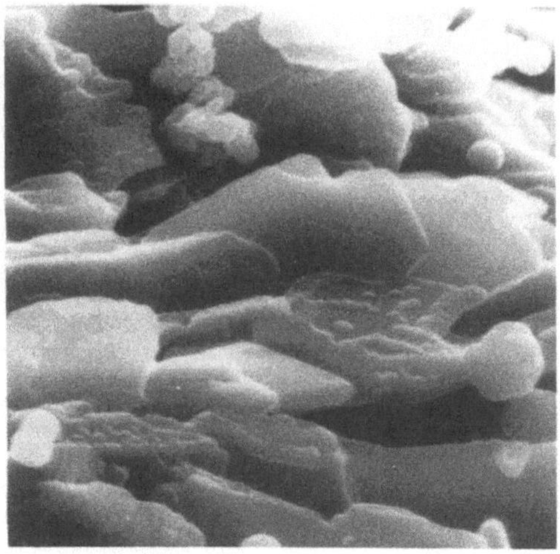

Fig. 3: (a) Talc sample TL 4 after a 30 days experiment at 50 °C with evident gel-like coatings covering the whole micrograph area
(b) Treated muscovite sample (30 days, 50 °C) with coatings following crystallographic orientations

Only small amounts of Si (<< 5 to 250 ppb) were removed from the talc samples. Long-term experiments suggest that the process of Si dissolution develops according to an exponential function. The corresponding saturation values depend on the experiment temperature and the fluid/sample ratio. An exponential model providing good results for fibrous minerals, e.g. chrysotile, crocidolite, and sepiolite (LARSEN, 1989), can be also applied for talc. The exponential function is given below:

$$c = c_{sat} \, (1 - e^{-kt})$$

where:
c - concentration [mg]
c_{sat} - saturation concentration [mg]
t - time [d]
k - exponential coefficient [1/d]

Figure 4 illustrates the Si dissolution for sample Tl 1.

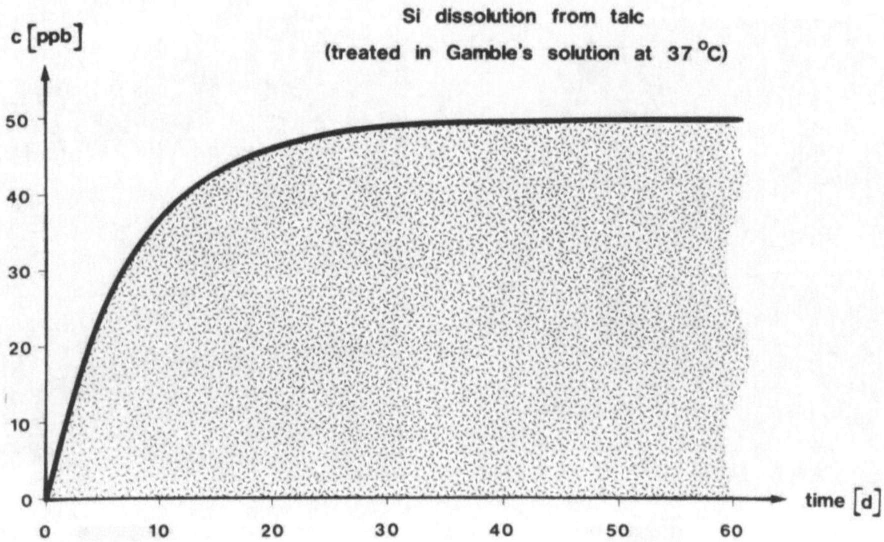

Fig. 4: Si dissolution of talc sample TL 1

XPS analyses of untreated and treated talcs confirm the observation of Si dissolution from the sample surface. The Mg/Si ratio (atomic concentration) increases from 0.88 (untreated talc sample) to 0.96 (treated talc sample).

Talc powders

Chemical analyses of the filtered fluids of the talc powders are comparable with those of the granulated talc samples.
For comparison purposes additional experiments were carried out in deionized water. Small amounts of Na, K, Ca, and Si are removed from the talc powders. In contrast to the tests in Gamble's solution the Mg concentration of the filtered fluids increases with the duration of the experiments and reaches values > 120 ppm.
TEM and electron diffraction analyses reveal that the talc powders - treated in the physiological fluid - exhibit a remarkable distortion of the diffraction patterns. This is

probably due to the influence of the physiological fluid. It can be suggested that the changes of the electron diffraction pattern are caused by the migration of cations into the interlayer positions resulting in interlayer tensions. The talc particles seem to swell up and flake off. In addition the crystal margins seem to be tilted. Very thin talc crystals tend to roll up into small tubes (Fig. 6b). In contrast diffraction patterns of talc powders treated in deionized water are not distorted.

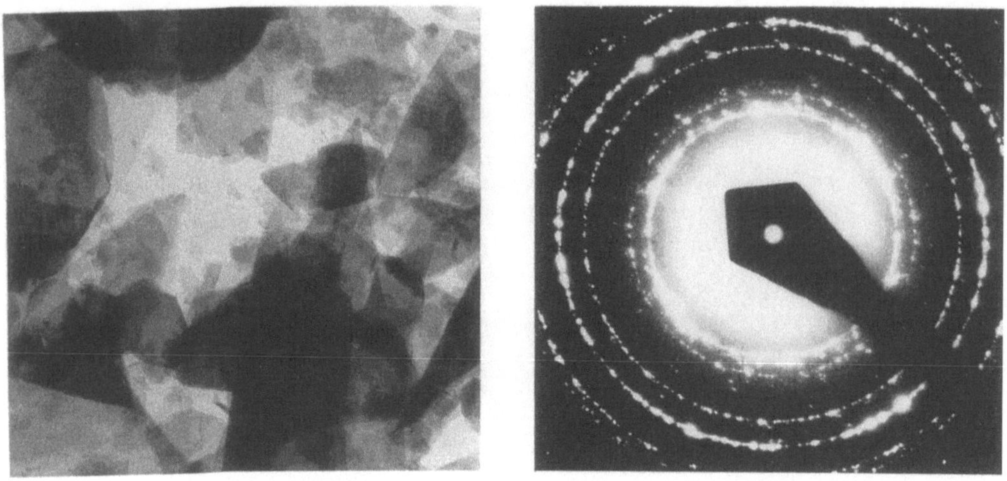

(a) 1 μm (b)

Fig. 5: TEM micrograph (a) and electron diffraction analysis (b) of untreated talc powder TLP 1

DISCUSSION

As result of the solubility experiments and surface properties investigations it can be stated that - under intermediate pH conditions - the crystal structure of talc is protected by surface coatings and cannot be attacked easily by the physiological fluid. Comparing the results with corresponding data for chrysotile and crocidolite asbestos there are no significant differences in solubility concerning the order of magnitude. But chrysotile and crocidolite, which have been treated under similar conditions, exhibit uncovered surfaces. Thus a different surface reactivity of talc towards the physiological fluid can be concluded.

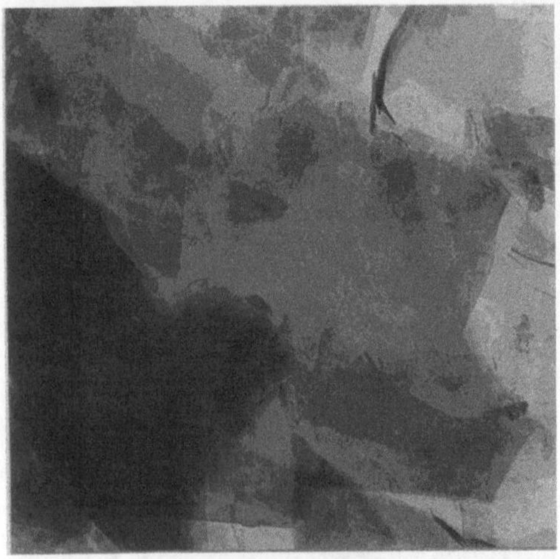

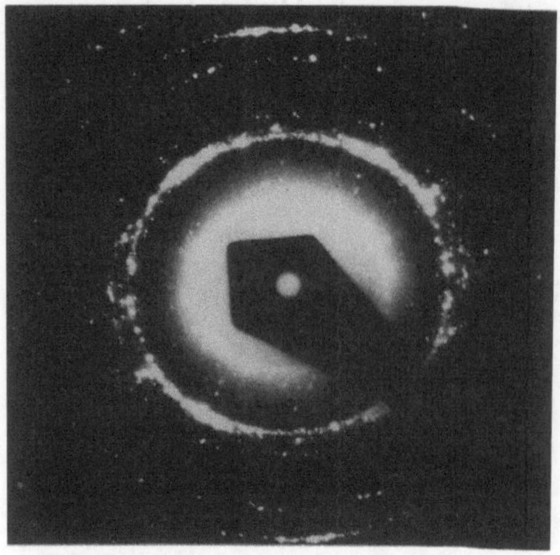

(a) 200 nm (b)

Fig. 6: TEM micrograph (a) and electron diffraction analysis (b) of talc powder TLP 1 treated in Gamble's solution for 30 days; some grains with thinning margins and few rolled up tubes can be observed. The diffraction pattern is slightly distorted.

REFERENCES

Deer WA, Howie RA, Zussmann J (1971) Rock-forming minerals. Vol 3. Sheet silicates. Longman Group Ltd., London

Feck G (1984) In vitro-Untersuchungen zur Löslichkeit von Glasfasern und einigen Mineralfasern in schwach alkalischer wäßriger Lösung. PhD thesis, RWTH Aachen

Förster H (1982) Behaviour of mineral fibres in physiological solutions. In: Biological effects of man-made mineral fibres. Vol 2. WHO regional office for Europe, Copenhagen, p 27-59

Larsen G (1989) Experimental data on in vitro fibre solubility. In: Bignon J, Peto J, Saracci R (eds) Non-occupational exposure to mineral fibres, IARC Scientific Publications No 90, p 135-139

Welton JE (1984) SEM petrology atlas. The AAPG Methods in Exploration Series No. 4, Tulsa, Oklahoma/USA

MINERALOGY OF TALC DEPOSITS

Jocelyne FERRET, Doctor in mineralogy
Philippe MOREAU, Doctor in Geology
Talc de Luzenac
B.P. 1162
31036 TOULOUSE CEDEX FRANCE

ABSTRACT

Synthesis of the mineralogy of products consumed by the world industry under the name of talc.
Nearly 4000 analyses were carried out in Talc de Luzenac's central laboratory, 1500 of which were conducted on products coming from other companies.

This poster shows the complexity of the notion of talc, mineral compositions varying considerably according to the deposit. The European picture is very different from the American one.
Tremolite is exceptional, always non asbestiform ; antophyllite, also non asbestiform, can only be found in one deposit (USA).

Talc is a hydrated magnesium silicate commonly used in numerous industrial sectors, because of its specific physical and chemical properties : inertia, whiteness, lamellarity, softness and low abrasiveness, highly hydrophobic (and lipophyllic) basal surfaces.

NATO ASI Series, Vol. G 21
Health Related Effects of Phyllosilicates
Edited by J. Bignon

The world production amounts to nearly 5.7 million tons, divided up as follows :

Western Europe :	1 150 000 T	(Turnover ≈ 130 M$)
Eastern Europe (included USSR) :	630 000 T	
Northern America :	1 300 000 T	(Turnover ≈ 125 M$)
Southern America :	300 000 T	
Asia :	1 800 000 T	
Australia :	350 000 T	
Others :	100 000 T	

Altogether, 600 indices and deposits are known, 250 of which are exploited. This industry generates about 35 000 direct jobs.

The concept "talc" has several meanings :

- the mineral : $Mg_3\ Si_4\ O_{10}\ (OH)_2$

- the ore : it can be associated in deposits with other minerals presenting properties that may be either complementary or prejudicial to the product for its industrial applications. These minerals can be eliminated by various techniques.

- the mineral product : it is generally a white or light grey powder, composed uniquely of talc or associated with minerals present in the deposit.

For the public, the term "talc" has generally a very precise meaning : perfumed white body dusting powder.

The poster is based on the study of about 140 deposits all over the world and on 4000 analyses of minerals and end-products carried out in the geology department and the Talc de Luzenac's laboratory over the last ten years.

I) GEOLOGY OF TALC DEPOSITS

1) General considerations

- Talc deposits were formed by transformation of other rocks of the earth's crust under the action of hydrothermal solutions.

- The quality of a talc deposit is dependent on :
  . the nature of the parent rock : direct influence on the mineralogy,
  . the formation conditions (temperature, pressure, duration, nature of hydrothermal solutions) ,

- The minerals associated with talc can be :
  . either inherited from the parent rock, generally as traces, e.g. : remains of dolomite, iron and nickel sulphides, etc...

  . either formed together with talc : in this case, they generally form a major component of the ore : e.g. : chlorite, magnesite (up to 50 %).

- These by-minerals :
  . can be processed with the talc and impart specific characteristics to the commercial product,
  . or can be removed by various techniques (selective mining, selective grinding and classification, flotation, magnetic separation, and so on),

2) The main types of talc deposits and related characteristics

We classify the talc deposits into 4 main types, according to the nature of their host rock :

1) Transformation of carbonates (dolomite, magnesite) :
   50 % of the world production.
   Talc is the dominant mineral.
   Product is often white.

2) Transformation of silico-aluminous rocks (granites, gneiss, schists)
   10 % of the world production.
   Chlorite is the dominant mineral.
   Product is generally green to grey.

3) Transformation of serpentines
   40 % of the world production.
   Talc or talc-carbonate mixture.
   Product is always grey (high iron content).

4) Transformation of magnesium clays
   No deposit of this type worked today.
   Generally mixture of talc and quartz.

Certain complex deposits may be a combination of two types. The enclosed table gives the mineralogical properties of these deposits, together with the transformation factors and some examples.

THE MAIN TYPES OF TALC DEPOSITS

EX. MAGNESIAN CARBONATES
$(Mg - Ca)CO_3$

. Major mineral : TALC
. Color : white or color of parent rock (grey to black)
. Main reaction :
Dolomite + SiO_2 + H_2O ==> Talc + CaO + CO_2
or Magnesite + SiO_2 + H_2O ==> Talc + CO_2

. Possible by-minerals :
- Inherited from host-rock : carbonates, graphite, pyrite

- Newly formed : at low temperature : traces of calcite, quartz
at high temperature (> 480°) : tremolite

EX SILICO ALUMINOUS ROCKS
(Schiste, Gneiss, Granite)

. Major mineral : CHLORITE
. Color : green to grey
. Main reactions :
Mica or Feldspath + MgO + Water ==> Chlorite + SiO_2 + x
Quartz + MgO + Water ==> Talc

. Possible by-minerals :
- Inherited from host-rock : traces of mica, quartz, calcium silicates, graphite, pyrite.
- Newly formed : traces of tremolite

PARENT ROCK

Motor of the transformation into talc

EX SERPENTINES (Mg,Fe, hydrous silicates)

CO_2 brought

. Major minerals : Talc + Carbonates
. Colors : grey (3 to 8% iron) (magnesite and/or dolomite)
. Main reaction :
Mg Serpentine + CO_2 ==> Talc + Magnesite
(Mg-Ca) Serpentine + CO_2 ==> Talc + Magnesite + Dolomite
. Possible by-minerals :
- Inherited from host-rock : iron oxide & sulphides
Ni traces, antigorite (rare)
Chrysotile is never present (first transformed into talc)
- Newly formed : chlorite.

SiO_2 brought

. Major mineral : Talc
. Color : grey (iron)
. Main reactions :
Mg Serpentine + SiO_2 ==> Talc + H_2O
(Mg-Ca) Serpentine + SiO_2 ==> Talc + Actinolite (1 to 5 %)
. Possible by-minerals :
- Inherited from host-rock : iron oxide or sulphide, antigorite.
- Newly formed : chlorite

EX MAGNESIAN CLAYS (rare type)

. Major minerals : Talc + Quartz
. Color : white to yellow (if iron traces)
. Main reaction :
Mg Clay ==> Talc + Quartz + H_2O
. Possible by-minerals :
- Inherited from host-rock : mica, clays, organic stuff, hydroxides
- Newly formed : chlorite, tourmaline

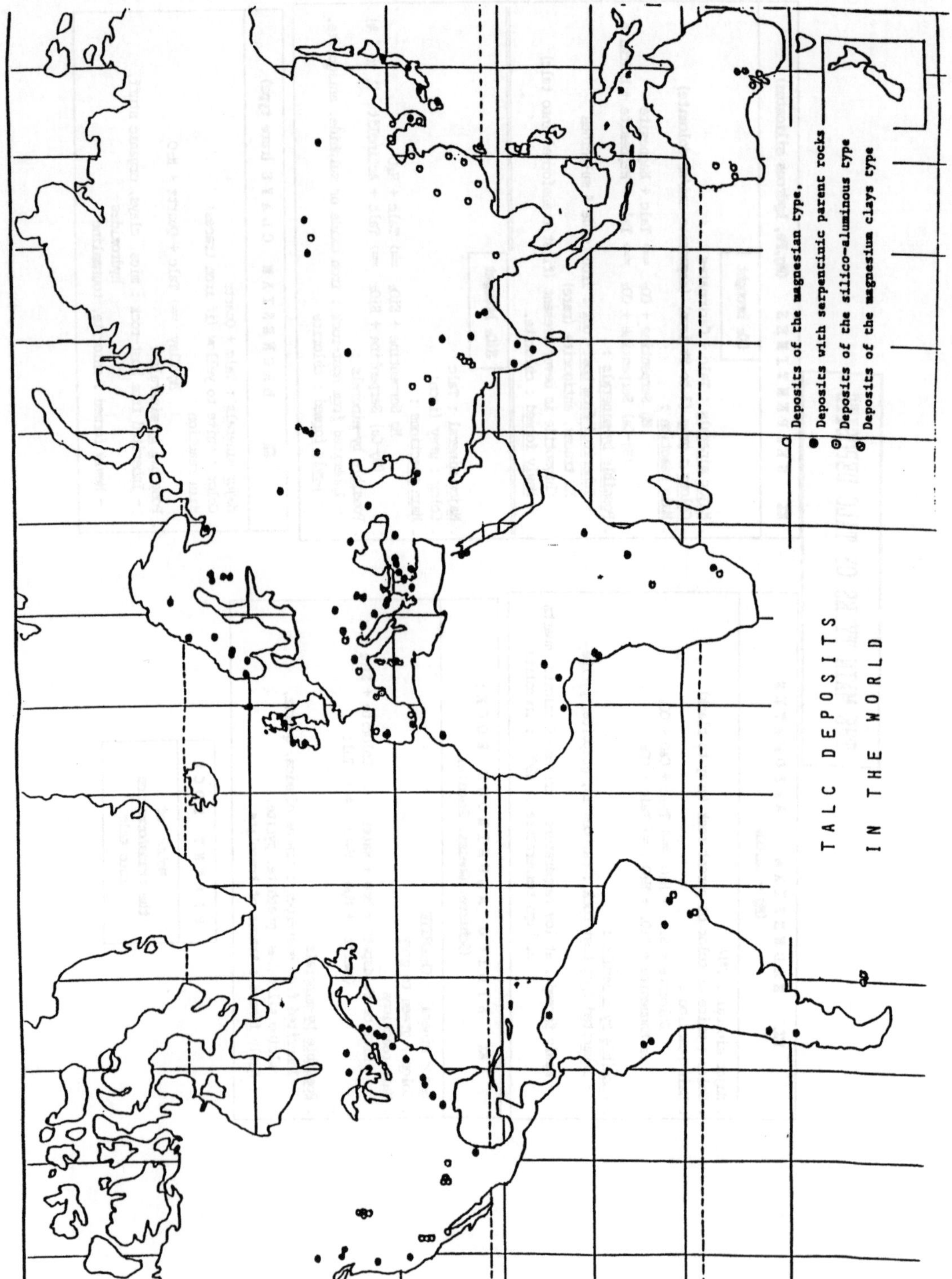

TALC DEPOSITS IN THE WORLD

CONCLUSIONS ON THE GEOLOGY OF TALC

- The 4 major types of talc deposit may be sub-divided into various types according to the possible variations of the parent rock or formation conditions.

In some deposits, selective mining enables a large number of different grades to be obtained : example : up to 13 different grades of ore from the Trimouns deposit, Luzenac France

In fact, each deposit is a specific case as regards mining and processing techniques.

These numerous types of deposits provide a variety of products called "talc" : more than 500 different products throughout the world.

- The deposits throughout the world may be divided up as follows :

. Deposits of the magnesian carbonates type, (talc is the dominant mineral) : aleatory situation : 7 to 8 deposits in Northern America, 3 to 4 in Europe, but a large number in Asia (India - China - Korea) and Australia.

. Deposits of the silico-aluminous type, (chlorite is the dominant mineral) this type is generally limited to Europe. Their exploitation is explained by the presence of a strong local market ; other deposits of this type may exist elsewhere but are not exploited because of their location.

. Deposits with serpentinic parent rocks (talc + carbonates, greyish) : located along the major ranges of different orogenic periods : the Andean range, the Rocky Mountains, the Alpine range (Europe, Asia), Caledonian range (East coast of Northern America and Scandinavia), Precambrian (Finland, Ontario, Brazil).

II) MINERALOGY OF PRODUCTS SOLD UNDER THE NAME TALC

Nearly 4000 analyses were carried out by Talc de Luzenac's central laboratory, 1500 of which were conducted on competitive talcs.

The two tables enclosed indicate the average mineralogical compositions of the analysed products, grouped according to their source, together with corresponding tonnages. For this study, the choice of products was limited to Europe and Northern America.

The mineralogical compositions are determined, qualitatively by X-ray diffractometric analysis and quantitively by thermogravimetric analysis.

These tables show the diversity in the mineralogical composition of products sold under the name talc and show that tremolite (fibrous or non-fibrous) is a rare occuring mineral.

They also underline the unique nature of the products from New-York state.

AVERAGE MINERALOGICAL COMPOSITION OF COMMERCIAL PRODUCTS SOLD UNDER THE NAME TALC IN NORTHERN AMERICA

	Canada			Vermont				California	Texas	Montana		New-York
Production in KT	40 floated	10	30	70	30 floated	200	12 floated	10	307	326	20	140
Mineralogy												
Talc (%)	92.5	64.5	60.5	55	90	52.5	94.5	54	80	94	8	25
Chlorite (%)	3	11.5	10.5	7	7	9	1.5	5	1	4.5	85.5	
Dolomite (%)	1	4	8	2	0.5	2	0.5	9	12.5	0.5	0.5	
Magnesite (%)	1.5	17	18	34	2	33.5	0.5	16		T	T	
Serpentine (%)			T									25
Quartz									T	T	T	
Mica	T	T							T	T	T	
Calcite	T											
Tremolite (%)												44
Anthophyllite (%)												5

T = identified mineral that can not be measured by our methods of analysis

AVERAGE MINERALOGICAL COMPOSITION OF COMMERCIAL PRODUCTS SOLD UNDER THE NAME TALC IN EUROPE

	Finland		Sweden	Norway	UK	France	Austria	Italy				Spain		
Production in KT	75 floated	250 floated	15	50	17	320	80	20	40	46	17	33	20	28
Mineralogy														
Talc (%)	93	88	64	55	54	59	51.5	51.5	86	51	47	89	80.5	53
Chlorite (%)	3.5	8.5	16.5	11	9	39	42	43	9.5	19.5	22.5	6	12	18.5
Dolomite (%)	0.5	T	11.5	2	2	1.5	1	2	1.5	12	14.5	2	1.5	6
Magnesite (%)	1.5	2		29	30.5		1		0.5	10	14.5			18.5
Serpentine										T	T			
Quartz			T	T	T		T	T		T			T	
Mica (%)			T				T	T				1.5	1.5	
Calcite			T	T				T		T		0.5	T	
Tremolite			T											

T = identified mineral that can not be measured by our methods of analysis

GENERAL CONCLUSIONS

The "concept talc" covers a very large range of industrial products adapted to various markets. The mineral talc can be either the dominant mineral or associated with other minerals. The analyses provided in this paper show the great qualitative variability covered by the word "talc".

For approximately 50 % of the tonnage produced in the world, talc is the dominant mineral (95 % or more).

For the second half, talc is associated with variable percentages (generally 10 to 50 %) of carbonates and/or chlorite.

Other minerals can be found, in minor amounts, either inherited from the parent rock or formed together with talc.

Tremolite is major component only in the deposits of New-York state. Traces of tremolite can be found in other deposits formed at high temperature ; it is very generally non-asbestiform.

We never observed in talc ores the presence of chrysotile (one of the first minerals transformed into talc) or crocidolite (talc deposits contain no sodium).

In fact, on the world market, more than 500 different products are sold under the name "talc", each having its own specific physical, chemical and mineralogical characteristics. This makes a fundamental difference with other industrial minerals, which are usually mono-minerals.

Therefore, when studying talc-containing products, one has to be extremely careful as regards the origin of products studied (deposits, producer) and in no case extend the observations or results obtained for one or several products to all marketed products. Such a lack of rigour would be detrimental to both the producers of a safe raw material and the credibility of the authors of these studies.

One has to be even more careful when those studies are the basis of regulations : the only way of working rigorously and scientifically is to establish a dialogue between geologist, mineralogists, producers, end-users, medecine doctors and regulation makers, each one having his/her place in the jigsaw puzzle and avoiding the other's excessive conclusions, interpretations or generalisations.

PART II

CLINICAL AND EPIDEMIOLOGICAL EVIDENCE OF HEALTH EFFECTS AFTER OCCUPATIONAL EXPOSURE

RESPIRATORY DISEASE FROM OCCUPATIONAL EXPOSURE TO NON-FIBROUS PHYLLOSILICATES

P Cullinan and JC McDonald
Department of Clinical Epidemiology
National Heart and Lung Institute
Brompton Hospital, London SW3, UK

1. INTRODUCTION

Many phyllosilicate minerals are non-fibrous in habit; some are usually fibrous (eg. sepiolites) and some are often found in association with fibrous forms (eg. serpentine with chrysotile). This review will consider only those which are characteristically non-fibrous:- kaolin, talc, vermiculite and mica.

Identification of pathological effects due to occupational exposure is hampered by the looseness with which some terms have been used (eg 'talc'), by the often great differences between past and current exposure levels and chiefly perhaps by the problem of coexposure. The most common contaminants are crystalline silica (particularly with kaolin and mica) and amphibole asbestos (with talc and vermiculite). These make difficult any study of the association between phyllosilicate exposure and lung cancer. The situation is further complicated by the uncertain relationship between early radiographic changes and pulmonary fibrosis.

Despite their very frequent occçurence and intensive mining, disease has seldom been attributed to the non-fibrous phyllosilicates. Where disease has been observed it has usually followed prolonged and heavy exposure. Inhalational and intratracheal studies of phyllosilicates in laboratory animals also suggest that their pulmonary fibrogenicity is low (Wagner et al, 1977; Goldstein and Rendall, 1970, King et al, 1940). A protective effect of kaolin and mica on the fibrogenic potential of quartz has even been suggested (King et al, 1950). Animal studies have failed to demonstrate

NATO ASI Series, Vol. G 21
Health Related Effects of Phyllosilicates
Edited by J. Bignon

convincingly the carcinogenicity of pure talc (Wagner et al, 1977, Ozesmi et al, 1985, Pott et al, 1976), mica (biotite) (Pott et al, 1974) or vermiculite (Hunter and Thompson, 1973). We are unaware of any evidence that pure kaolin is carcinogenic to animals.

2. KAOLIN

Kaolin, a non-fibrous hydrated aluminium silicate derived from the metamorphic alteration of feldspars, has been mined for over 3000 years in China. Its diverse physical properties and its low cost account for its very widespread use in the paper and ceramic industries, in the manufacture of plastics, paints and fertilisers, and in certain medicines. The mineral is frequently found in association with quartz and mica (as in the 'china-clay' deposits in Cornwall, UK) but some ores are remarkably pure, as in the sedimentary deposits of central Georgia, US, where the free quartz contamination is less than 1%.

The mining and processing of kaolin involves both wet and dry phases but significant occupational dust exposure occurs only during the latter; during kiln drying, bagging and loading. Processed kaolin particles vary from less than two microns to about ten microns across. With better control methods dust exposures are considerably lower than in previous years.

MORBIDITY Middleton (1936) is credited with the first suggestion of an association between lung disease and occupational kaolin exposure. He reported two cases of pneumoconiosis in Cornish china-clay workers, one of which was probably silicosis. Twenty years later, Lynch and McIver (1954) reported the post mortem findings of two US kaolin miners with progressive massive fibrosis, and subsequently coined the term 'kaolinosis'. A number of case reports since have described fine nodular radiographic changes in kaolin workers with high exposures but these have been frequently complicated by coexposures (particularly to crystalline silica) and by past exposures to other dusts.

Ten cross-sectional surveys of kaolin workers mainly in Georgia or Cornwall are summarised in Table 1. Radiological evidence of simple pneumoconiosis was reported in between 0.6% and 22.6% of subjects. Opacities were usually of the small rounded type. Complicated pneumoconioses were much less common and pleural involvement very unusual. Radiographic changes were associated with high or long dust exposure and complicated pneumoconiosis only after very long exposure.
In the study of Cornish miners by Oldham, (1983), the presence of radiographic abnormalities was associated with time spent in dusty jobs, regardless of age and smoking history. The effect was greater for more recent exposures, suggesting to the author that X ray changes reflect alveolar and lymphatic deposition of dust rather than fibrosis. Similarly, in a survey of miners in Georgia the radiographic appearances of fine nodular shadowing had not changed when x-rays were repeated between 3 and 12 years later. In the study of Cornish miners by Ogle et al, (1989), regression analysis showed that exposure to china stone dust (containing crystalline silica) had the greatest effect on radiographic appearance.

In a small survey of 65 kaolin miners and millers in Georgia (Sepulveda et al, 1983), FEV1 and FVC were significantly reduced in comparison to an unexposed referent group, but the changes were seen only in past or current cigarette smokers. Two larger studies of miners from the same area (Kennedy et al, 1983; Morgan et al, 1988) have shown minor restrictive changes, confined to men with complicated pneumoconiosis. Oldham (1983) and Ogle et al (1989) have also demonstrated small reductions in vital capacity in men with simple pneumoconiosis, but in the Cornish mines there was probably coexposure to free silica.

MORTALITY There has been the occasional mention(Sheers, 1964; Lapenas et al, 1984) of respiratory cancers in workers exposed to kaolin but no cohort study of mortality has been reported.

3. TALC

Pure talc is an hydrated magnesium silicate of a loosely bound triclinic structure. Since industrial talc is derived from the metamorphic alteration of several different rocks, it is commonly found in association with a variety of other 'contaminating' minerals, commonly fibrous tremolite and quartz and occasionally chrysotile.

About 5m tonnes of talc are mined annually, largely in China, Western Europe and the United States. Significant occupational exposure to talc dusts has occured during mining, processing and bagging and in secondary industries such as rubber and ceramic manufacture and in dusting. Talc ores from different sources are variously contaminated; those from Vermont, France and Italy are said to contain very low concentrations of quartz and asbestiform fibres, while those from New York State are frequently contaminated with these minerals (Dement and Zumwalde, 1979).

MORBIDITY Pneumoconiosis following occupational talc exposure was first reported as long ago as 1896 (Thorel, 1986). There have been several case reports and case series since, which have described fine reticular infiltrates, pulmonary fibrosis and pleural thickening in workers heavily exposed to allegedly pure talc. Post mortem findings are not typical of silicosis.

Six cross-sectional studies of respiratory morbidity in talc workers are listed in Table 2. Radiographic pulmonary infiltrates (small opacities) were reported in between 0.7% and 37.2% of subjects. This last figure was in 46 selected New York State miners with an average of 19 years of high exposure (Kleinfield et al, 1965). In a later study of miners from the same area (Gamble et al, 1979) radiographic abnormalities were found only in 3.4% workers with more than 15 years exposure. In neither study can coexposures be excluded.

In two studies of workers exposed to talc (Fine et al, 1976, Gamble et al, 1982) with less than 2.2% respirable silica and very low fibre concentrations, radiographic pneumoconiosis was found in only 0.8% of 359 workers. On the other hand Wegman et al, (1982), found small rounded opacities in 10.3% of a selected group of 116 Vermont miners and millers exposed to allegedly pure talc. In only one survey, has complicated pneummoconiosis been reported, in 5 of 22 talc workers from Piedmont Italy (Scansetti et al, 1963). It is not clear how these workers were selected or whether they were also exposed to silica.

In 121 miners from St Lawrence county New York (Gamble et al, 1979), there were small but significant decreases in FEV1 and FVC (but not in their ratio) in talc workers compared with a referent group of potash miners, after correction for smoking habit. There were significant associations between these parameters and talc exposure estimates. In a one year follow up (Wegman et al, 1982) of 103 workers exposed to allegedly uncontaminated talc, the annual rates of loss of FEV1 and FVC were not significantly greater than predicted nor correlated with exposure. Gamble et al, (1982) did not find any increase in symptoms or reduction in lung function in 299 talc miners and millers when compared with two unexposed referent groups.

Pleural thickening (usually bilateral) has been demonstrated in talc workers. Prevalences range from 4% in Montana,(Gamble et al, (1982) to 31% in New York,(Gamble et al, 1979) and have tended to be higher in workers exposed to amphibole contaminated talcs. In two studies involving uncontaminated talc pleural thickening was seen in 3.4% (Wegman et al, 1982) and 4% (Gamble et al, 1982) of exposed workers; the significance of these pleural changes is unclear.

<u>Mortality</u> The occasional presence of asbestos contaminants in talc has prompted the question of whether occupational talc exposure is associated with an increased risk of malignancy and, if so, whether talc itself might be partially responsible. Table 3 summarises seven cohort studies of mortality in talc workers. Four of the studies were of talc workers in New York State where the exposure may or may not have included mineral fibres. In the study by Kleinfeld <u>et al</u> (1974) of 108 deaths in 260 New York talc workers the proportional mortality from respiratory cancers was over three times that expected; the excess was seen only in men who died between the ages of 60 and 79 years. In later studies in the same mines and to some extent in overlapping cohorts, SMRs for respiratory cancers ranged from 163 to 290. These findings are difficult to interpret and have been subject to considerable controversy.

In the study by Stille and Tabershaw (1982), the small excess (11 vs 6.7) could have been explained by previous employment elsewhere. Lamm et al (1988), in a reanalysis, reported that the lung cancer excess was concentrated in men who had either been employed for less than one year at the mine in question or had prior employment which could have been responsible.

In three of the four cohort studies in the New York talc industry, a total of 4 deaths from malignant mesothelial disease were reported. One was described as a pleural fibrosarcoma, one as a peritoneal mesothelioma and in the remaining two the site was unspecified. It is not clear whether these represent four different cases.

Three of the studies listed in Table 3 are of workers exposed to talc apparently free from mineral fibre. In only one, of Vermont talc miners and millers (Selevan <u>et al</u>, 1979), was there an excess of respiratory malignancies. This was apparent only in the miners where the possibility of exposure to radon daughters or tremolite fibres could not be excluded. In a further study of unspecified size from the USSR

(Katsnelson and Mokronosova, 1979), very high death rates from lung cancer were found in men occupied in mining and processing talc. Details of the methodology and mineral analysis in this study were insufficient for interpretation.

In all three studies concerned with non-asbestiform talc there were excesses of deaths due to non-malignant respiratory disease (NMRD), including pneumoconiosis. In the Italian study (using a national referent population) the excess in deaths from NMRD (SMR=279 based on 127 deaths) was seen only in miners, in whom there was a clear relationship with cumulative exposure. This was attributed to exposure to free silica during mining. In the Vermont study, on the other hand, the excess in deaths from NMRD (SMR 300, on 11 observed deaths) was confined to millers presumed to have had a greater cumulative dust exposure. In five out of the six deaths attributed to pneumoconiosis there were radiographic changes of pneumoconiosis (2/1 or more) on chest x-rays taken during employment.

4. VERMICULITE

Vermiculite, an hydrated aluminium-magnesium phyllosilicate has the unique property of exfoliating when flash heated to high temperatures. The resultant product is fire resistant, highly insulating and of very low density. The mineral is mined chiefly in the United States and in the Republic of South Africa. In its expanded phase it is used as a thermal and acoustic insulator in the construction industry and as a horticultural and agricultural soil filler. The natural unprocessed mineral is frequently contaminated with amphibole fibres in the tremolite series, in varying concentrations. Dust exposure occurs during mining, dry processing, bagging and loading of the mineral. During mining, there may be exposure to free silica also.

MORBIDITY There have been three cross sectional studies of respiratory morbidity in vermiculite miners and millers and one in factory workers, all in the United States (see Table 4). Pneumoconiotic changes (usually small irregular opacities) were found in up to 50% of exposed workers, the prevalence related to cumulative fibre exposure. Pleural thickening and calcification were also frequent and directly related to fibre exposure. Twelve cases of benign pleural effusion in factory workers exposed to Montana and South African vermiculites were described by Lockey et al, (1984).

MORTALITY Three cohort studies of respiratory mortality in vermiculite workers are summarised in Table 5. In the studies at the Libby mine in Montana, where there was exposure to amphibole-contaminated vermiculite, the SMRS from malignant and non malignant respiratory diseases were significantly increased in direct relation to estimates of cumulative fibre exposure. In the smaller study at Enoree, South Carolina, where amphibole exposure was minimal, SMRs for malignant and non malignant respiratory disease were not significantly raised. Four deaths from mesothelioma were reported in the Libby studies; three were pleural and one peritoneal. No mesothelioma death was reported from Enoree.

5. MICA

The micas are complex aluminium silicates which are used in paint and construction industries, in the electrical industry and in oil-drilling. Its fire resistant properties make it an increasingly used substitute for asbestos. Contamination of mica dust with other minerals (chiefly quartz) is very common, particularly during mining. It is only in the grinding and processing of the mineral, and in secondary industries, that exposure to relatively pure mica occurs.

Morbidity Ferguson, in an unpublished memorandum (Middleton, 1936) described 5 cases of pneumoconiosis in 12 mica workers. Davies and Cotton, (1983) have reported disabling pneumoconiosis in two workers in a mica (muscovite) processing plant. Neither had previous dust exposure and in both cases the lung disease was attributed to pure mica dust. Pimentel and Menexes (1978) described severe restrictive pneumoconiosis and pulmonary granulomas in a woman exposed to uncontaminated ground mica. A case of mild interstitial fibrosis has been described in a young woman exposed to pure mica dust through her husband's occupation (Lapenas et al, 1982).

Table 6 summarises three epidemiological surveys of occupational exposure to mica. The first two were in groups of workers who were probably exposed to varying concentrations of pure mica dust. Workers in the third survey were exposed to quartz-contaminated mica. In the survey by Dreessen et al (Dreesen et al, 1940) the prevalence of pneumoconiosis correlated with length and intensity of exposure. In a survey of 302 Chinese mica workers (Li Weizu, 1980) no case of mica pneumoconiosis was found (Skulberg et al, 1985). Pleural thickening has been reported in persons exposed to mica for long periods (Smith, 1952; Green and Churg, 1988); the significance of this is unclear.

Mortality There has been no cohort or other epidemiological survey of respiratory cancer in mica exposure; however a single case of a peritoneal mesothelioma in a mica worker has been reported (Chahinian et al, 1982).

6. CONCLUSIONS

There is little evidence that occupational exposure to pure kaolin, talc, vermiculite or mica carries any important risk to health. Some studies suggest that long heavy exposure to kaolin and mica may give rise to low grade radiographic changes but very seldom to large opacities. Pleural changes are common in talc workers but may be due to mineral fibre impurities. It appears unlikely that lung cancer or mesothelioma have been caused by any of the four phyllosilicates reviewed. Further well designed studies are still needed, nevertheless, to confirm these conclusions;

these should be on a sufficient scale with detailed qualitative and quantitative characterization of exposure.

References

Altekruse EB, Chaudhary BA, Pearson MG, Morgan WKC (1984). Kaolin dust concentrations and pneumoconiosis at a kaolin mine. *Thorax*, 39, 436-441.

Amandus HE and Wheeler R (1987). The morbidity and mortality of vermiculite miners and millers exposed to tremolite - actinolite. Part II. Mortality. *Am. J. Ind. Med.*, 11, 15-26.

Amandus HE and Wheeler R (1987). The morbidity and mortality of vermiculite miners and millers exposed to tremolite - actinolite. Part III. Radiographic findings. *Am. J. Ind. Med.* 11, 27-37.

Brown DP, Dement JM and Wagoner JK. (1979). Mortality patterns among miners and millers occupationally exposed to asbestiform talc. *In: Lemen R and Dement JM, eds. Dusts and Diseases.* pp317-324. Pathotox. Inc., Illinois.

Chahinian P, Pajak TF, Holland JF et al 1982. Diffuse malignant mesothelioma. Prospective evaluation of 69 patients. *Ann. Intern. Med.* 96, 746-755.

Davies D and Cotton R (1983). Mica pneumoconiosis. *Br. J. Ind. Med.*, 40, 22-27.

Dement JH, Zumwalde R (1979). Occupational exposures to talcs containing asbestiform minerals *In: Lemen R and Dement JM eds. Dusts and Diseases.* pp 287-305. Pathotox Inc. Illinois.

Dreesen WC, Dallavalle JM, Edwards T et al (1940). Pneumoconiosis among mica and pegmatite workers. *US Public Health Bulletin* No 250. Washington.

Edenfield RW (1960). A clinical and roentgenologic study of kaolin workers. *Arch. Environ. Health*, 5, 392-403.

Fine LJ, Peters JM, Burgess WA and Di Berardinis LJ (1976). Studies of respiratory morbidity in rubber workers. *Arch. Envir. Health*, 31, 195-200.

Gamble J, Greife A and Hancock J (1982). An epidemiological-industrial hygiene study of talc workers. *Ann Occup. Hyg.*, 26, 841-859.

Gamble JF, Fellner W and DiMeo MJ (1979). An epidemiologic study of a group of talc workers. *Am. Rev. Respir. Dis*, 119, 741-753.

Goldstein B and Rendall REG (1970). The relative toxicities of the main classes of minerals. In: *Pneumoconiosis, Proceedings of the International Conference, Johannesburg 1969, ed., HA Shapiro*, pp429-434. Oxford University Press, Cape Town.

Green FH and Churg A (1988). Diseases due to nonasbestos silicates. *In: Churg A and Green FH, eds, Pathology of Occupational lung disease.* pp.177-211. Iga Ku-Shon, New York, Tokyo.

Heimann H, Moskowitz S, Iyer C et al (1953). Silicosis in mica mining in Bihar, India. *Arch Ind. Hyg. Occup. Med*, 8, 420-435.

Hunter B and Thompson C (1973). Evaluation of the tumorigenic potential of vermiculite by intrapleural injection in rats. *Br. J. Ind. Med.*,30, 167-173.

Katsnelson BA and Mokronosova KA (1979). Non fibrous mineral dusts and malignant tumors. *J. Occup. Med*, 21, 15-20.

Kennedy T, Rawlings W, Baser M and Tockman M (1983). Pneumoconiosis in Georgia kaolin workers. *Am. Rev. Resp. Dis*, 127, 215-220.

King EJ, Harrison CV and Nagelschmidt G (1940). The effects of kaolin on the lungs of rats . *J. Path. Bact.*, Vol LX, 435-440.

King EJ, Ray SC and Harrison CV (1950). The effects of granite on the lungs of rats. *Br. J. Ind. Med*, 7,37-41.

Kleinfield M, Messite J and Zaki MH (1974). Mortality experiences among talc workers: a follow up study. *J. Occup. Med.*, 16, 345-349.

Kleinfield M, Messite J, Shapiro J and Snencicki R (1965). Effect of talc dust inhalation on lung function. *Arch Environ. Health*, 10, 431-437.

Lamm SH, Levine MS, Starr JA and Tirey SL (1988). Analysis of excess lung cancer risk in short term employees. *Am. J. Epidem.*, 127, 1202-1209.

Lapenas D, Gale P., Kennedy T et al. 1984. Kaolin pneumoconiosis. *Am Rev Respir Dis*, 130, 282-288.

Lapenas DJ, Davis GS, Gale PN and Brody AR (1982). Mineral dusts as etiologic agents in pulmonary fibrosis. *Am. J. Clin. Pathol.*, 78, 701-706.

Léophonte P, Basset MF, Pincemin J et al (1983). Mortalité des travailleurs du talc en France. *Rev. Jr. Mal. Resp.*, 11, 489-490.

Li Weizu (1980). A preliminary report on the conditions of the Mou mica mine dusts causing lung disease. *Chin, J. Prev. Med*, 14, 177-178.

Lockey JE, Brooks SM, Jarabek AM et al (1984). Pulmonary changes after exposure to vermiculite contaminated with fibrous tremolite. *Am. Rev. Respir. Dis.*, 129, 952-958.

Lynch KM and McIver FA (1954). Pneumoconiosis from exposure to kaolin dust: kaolinosis. *Am J Pathol.*, 30, 1117-1127.

McDonald JC, McDonald AD, Armstrong B and Sébastien P (1986). Cohort study of mortality of vermiculite miners exposed to tremolite. *Br. J. Ind. Med.*, 43, 436-444.

McDonald JC, McDonald AD, Sébastien P, May K (1988). Health of vermiculite miners exposed to trace amounts of fibrous tremolite. *Br. J. Ind. Med.*,45, 630-634.

McDonald JC, Sébastien P and Armstrong B (1986). Radiological survey of past and present vermiculite miners exposed to tremolite. *Br. J. Ind. Med*, 43, 445-449.

Middleton EL, (1936). Industrial pulmonary disease due to the inhalation of dust. *Lancet*, 2, 59-64.

Morgan WKC, Donner A, Higgins IT et al (1988). The effects of kaolin on the lung. *Am. Rev. Respir. Dis*, 138, 813-820.

Ogle CJ, Rundle EM and Sugar ET (1989). China clay workers in the south west of England: Analysis of chest radiograph readings ventilating capacity and respiratory symptoms in elation to type and duration of occupation. Br. J. Ind. Med., 46, 261-270.

Oldham PD (1983). Pneumoconiosis in Cornish china clay workers. *Br. J. Ind. Med.*, 40, 131-137.

Ozesmi M, Patiroglu TE, Hillerdal G and Ozesmi C (1985). Peritoneal mesothelioma and malignant lymphoma in mice caused by fibrous zeolite. *Br. J. Ind. Med*,42, 746-749.

Pimentel JC and Menezes AP (1978). Pulmonary and hepatic granulomatous disorders due to the inhalation of cement and mica dusts. *Thorax*, 33, 219-227.

Pott F, Dolgner R, Friedrichs KH and Huth F (1976). Oncogenic effect of fibrous dusts. Animal experimentation and its relation to human carcinogenesis. *Ann Anat. pathol.*,21, 237-246.

Pott F, Huth F, and Friedrichs KH (1974). Tumorigenic effect of fibrous dusts in experimental animals. *Environ. Health Perspec.*,9, 313-315.

Rawlings W, Kennedy T., Menkes H and Morgan R. (1981). Pulmonary function and respiratory symptoms in a population of Georgia kaolin workers (abstract). *Am. Rev. Resp. Dis*, 123, 4, 146.

Rubino GF, Scansetti G, Piolatto G and Gay G (1979). Mortality and morbidity among talc miners and millers in Italy. *In: Lemen R and Dement JM eds, Dust and Diseases*, pps 357-363. Pathotox Inc, Illinois.

Scansetti G. Rasetti L and Ghemi F. (1963). Evoluzione clinica e radiologica delle pneumoconiosi nell' industria estrattiva del talco. *Med. Lavora*, 54, 746-749.

Selevan S, Dement JM, Wagoner JK and Froines JR (1979). Mortality patterns among miners and millers of non-asbestiform talc. *In: Lemen R and Dement JM eds. Dust and Diseases*, pps 379-388. Pathotox Inc., Illinois.

Sepulveda M-J, Vallyathan V, Attfield MD et al (1983). Pneumoconiosis and lung function in a group of kaolin workers. *Am Rev Respir Dis.*, 1983, 127, 231-235.

Sheers G (1964). Prevalence of pneumoconiosis in Cornish kaolin workers. *Br. J. Ind. Med.*, 21, 218-225.

Skulberg KR, Gylseth B, Skaug V and Hanoa R. 1985. Mica pneumoconiosis - a literature review. *Scand. J. Work Environ Health.*, 11, 65-74.

Smith AR 1952. Pleural calcification resulting from exposure to certain dusts. *AJR*, 67, 375-382.

Stille WT and Tabershaw IR (1982). The mortality experience of upstate New York talc workers. *J. Occup. Med.*,24, 480-484.

Thorel C (1986). Talc lung. A contribution to the pathological anatomy of pneumoconiosis. Beiträge zur Pathologischen Anatomie und zur Allgemeinen Pathologie, 20, 85-101.

Vestal TF, Winstead JA and Joliet PV (1943). Pneumoconiosis among mica and pegmatite workers. *Ind. Med* 12, 11-14.

Wagner JC, Berry G, Cooke TJ et al (1977). Animal experiments with talc. In: *Inhaled Particles* vol IV, *eds, Walton WH and McGovern B,* Part 2, pp647-652, Pergamon Press, Oxford.

Warraki S and Herant Y (1963). Pneumoconiosis in China Clay workers. *Br. J. Ind. Med.*, 20, 226-230.
Wegman DM, Peters JM, Boundy MG and Smith TJ (1982). Evaluation of respiratory effects in mines and miners exposed to talc free of asbestos and silica. *Br. J. Ind. Med.*, 39, 233-238.

OCCUPATIONAL EXPOSURE TO KAOLIN AND RESPIRATORY MORBIDITY
CROSS SECTIONAL SURVEYS

TABLE 1

SURVEY	PLACE	NUMBER OF SUBJECTS	EXPOSURE	CONTAMINATION silica	RADIOLOGICAL PNEUMOCONIOSIS	OTHER INFORMATION
Edenfield (1960)	Georgia US	1130	NS	0	4%	95% of subjects with pneumoconiosis had >10 years exposure.
Rawlings et al (1981)	Georgia US	404	NS	0	9%	Pneumoconiosis correlated with age.
Kennedy et al (1983)	Georgia US	459	>1yr (mean 12 yrs)	0	9%	Pneumoconiosis independently associated with length of exposure.
Sepulveda et al (1983)	Georgia US	47	Mean 16 yrs	0	11%	No pneumoconiosis in those exposed <5 yrs. Reduced FVC and FEVI in ex-smokers.
Altekruse et al (1984)	Georgia US	65	NS	0	8%	No pneumoconiosis in those exposed <5yrs. FVC and FEV1 reduced in subjects with pneumoconiosis.
Morgan et al (1988)	Georgia US	2022	>3yrs	0	4%	Pneumoconiosis associated with "dusty" jobs.
Warraki and Herant (1963)	Egypt	914	Mean ~15yrs	(+)	<1%	Initial examination by mass miniature radiography.
Sheers (1964)	Cornwall UK	553	>5yrs	+	9%	Pneumoconiosis related to duration of exposure
Oldham (1983)	Cornwall UK	1676	NS	+	23%	Pneumoconiosis associated with "dusty" jobs.Restrictive changes in lung function related to radiographic appearance.
Ogle et al (1989)	Cornwall UK	3689	All	+	9%	Pneumoconiosis related to dusty jobs and to silica exposure.

* NS = not stated

OCCUPATIONAL EXPOSURE TO TALC AND RESPIRATORY MORBIDITY
CROSS SECTIONAL SURVEYS

TABLE 2

SURVEY	PLACE	NUMBER OF SUBJECTS	EXPOSURE (Mean)	CONTAMINATION		RADIOLOGICAL PNEUMOCONIOSIS	OTHER INFORMATION
				Amphiboles	Silica		
Scansetti et al (1963)	Piedmont Italy	236	4.9yrs	(0)	(0)	36%	100% (18% complicated) pneumoconiosis in 22 workers exposed a mean of 19.5 years.
Fine et al (1976)	Rubber Factory US	80	8.9yrs	0	0	1%	Increased respiratory symptoms in subjects (vs.controls). FEV_1 reduced in subjects with >10 years exposure.
Wegman et al (1982)	Vermont US	116	8.5yrs	0	0	10%	Reduced FEV_1 (vs. predicted) 3.4% pleural thickening/plaques
Gamble et al (1982)	Montana S. Carolina Texas	279	7-10yrs	0/+	0	1%	No reduction in FEV_1, FVC (vs.controls) 9% pleural thickening.
Kleinfeld et al (1965)	New York US	43	>10 yrs	+	+	37%	Reduced FEV_1 and FVC in subjects (vs controls)
Gamble et al (1979)	New York US	121	10.2yrs	+	+	1%	Radiographic changes confined to subjects with > 15 years exposure. 31% pleural thickening in subjects with > 15 years exposure.

TABLE 3

OCCUPATIONAL EXPOSURE TO TALC AND RESPIRATORY MALIGNANCY
COHORT STUDIES

SURVEY	PLACE	NUMBER OF SUBJECTS	EXPOSURE	CONTAMINATION Amphiboles	Resp. Malig. SMR (no.)	Mesothelioma (no.)
Rubino *et al* (1979)	Piedmont Italy	1678	>1yr	0	52 (12)	0
Selevan *et al* (1979)	Vermont US	392	>1yr	0	163 (6)	0
Léophonte *et al* (1983)	France	470	>1yr	0	No change	0
Kleinfeld *et al* (1974)	New York US	260	>15yrs	+	320+ (13)	2
Brown *et al* (1979)	New York US*	398	All	+	290 (10)	1
Stille and Tabershaw (1982)	New York US*	655	All	+	163 (11)	0
Lamm *et al* (1988)	New York US*	741	All	+	246 (13)	1

\+ From proportional mortality ratio of 3.2
* Same mine

TABLE 4

OCCUPATIONAL EXPOSURE TO VERMICULITE AND RESPIRATORY MORBIDITY
CROSS SECTIONAL SURVEYS

SURVEY	PLACE	NUMBER OF SUBJECTS	MEAN EXPOSURE	CONTAMINATION Amphiboles	RADIOGRAPHIC PNEUMOCONIOSIS	PLEURAL THICKENING
Lockey et al (1984)	Montana* US	501	0.4-7.6 f-yr	++	0.2%	3%
McDonald et al (1986)	Montana** US	164	40f-yr	++	18%	28%
Amandus & Wheeler (1987)	Montana** US	184	119f-yr	++	10%	13%
McDonald et al (1988)	S. Carolina US	86	1f-yr	+	5%	8%

* Processing plant, ** Same mine

TABLE 5

OCCUPATIONAL EXPOSURE TO VERMICULITE AND RESPIRATORY MALIGNANCY
COHORT STUDIES

STUDY	PLACE	NUMBER OF SUBJECTS	EXPOSURE	CONTAMINATION Amphiboles	Resp. Malig. SMR (no.)	Mesothelioma (no.)
McDonald et al (1986)	Montana* US	406	>1yr	++	245 (2)	4
Amandus & Wheeler (1987)	Montana* US	575	>1yr	++	223 (20)	2
McDonald et al (1988)	S.Carolina US	193	>6m	+/=	121 (4)	0

* Same mine

TABLE 6

OCCUPATIONAL EXPOSURE TO MICA AND RESPIRATORY MORBIDITY
CROSS SECTIONAL SURVEYS

SURVEY	PLACE	NUMBER OF SUBJECTS	EXPOSURE	CONTAMINATION Silica	RADIOGRAPHIC PNEUMOCONIOSIS	OTHER INFORMATION
Dreessen _et al_ (1940)	N.Carolina US	109	~2.5 m particles/f^3	0	1%	Sheet mica (Single worker with previous heavy exposure to ground mica).
		57	> 10 m particles/f^3	0	18%	Ground mica dust
Vestal et al (1943)	N.Carolina US	79	0-30 yrs (mean 5 yrs)	0	11%	Mica grinding + milling
Heimann _et al_ (1953)	Bihar India	61	< 5 yrs.	++	44%	Probably silicosis

PATHOLOGICAL FEATURES OF PULMONARY DISEASE DUE TO SILICATE DUST INHALATION

John E. Craighead, M.D.
Department of Pathology
University of Vermont College of Medicine
Burlington, VT 05405
U.S.A.

Clinical and pathological information is limited on diseases of the lung caused by inhalation of plate silicates. Many reports and descriptions of the pathology of these diseases depict postmortem studies conducted on workers with exceedingly heavy pulmonary dust burdens. While these observations are important, the pathological changes seen in severe, advanced cases of disease due to occupational exposure do not accurately reflect the pathological changes occurring in the lungs of those with lesser degrees of exposure (Hale et al. 1956, Vallyathan and Craighead 1981, Lapenas et al. 1984, Davies and Cotton 1983, Seaton et al. 1986). Countless numbers of us are exposed to silicates avocationally and in industries where particulate concentrations in the ambient air are controlled. And, community hospital pathologists occasionally see lesions due to silicates when no documented exposure exists. When a history of silicate exposure is present in those with in chronic obstructive lung disease, definitive evaluation of the functional effects of the silicates is difficult. Since analytical studies to determine the nature of the dust in the lungs are expensive and often not feasible, recognizing the pathologic pattern of lesions due to silicate exposure may be critical to the diagnostic pathologist. Assembled here are observations on sporadic cases occurring in workers exposed to varying degrees of dust.

Contributors to this symposium have described in detail the mineralogical features of the plate silicates; others have documented experiments which examine the toxicity of plate silicates <u>in vitro</u> (Wallace et al. 1988,

NATO ASI Series, Vol. G 21
Health Related Effects of Phyllosilicates
Edited by J. Bignon

Vallyathan et al. 1988). Yet we still know very little regarding the pathogenetic processes involved, and the basis for the pathological features of the lesions has not been defined. A number of questions require resolution. Firstly, the effects of plate silicate particle size and configuration on pulmonary clearance have not been fully evaluated. Secondly, the normal clearance capability of the human lung for plate silicates is unknown. Under what circumstances does exposure exceed clearance and result in increasing lung burdens? Does smoking inhibit the elimination of these particulates from the lung and, therefore, predispose to development of silicate pneumoconiosis? There are other questions as well. Do excessive numbers of plate silicate particles influence the development of the lesions of silicosis in workers exposed to a variety of particulates, including both silicates and silica (Walton et al. 1977, Morgan and Lapp 1976)? Are the reactive surfaces of the particles altered _in vivo_, reducing their ability to promote proliferation of fibrous tissue? Why are particles that are cytolytic in _in vitro_ assays not particularly fibrogenic in tissue?

Perivascular and peribronchiolar accumulations of dust in macrophages, associated with modest deposits of reticulin and collagen, are the characteristic pulmonary lesions associated with long-term exposure to silicates. The extent of these changes in the lungs and their distribution in relation to dust types and burdens have not been systematically evaluated. Indeed, carefully conducted systematic anatomic studies are not recorded in the literature. Thus, our knowledge of these disease processes is scant in comparison to the well studied conditions of asbestosis (due to fibrous silicates) and silicosis (caused by silica dioxide particles).

The early lesions of plate silicate disease in man have not been identified. Observations that we and others have made on feral animals may be relevant, however (Brambilla et al 1979, Schwartz et al. 1981). Figures 1 and 2 illustrate localized lesions in the individual acinar units of a zoo-encaged animal. Macules comprised of dust-laden macrophages are present in the region of the distal airway branch of this lung. Presumably the accumulations of dust result from particulates inhaled on a daily basis as this animal browses dry over clay soil for food. This is a silicate-associated lesion seen relatively commonly by veterinary pathologists, but only rarely by pathologists studying human disease (Gylseth et al. 1984).

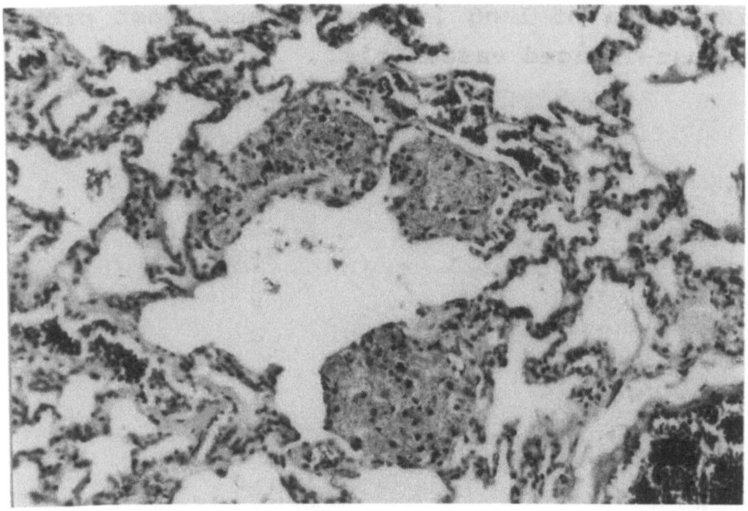

Figure 1. Respiratory bronchioles of a wolf who died in a zoo from causes not related to the lung lesion. Note the macules containing dust particulates which are a mixture of silicates and small amounts of silica. Polarization microscopy demonstrates abundant highly refractile particulates. These lesions contain minimal amounts of collagen.

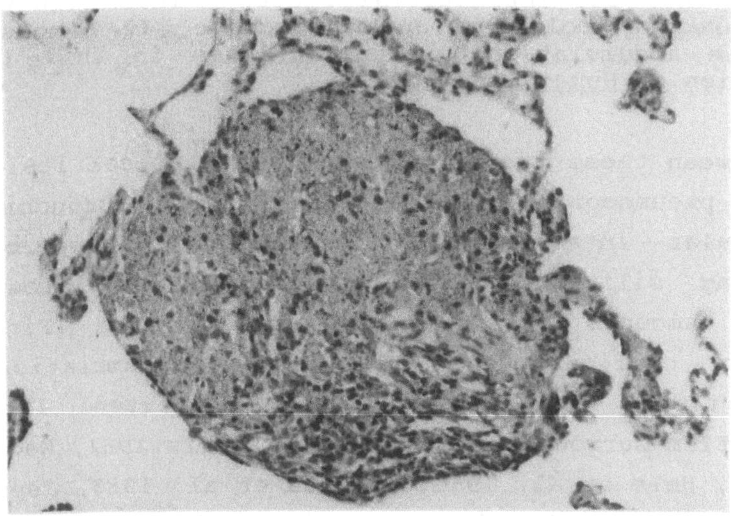

Figure 2. A high resolution microscopy illustration of a macular lesion in the lung of the animal illustrated in Fig. 1. The cellular accumulation is nonspecific.

Progressive massive fibrosis is the extreme of lesions which result from silicate exposure (Fig. 3) (Kleinerman et al. 1979, Leibowitz and Goldstein 1987). It occurs in workers exposed to massive amounts of dust over long periods of time. While the pathogenesis of this lesion is unclear, it is plausible to propose that the circulation to a localized area of lung is compromised, most probably due to focal dust-induced vasculitis.

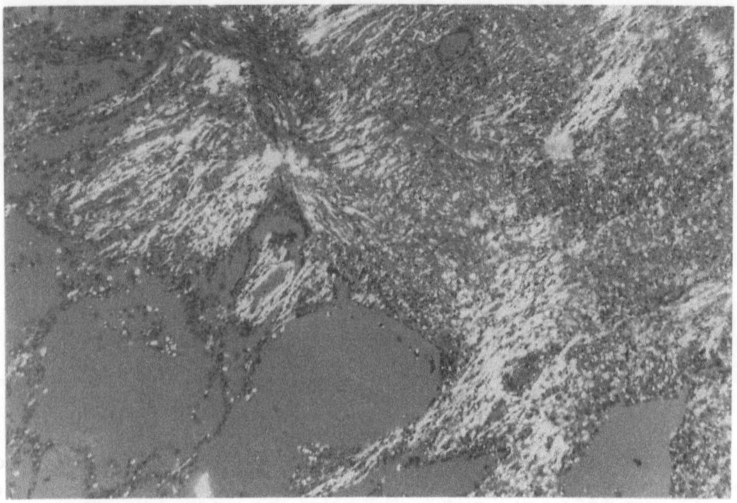

Figure 3. Progressive massive fibrosis in a long-term worker in the Vermont (U.S.A.) talc industry. The talc does not contain asbestiform minerals and the silica concentrations are low. Note the diffuse fibrosis which radiates into the interstitial spaces. This illustration is prepared by polarization microscopy. It demonstrates the dense accumulations of particulates in this lesion. (Permission of Human Pathology).

Between these two extremes are the typical lesions of silicate pneumoconiosis. They consist of peribronchial and perivascular interstitial accumulations of macrophages containing silicate and carbon dust, associated with variable amounts of reticulin and collagen (Fig. 4). Irregular configurations of whorled accumulations of fibrous tissue form which occasionally extend into the interstitium surrounding air spaces (Sheers 1964, Wagner et al. 1986, Hale et al. 1956, Lapenas et al. 1983, Jepson et al. 1985, Craighead and Emerson 1984, Gibbs, et al. 1986, McConnochie et al., Gloyne et al. 1949, Musk et al. 1980, Sherwin et al. 1979).

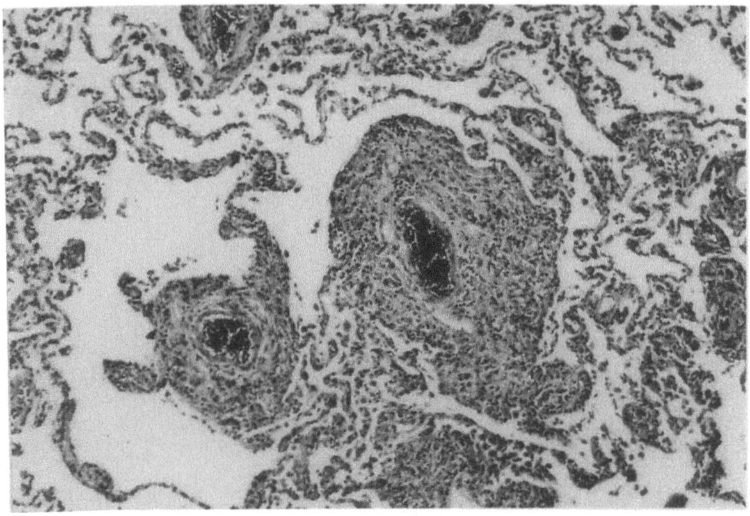

Figure 4. Lung of a Vermont slate worker dying of causes unrelated to dust exposure. Note the accumulations of fibrous tissue and the cellular infiltrate around the vessels. Bronchi and bronchioles show similar lesions at scattered sites. Occasional workers exposed to slate often have within their lungs the nodular lesions of silicosis in addition to those caused by silicates. Interestingly enough, the slate and silica lesions do not coalesce, but form separately.

Plate silicate dust in lesions is highly refractile when examined by polarized light (Figs. 3,5) (Craighead et al. 1988). The particles exhibit a diversity of forms; in a single lesion, round, square, lancet-shaped, rhomboid or granular particulates may be found. In contrast, silica particles customarily are small, granular and only slightly refractile. And, in comparison to the plate silicates, silica often is difficult to see when poor quality polarizing lenses are used and the ambient light is intense. All too often pathologists identify the particles of silicates as silica and convey the impression that a lesion is silicosis when, in fact, it is due exclusively or in large part to silicates.

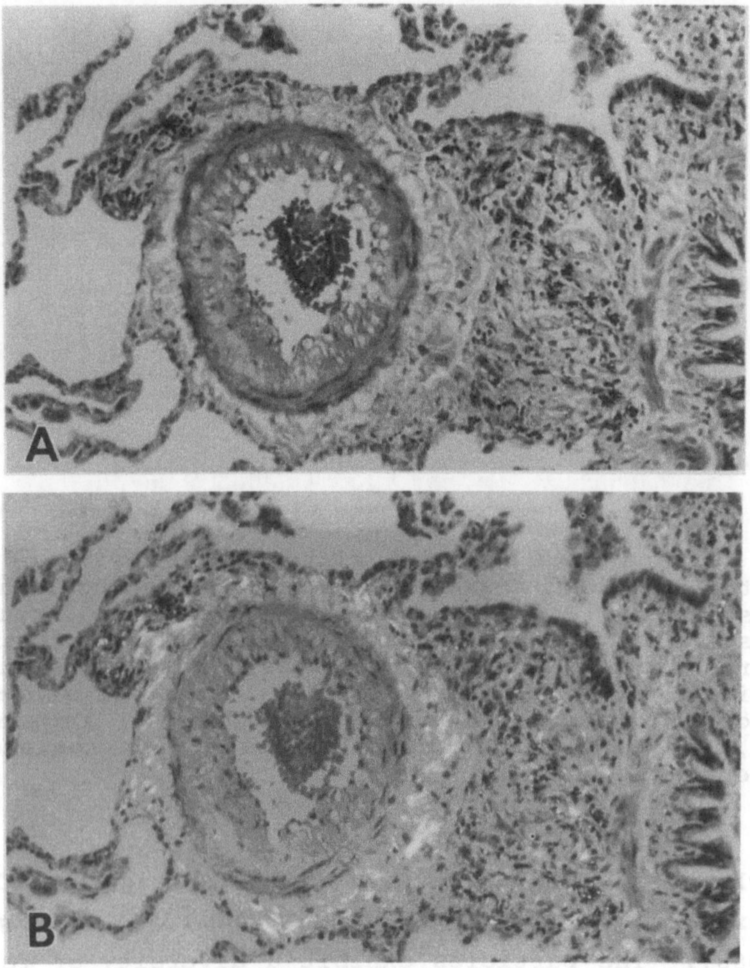

Figure 5. Perivascular accumulations of Vermont talc by light (A) and polarization microscopy (B). Note the fibrotic lesion bridging between the bronchiole (right) and the vessel (center) of the photomicrograph. (Permission of Human Pathology)

Occasionally one observes multinucleate foreign body giant cells in the lungs of workers exposed to coarse particles of silicates (Fig. 6). In contrast, when fine particulates such as pharmaceutical or cosmetic talc are responsible for the lesions, the giant cells and associated tissue reaction can strikingly resemble the lesions of Boeck's sarcoid (Figs 7,8) (Factor 1975, Terzakis et al. 1978). Inhaled plate silicate particles occasionally serve as the core of ferruginous bodies (Fig. 9) (Sheers 1964). These structures exhibit a great variety of configurations, and can resemble the asbestos body seen in disease caused by the fibrous silicate.

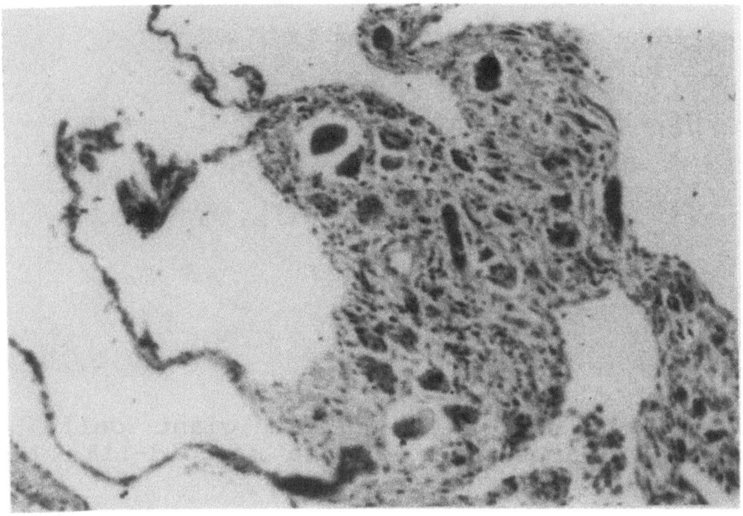

Figure 6. Multinucleate giant cells in the interstitial fibrotic lesion of a Vermont talc worker.

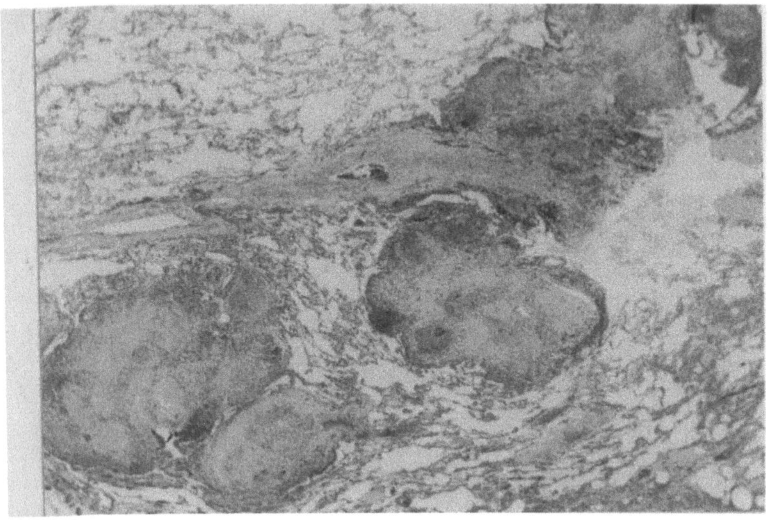

Figure 7. Nodular sarcoid-like lesions in the lung of a middle-aged man vicariously exposed to cosmetic talc.

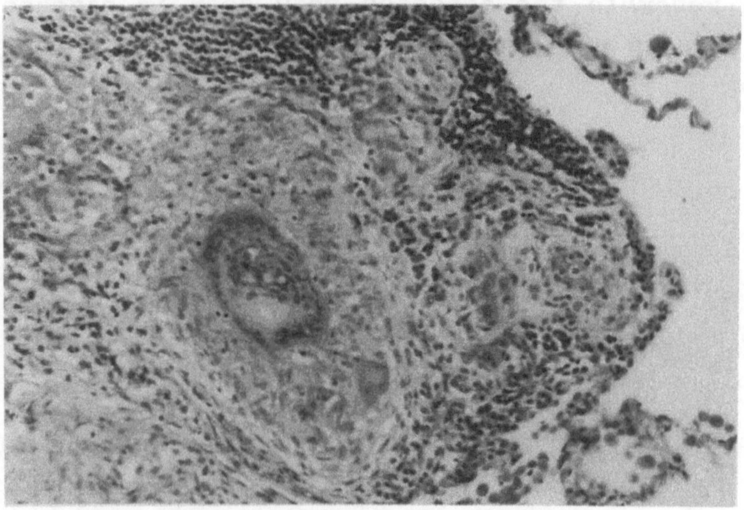

Figure 8. Note the multinucleate giant cell, and the whorled nodular appearance of the sarcoid-like lesion of the case illustrated in Fig. 7.

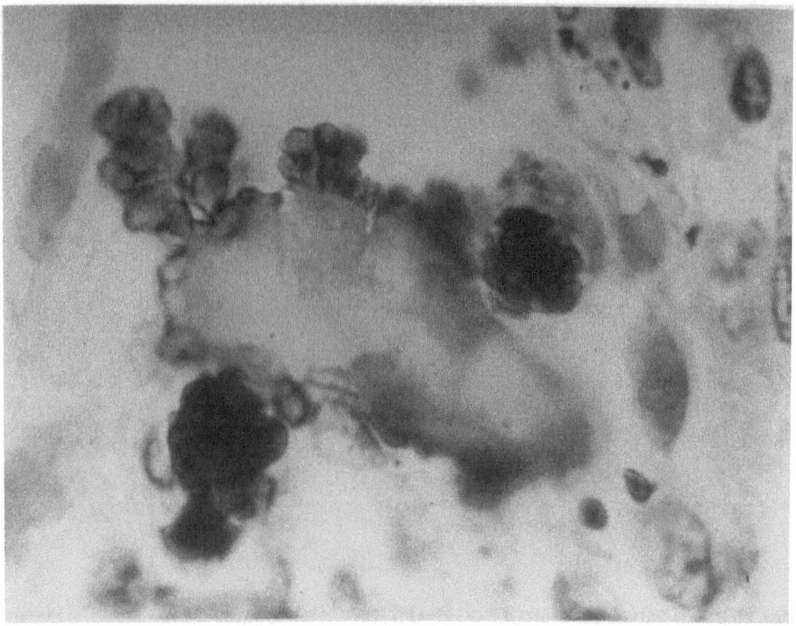

Figure 9. A mica ferruginous body. Note the irregular plate-like structure encompassed by the adherent globules of iron-protein containing material. (Permission of the Archives of Pathology and Laboratory Medicine))

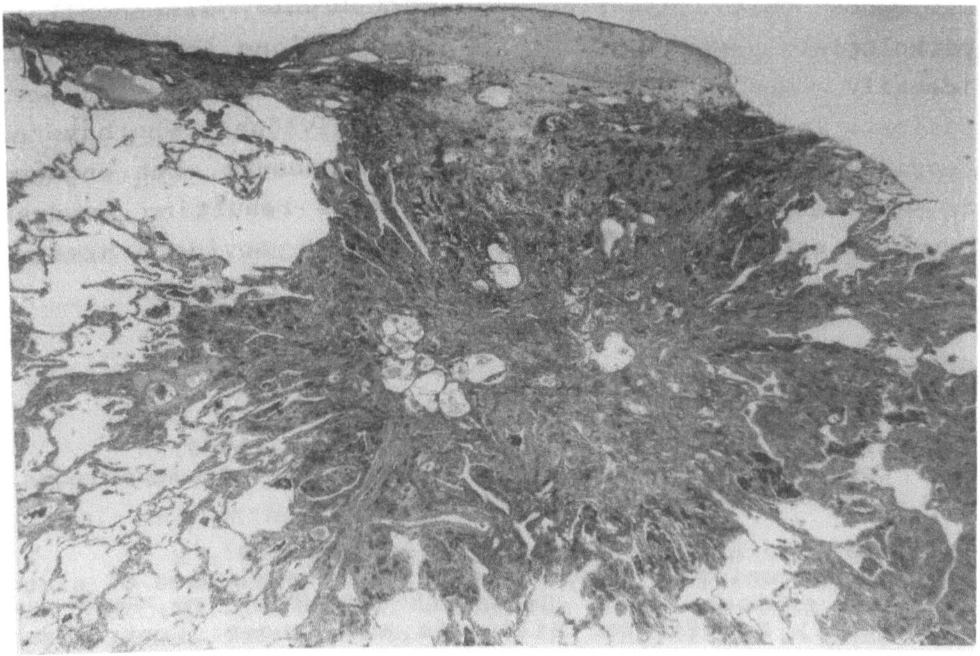

Figure 10. A mixed dust pneumoconiosis lesion adjacent to the pleura. Note the stellate extensions of fibrotic tissue into the adjacent interstitial spaces. These lesions contain a mixture of silica and silicates as well as carbonaceous material. Their pathogenesis is not defined.

McLaughlin and Harding (1956) and McLaughlin (1957) described stellate nodules histologically in the lungs of workers in several different industries (Fig. 10). Analysis revealed mixtures of silica and silicates accompanied by variable amounts of carbon. The pathogenesis of this unique mixed dust lesion is uncertain. More frequently, one observes isolated silicotic nodules in the lungs of workers, associated with, but anatomically separate from typical silicate lesions and carbon-containing dust macules. Nodules of the latter type occur frequently in those with coal workers pneumoconiosis (Kleinerman et al. 1979).

Plate silicate lesions are unique, but not strikingly different from those caused by non-platy, nonfibrous silicates. Their appearance is sufficiently distinct that pathologists acquainted with the lesions can readily identify them in tissue sections. In my experience, silicate lesions often are found in individuals who have no readily apparent history of silicate exposure, but engaged in vocational or avocational activities resulting in dust exposure early in life. Thus, the lungs provide a history of dust inhalation during an individual's lifetime.

REFERENCES

Brambilla C, Abraham J, Brambilla E, et al (1979) Comparative pathology of silicate pneumoconiosis. Amer J Pathol 96:149-163

Craighead JE, Emerson RJ (1984) Slateworkers' pneumoconiosis: Lung disease due to a mixture of slate and silicate dust. Goldsmith DF, Winn, DM, Shy, CM (eds.) In: Silica, Silicosis and Cancer: Controversy in Occupational Medicine. Cancer Research Monographs 2:533

Craighead JE, Kleinerman J, Abraham JL, et al (1988) Diseases associated with exposure to silica and nonfibrous silicate minerals. Arch Pathol Lab Med 112:673-720

Davies D, Cotton R (1983) Mica pneumoconiosis. Br J Ind Med 40:22-27

Factor SM (1975) Granulomatous pneumonitis: A result of intrapleural instillation of quinacrine and talcum powder. Arch Pathol Lab Med 99:499-502

Gibbs AR, Craighead JE, Pooley FD, et al (1988) The pathology of slateworkers' pneumoconiosis in Wales and Vermont. Ann Occup Hyg 32:273-278

Gloyne SR, Marshall G, Hoyle C (1949) Pneumoconiosis due to graphite dust. Thorax 4:31-38

Gylseth B, Stettler L, Mowe G, et al (1984) A striking deposition of mineral particles in the lungs of a farmer: A case report. Am J Ind Med 6:231-240.

Hale LW, Gough J, King EJ, et al (1956) Pneumoconiosis of kaolin workers. Brit J Ind Med 13:251-259

Jepson WB, Comyns RA, Sewart JH (1985) Pulmonary implications of the inhalation of china clay (kaolin) dust. Sixth Inter. Symposium on Inhaled Particles. Br Occup Hyg Soc 108-109

Kleinerman J, Green F, Laqueur W, et al (1979) Pathology standards for coal workers' pneumoconiosis: Report of the Pneumoconiosis Committee of the College of American Pahtologists to the National Institute for Occupational Safety and Health. Arch Pathol Lab Med 103:373-432

Lapenas D, Gale P, Kennedy T, et al (1984) Kaolin pneumoconiosis: radiologic, pathologic, and mineralogic findings. Am Rev Respir Dis 4:282-288

Lapenas D, Gale P, Kennedy T, et al (1984) Kaolin pneumoconiosis. Am Rev Respir Dis 130:282-288

Leibowitz MC, Goldstein B (1987) Some investiations into the nature and cause of massive fibrosis (MP) in the lungs of South African gold, coal and asbestos mine workers. Am J Ind Med 12:129-143

McConnochie K, Gibbs AR, Pooley FD, et al (in press) Pneumoconiosis in Welsh Slateworkers. Correlation Study

McLaughlin AIG (1957) Pneumoconiosis in foundry workers. Br J Tuber and Dis Chest 4:297-308

McLaughlin AIG, Harding HE (1956) Pneumoconiosis and other causes of death in iron and steel foundry workers. AMA Arch Ind Health 350-378

Morgan, WKC, Lapp, NL (1976) Respiratory disease in coal miners. Am Rev Respir Dis 113:531-559

Musk AW, Greville HW, Tribe AE (1980) Pulmonary disease from occupational exposure to an artificial aluminum silicate used for cat litter. Br J Ind Med 37:367-372

Schwartz LW, Knight HD, Whittig LD, et al (1981) Silicate pneumoconiosis and pulmonary fibrosis in horses from the Monterey-Carmel peninsula. Chest 80:82-85

Seaton A, Louw SJ, Cowie HA (1986) Epidemiologic studies of Scottish oil shale workers: 1. Prevalence of skin disease and pneumoconiosis. Am J Ind Med 9:409-421

Sheers G (1964) Prevalence of pneumoconiosis in Cornish kaolin workers. Br J Ind Med 21:218-225

Sherwin RP, Barman ML, Abraham JL (1979) Silicate pneumoconiosis of farm workers. Lab Invest 40:576-582

Terzakis JA, Shustak SR, Stock EG (1978) Talc granuloma identified by x-ray microanalysis. JAMA 239:2371-2372

Vallyathan NV, Craighead JE (1981) Pulmonary pathology in workers exposed to nonasbestiform talc. Human Pathol 12:28-35

Vallyathan V, Schweglet D, Reasor M, et al (1988) Comparative in vitro cytotoxicity and relative pathogenicity of mineral dusts. Ann Occup Hyg 32:279-289

Wagner JC, Pooley FD, Gibbs AR et al (1986) Inhalation of china stone and china clay dusts: Relationship between the mineralogy of dust retained in the lungs and pathological changes. Thorax 41:190-196

Wallace WE, Keane MJ, Vallyathan V, et al (1988) Suppression of inhaled particle cytotoxicity by pulmonary surfactant and re-toxification by phospholipase: Distinguishing properties of quartz and kaolin. Ann Occup Hyg 32:291-298

Walton WH, Dodgson J, Hadden GG, et al (1977) The effect of quartz and other non-coal dusts in coalworkers' pneumoconiosis. Walton WH (ed) In: Inhaled Particles IV, Pergamon Press, New York 4:669-690.

CLINICAL ASPECTS OF KAOLIN PNEUMOCONIOSIS

W.K.C. Morgan, A. Donner*, I.T.H.Higgins, M.G. Pearson***,
W. Rawlings, Jr.+
University Hospital,
London, Ontario,
N6A 5A5
Canada

China clay or kaolin is mined extensively in east central Georgia and is the major industry in that region. The mining and processing of this mineral is known to lead to the development of a pneumoconiosis (Middleton 1936). Several limited studies have been published, but no well designed large investigation that has included most of the work force has been carried out until recently (Edenfield 1960, Kennedy et al. 1983, Oldham 1983, Altekruse et al. 1984). The present communication summarizes a recent comprehensive study that was carried out and adds some additional information to that which has been published (Morgan et al. 1988).

MATERIAL AND METHODS

Subjects

Twelve companies were included, the work force of which constituted over 95% of those currently employed in this industry in the U.S.A. All workers over 40 who had been employed for more than three years were asked to participate. In addition, a 50% sample of those aged below 40 were asked if they would agree to be included in the study. Women were excluded since less than half a dozen subjects had a significant exposure to dust. The protocol was explained to all participants who signed an appropriate consent form.

---

* A. Donner, Department of Epidemiology, University of Western Ontario, London, Ontario, Canada.
** I.T.H. Higgins, Betheseda, Maryland, U.S.A.
*** M.G. Pearson, Fazakerly Hospital, Liverpool, England.
+ W. Rawlings Jr., Sandersville, Georgia, U.S.A.

NATO ASI Series, Vol. G 21
Health Related Effects of Phyllosilicates
Edited by J. Bignon

The Processing of Kaolin

Kaolin is a constituent of granitic rocks and occurs as microscopic plate like crystals. It is removed from the ground by open cast mining and is then processed by either wet or dry method. In the former, water is added to it and a slurry formed. Subsequently, sand, sludge and other contaminants are removed by differential sedimentation using either vacuum filtration or spray drying. In dry processing the clay is first dried and then pulverized. The sand and other contaminants are later removed by differential air flotation.

Processed kaolin has a particle size of around 2 µm with over 90% of the finer grades of particles having a MMD of below 2 µm. Over the recent years an increasing proportion of kaolin mined has been calcined. This process involves heating the kaolin to 1000° C for 20 to 30 minutes. Processed kaolin contains less than 1 % free silica. Prior industrial hygiene measurements have shown the greatest dust exposures to occur in those workers who are baggers, processors or loaders (Kennedy et al 1983).

Medical Examinations

All participants had a respiratory questionnaire administered to them. This was based on the MRC questionnaire on chronic bronchitis (Higgins and Whittaker 1981). In addition, a complete smoking history was taken, along with an occupational history. Every job from the time the subject left school to the time he was interviewed was recorded. Subsequently, each subject had his ventilatory capacity assessed using a rolling seal spirometer (Spirotech, Atlanta, Georgia). The forced vital capacity (FVC) and forced expiratory volume in one second (FEV_1), as well as the forced expiratory flow between 25% and 70 % of the FVC (FEF_{25-75}) were measured. Morris' predicted values were used and a 15% correction for black subjects was included. For the FVC and FEV_1, a value below 80% of the predicted was assumed to be abnormal. The ATS/DLD criteria for acceptability and reproducibility were applied to all spirometric tracings, i.e. the two largest values had to be within 100 ml or 5% of each other (Epidemiology Standardization Project, 1978).

Radiographic Examinations

All participants had a recent (within two years of the examination) PA chest film interpreted according to the 1980 ILO classification (1980). Three experienced readers independently interpreted each film and the median reading was accepted as the patient's value.

Composition of the Cohort

Of the 2,379 subjects selected for study, 2,111 agreed to participate (89%). Of those who did not participate, 67 were either sick or on holiday at the time of the examination. Of those participating, 2,020 had satisfactory investigations carried out. There were 1,215 white subjects and 805 black subjects. Of the total, 41% were cigarette smokers, 27% ex-smokers, 27% nonsmokers while the remaining 5% were cigar or pipe smokers. Of the white subjects, 39% were current smokers as compared to 44% amongst the blacks. However, 57% of the current white smokers smoked more than 20 cigarettes a day, whilst the comparable figure for blacks was 14%. Life time smoking was graded according to those with less than 20 pack years, those with 21 to 40 pack years, and those with more than 40 pack years.

RESULTS

Radiographic Findings

In all, 2,069 subjects out of 2,379 subjects had satisfactory chest x-rays available for interpretation. Thirty-one of the radiographs could not be interpreted because of inadequate or poor technique. A total of 90 subjects had simple pneumoconiosis and of this total, 64 had category 1, 21 had category 2 and 5 had category 3. Eighteen subjects had complicated pneumoconiosis, 5 having stage A, 8 stage B, and 5 stage C.

TABLE 1

PREVALENCE OF KAOLINOSIS

	AGE	RADIOGRAPHIC CATEGORY 0	1	2	3	COMP	TOTAL KAOLINOSIS	TOTAL WORKERS
WHITES	20-39	507	2	-	-	-	2	509
	39-60+	674	25	4	-	3	32	706
BLACKS	20-39	320	-	-	-	-	-	320
	39-60+	411	37	17	9	13	76	487

The prevalence of simple and complicated pneumoconiosis were 3.2% and 0.89% respectively. It will be observed that a far greater percentage of black subjects had pneumoconiosis. The difference was explained by the greater proportion of whites employed in less dusty jobs (49% versus 15%). Agreement within one subcategory was present in 96% of the films interpreted.

56 men were involved in dry processing and 701 men in wet processing. The percentages of men who had either simple or complicated pneumoconiosis according to the type of processing are shown in Table 2. It is apparent that dry

processing led to a significantly increased prevalence rate of pneumoconiosis.

TABLE 2

RADIOGRAPHIC CATEGORY IN WET AND DRY PROCESSORS BY YEARS WORKED

		RADIOGRAPHIC CATEGORY							
		0		SIMPLE		COMP		% WITH KAOLINOSIS	
	YEARS WORKED	DRY	WET	DRY	WET	DRY	WET	DRY	WET
WHITES	3-19	14	164	-	2	-	-	-	1.2
	20-30+	12	226	2	13	-	2	14.3	6.2
BLACKS	3-19	21	115	1	-	1	-	4	-
	20-30+	26	152	6	24	3	3	25.7	15

Ventilatory Capacity and Pneumoconiosis

The number and percentage of subjects with an FEV_1 or FVC of less than 80% of the predicted value are shown in Table 3. There is a significant difference between normal subjects, and those with simple and complicated pneumoconiosis. The impairment in those with complicated pneumoconiosis is immediately apparent, and for the most part was restrictive in character with both the FEV_1 and FVC being reduced. When compared to those without pneumoconiosis, subjects with simple pneumoconiosis had a small reduction in ventilatory capacity.

TABLE 3

RESPIRATORY FUNCTION AND CATEGORY OF PNEUMOCONIOSIS
(ALL SUBJECTS)

	FEV_1 (% pred)			FVC (% pred)		
Category	≥80	<80	Total	≥80	<80	Total
Normal	1,656	135(7.5%)	1,791	1,631	141(8.0%)	1,772
Simple	75	11(12.8%)	86	77	9(10.5%)	86
Complicated	10	5(33.3%)	15	10	5(33.3%)	15
Total	1,741	151	1,892	1,718	155	1,873
		$\chi^2 = 16.32$ (p = 0.003)			$\chi^2 = 13.19$ (p = 0.0014)	

When smoking was taken into account this was not statistically significant. The relationship of ventilatory capacity to the predicted FVC and FEV_1 in subjects with normal films and those with simple and complicated films are shown in Table 4. Here again, those with complicated pneumoconiosis have an obvious reduction.

TABLE 4

RELATIONSHIP OF VENTILATORY CAPACITY TO RADIOGRAPHIC
CATEGORY (ALL SUBJECTS)

Category	FEV_1 n	FEV_1 Mean Age	FVC n	FVC Mean Age	FEV_1/FVC% n	Mean Value (% pred) FEV_1*	FVC*	FEV_1/FVC%*
Normal	1,895	41.46	1,872	41.41	1,872	99.73	96.06	78.84
Simple	89	52.26	89	52.26	89	98.08	94.57	75.80
Complicated	16	54.75	16	54.75	16	86.43	85.25	72.69

* Normal and simple pneumoconiosis mean values differ significantly from mean value for those with complicated pneumoconiosis ($p<0.001$).

The Relationship of Pneumoconiosis to Lung Function in Nonsmokers

The confounding effects of cigarette smoking were

excluded by considering only nonsmokers. The number and percentage of nonsmoking subjects with an FEV_1 or FVC less than 80% of the predicted are shown in Table 5. There was a total of 549 life long nonsmokers who participated in this study and of this total 542 had a normal FEV_1 or FVC. This constitutes 7 out of 542 subjects, i.e. less than expected using a 95% confidence limit.

TABLE 5

NUMBER AND PERCENTAGE OF NONSMOKERS WITH AN FEV_1 AND FVC BELOW 80% OF PREDICTED ACCORDING TO RADIOGRAPHIC CATEGORY

	Normal		Simple		Complicated		Total	
(%pred)	FEV_1	FVC	FEV_1	FVC	FEV_1	FVC	FEV_1	FVC
< 80	12(2.3%)	23(4.5%)	0(-)	1(5.3%)	4(57.1%)	4(57.1%)	16(3%)	28(5.2%)
$\geqslant 80$	504	488	19	18	3	3	526	509
TOTAL	516	511	19	19	7	7	542	537

$FEV_1: \chi^2 = 73.04$ ($p < 0.001$)
$FVC: \chi^2 = 38.72$ ($p < 0.001$)

The relationship of lung function to pneumoconiosis was also examined by multiple logistic regression analysis using an abnormality of respiratory function as the dependent variable. The analysis showed that complicated, but not simple pneumoconiosis was associated with a significantly lower ventilatory capacity. Similarly, we compared the relationship of radiographic category to the FEV_1 and FVC when the latter were expressed as a percentage of the predicted values, i.e. by treating them as continuous variables (Table 6). Once more it was apparent that complicated, but not simple pneumoconiosis was associated with a reduced ventilatory capacity.

TABLE 6

MEAN FEV_1 AND FVC OF NONSMOKERS ACCORDING TO RADIOGRAPHIC CATEGORY

	FEV_1		FVC		$FEV_1/FVC\%$	Mean Value (% pred)		
Category	n	Mean Age	n	Mean Age	n	FEV_1	FVC	$FEV_1/FVC\%$
Normal	516	39.01	511	38.93	511	103.3	97.3	81.3
Simple	19	49.50	19	49.50	19	104.9	98.8	78.4
Complicated	7	52.93	7	52.93	7	81.1	77.0	74.6

Note: FEV_1: $p < 0.001$, FVC: $p < 0.001$, and $FEV_1/FVC\%$: $p = 0.24$

Dust Exposure and Lung Function

The effects of dust exposure on the lung function of nonsmokers with normal radiographs were determined. 522 men fulfilled these criteria; of them 516 had a FEV_1 and 511 a FVC which were within normal limits. Normal respiratory function was defined in the same way as previously described. There was no difference between the wet and dry processors, however, those who were employed with calcined clay had a greater prevalence of lung function abnormalities.

TABLE 7

RELATIONSHIP OF JOB TO FEV_1 AND FVC IN NONSMOKERS

WITH NO PNEUMOCONIOSIS

	FEV_1			FVC		
	n	n < 80%		n	n < 80%	
	normal	abnormal	% < 80%	normal	abnormal	% < 80%
Dry processors (1.1 or 2.1)	13	0	0	13	0	0
Wet processors (1.2 or 2.2)	178	2	(1.1%)	177	6	(3.4%)
Calciners (1.3 or 2.3)	47	5	(10.6%)	45	4	(8.9%)
Non-dusty jobs (Only 3+)	202	4	(2.0%)	200	10	(5.0%)
Total	440	11		435	20	

$\chi^2 = 16.14 (p = 0.003)$ $\chi^2 = 3.29 (p = 0.51)$

Similar findings resulted from treating the FEV_1 and FVC as continuous variables. The results make is apparent that nonsmoking calciners have a higher rate of abnormal function than other nonsmoking men doing other types of processing. This persisted after adjustments for age, race and number of years worked in the work force.

TABLE 8

RELATIONSHIP OF FEV_1, FVC, AND FEV_1/FVC % TO PROCESS TYPE AND JOB IN NONSMOKERS WITH NO PNEUMOCONIOSIS

	FEV_1 (% pred)			FVC (% pred)			(FEV_1/FVC)%		
	n	Mean	S.D.	n	Mean	S.D.	n	Mean	S.D.
Dry processors (1.1 or 2.1)	13	100.93	7.63	13	93.99	5.56	13	82.69	5.04
Wet processors (1.2 or 2.2)	178	103.78	12.11	177	97.30	10.07	177	81.60	5.60
Calciners (1.3 or 2.3)	47	100.08	15.60	45	95.69	14.49	45	82.09	6.20
Non-dusty jobs (Only 3+)	202	102.82	12.07	200	97.70	10.79	200	80.35	6.54
Total	440			435			435		
		$p = 0.14$			$p = 0.64$			$p = 0.20$	

It was apparent that the effect of increased smoking on respiratory function was greater among subjects with normal chest films than it was among subjects with pneumoconiosis. The effect of increased cigarette smoking was greater on the other work categories than it was in the calciners. The apparent association of abnormal lung function in the calcined clay workers could not be explained by an increased prevalence of pneumoconiosis.

Type of Radiographic Opacity

The predominant type of opacity found in subjects with definite kaolin pneumoconiosis was rounded, and was the only type of opacity found in 60% of subjects with simple kaolinosis. In the remainder, the secondary type of opacity seen was irregular, but only in once instance was an irregular opacity found to be the primary opacity in a subject with greater than category 1/1. This was of the u type. It was apparent, however, that the rounded opacities were sometimes less well defined than those present in classical silicosis or in conditions such as welders' siderosis. All sizes of rounded opacities were seen, viz. p, q and r. In our study, the presence of irregular opacities was weakly associated with cigarette smoking and age, but no obvious relationship to dust exposure as measured by years worked was apparent.

Summary

Ventilatory capacity, the prevalence of chest symptoms, and radiographic abnormality was assessed in a randomly selected sample of over 2,000 kaolin workers from east central Georgia. Ventilatory impairment was related to the presence of complicated pneumoconiosis, employment as clay calciners, and in those who were cigarette smokers. By far the greatest decrement in ventilatory capacity was seen in those were cigarette smokers. In calcined clay workers, there was an increased prevalence of abnormality of the FEV_1, but not of the FVC, although a trend in the latter was apparent. The increased prevalence could not be increased by cigarette smoking or the presence of pneumoconiosis. In workers with more than three years tenure, there were 90 subjects with simple pneumoconiosis and 18 with complicated, yielding an overall prevalence of 3.2% and 0.63% respectively in the sample examined. Dry processors were at a greater risk of kaolinosis than were wet processors.

Acknowledgement

The tables are based on data that has previously been published in the American Review of Respiratory Diseases (138, 813-820).

REFERENCES

Altekruse EB, Chaudhary BA, Pearson MG, Morgan WKC (1984) Kaolin dust concentrations and pneumoconiosis at a kaolin mine. Thorax 39:436-441

Edenfield RW (1960) A clinical and roentgenological study of kaolin workers. Arch Environ Health 5:28-39

Epidemiology Standardization Project (1978) Am Rev Respir Dis 118(Suppl 11):1-88

Higgins ITT, Whittaker PE (1981) Chronic respiratory disease in two mining communities in West Virginia. NIOSH Technical Report, Washington D.C., U.S. Government Printing Office

International Labour Office (1980) Guidelines for the use of the ILO International classification of radiographs of pneumoconiosis: Revised Edition. Geneva:Geneva International Labour Office

Kennedy T, Rawlings W, Baser M, Tockman M (1983) Pneumoconiosis in Georgia kaolin workers. Am Rev Respir Dis 127:215-220

Middleton EL (1936) Industrial pulmonary disease due to the inhalation of dust. Lancet 2:59-64

Morgan WKC, Donner A, Higgins ITT, Pearson MG, Rawlings Jr W (1988) The effects of kaolin on the lung. Am Rev Respir Dis 138:813-820

Morris JF, Koski A, Johnson LC (1971) Spirometric standards for healthy nonsmoking adults. Am Rev Respir Dis 103: 57-67

Oldham PD (1983) Pneumoconiosis in Cornish clay workers. Brit J Ind Med 40:131-137

FRENCH TALC PNEUMOCONIOSIS

P. LEOPHONTE, A. DIDIER
Service de Pneumologie et Allergologie
C.H.U. Rangueil
31054 Toulouse Cedex, France

ABSTRACT

French talc (talc from LUZENAC) is associated with various amounts of chlorites, small quantities of carbonates ; there is no asbestos and quartz is below 3 %. The clinic and radiologic features are not specific : micronodular pneumoconiosis, generally without clinical symptoms. In some cases, confluent opacities and a pseudo-tumoral aspect with emphysema may be observed. In these cases, restrictive or mixed ventilatory insufficiencies are noticed. Talc and talc-bodies are always found in the bronchio-alveolar lavage of such patients. The histologic lesions are composed of macrophages containing particles with slight fibrosis. Radiologic signs, clinical and functional symptoms progressed in some cases several years after dust exposure has been stopped. Broncho-pulmonary infectious factors can sometimes explain a more severe evolution. Mortallty by cancer is not significantly higher than in general population. We did not observe any case of mesothelioma. Morbidity by non malignant respiratory diseases is significantly higher compared with control subjects.

The French talc

The only French talc deposit currently exploited is located in LUZENAC (Ariège) in the South of France. Talc is extracted in open air with no dust, and then processed in a plant (sorting, drying, milling, packaging). In the factory very high dust levels were observed in the past but had gradually decreased during the last years (Fig 1)

NATO ASI Series, Vol. G 21
Health Related Effects of Phyllosilicates
Edited by J. Bignon

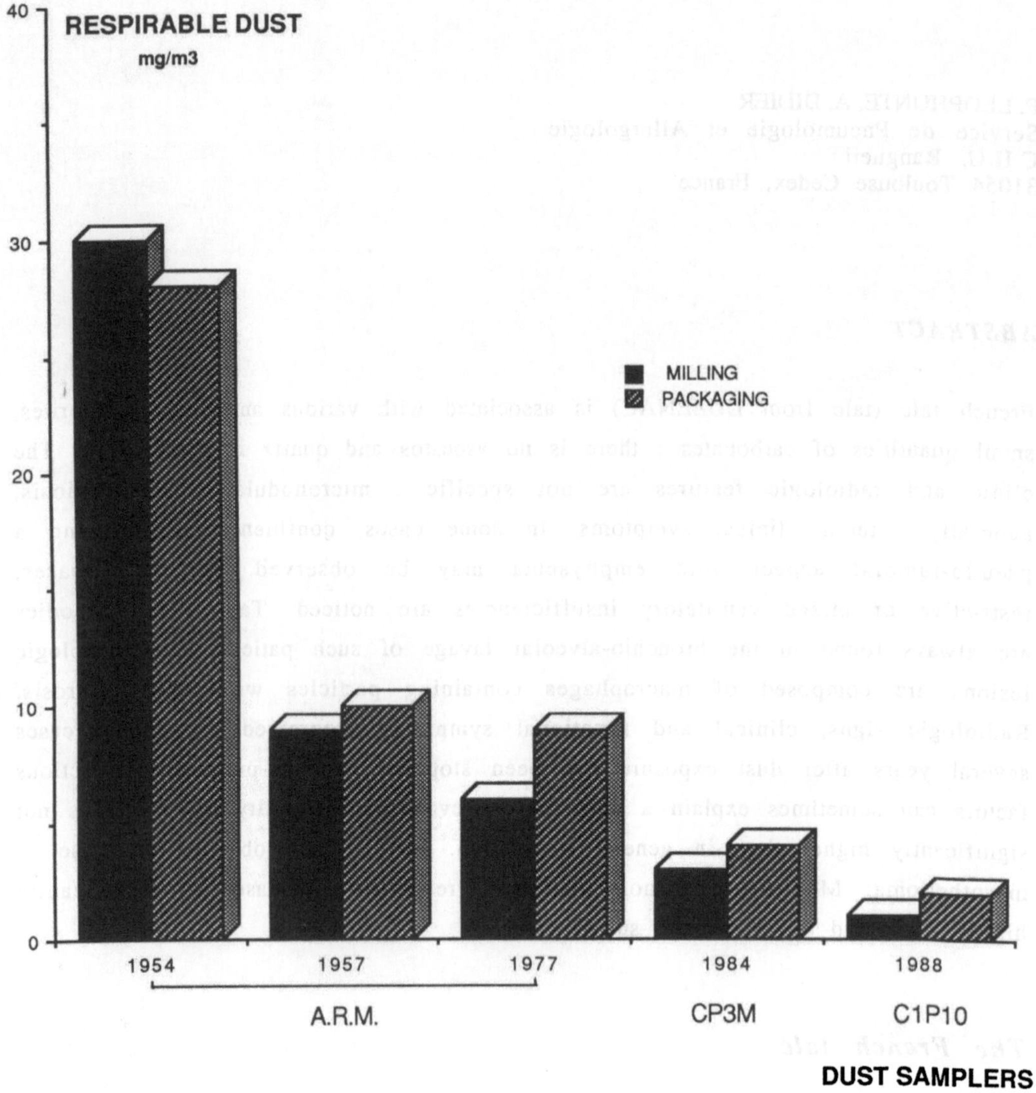

Fig. 1 - STATIC DUST MEASUREMENTS INSIDE THE FACTORY

The French talc contains talc and chlorite with a small quantity of dolomite ; there are traces of pyrite ; the amount of free silica varies from 0.5 % to 3 % (less than 1 % for a particle size distribution of less than 10 microns). This talc does not contain any asbestos (table I).

Table I - FRENCH TALC MINERALOGICAL COMPOSITION (Talc de Luzenac)

MINERALS	TALC 1 (%)	TALC 2 (%)	TALC 3 (%)
TALC	56	41	90
CHLORITE	42	58	9
DOLOMITE	2	1	1

Small traces of pyrite (150 ppm)
No asbestos. Free silica : 0.5-3 % (< 1 % dust < 10 μ)

Material and methods

Our experience is based on the following data :
- a cross sectional study carried out in 1978 on 176 millers with a ECSC (European Coal and Steel Community) questionnaire, a clinical examination, a chest X-ray, a vitalographic spirometry ;
- a retrospective mortality study from 1945 to 1981 completed by a prospective study until 1988 ;
- a functional respiratory study on 39 pneumoconiotic workers matched to 39 non dust exposed control subjects with similar age and smoking habits ;
- a more detailed study of 8 hospitalized pneumoconiotic workers. A bronchioalveolar lavage was carried out on all of them and a surgical lung biopsy on three.

Results

The prevalence of pneumoconiosis is highly related with dust levels observed in the past years (Fig. 1). In the cross sectional study on 176 workers having worked exclusively in the talc industry, 46 suffered from pneumoconiosis (27 %). Among them 36 had a slight pneumoconiosis (small opacities in categories 1/0, 1/1, 1/2), 10 showed signs of pneumoconiosis with higher profusion (categories 2 and 3) or large opacities. Pleural thickenings were rare (three cases) and never calcified. In two cases not included in the study, a progressive massive fibrosis with emphysema was observed. Intensity and duration of dust exposure were linked to the radiologic signs of pneumoconiosis.

The functional respiratory symptoms (cough, phlegm, dyspnea) were significantly higher among pneumoconiotic workers than in non pneumoconiotic patients.

The ventilarory function of pneumoconiotic workers was significantly impaired (Table II). This observation was based on a study carried out on 39 pneumoconiotic workers (33 cases with small opacities with profusion equal to 1, and 6 cases with small opacities with profusion equal to 2 and 3 and/or large opacities). They were matched to 39 non exposed control subjects with similar age and smoking habits. The TCO (single breath method) was decreased and a moderate restrictive ventilatory syndrom was observed among the 6 most impaired pneumoconiotic workers.

In both group V'max 25 % and V'max 50 % were lower than reference workers, in relation with smoking habits (29 smokers).

Table II - TALC PNEUMOCONIOSIS FUNCTIONAL RESPIRATORY STUDY

FUNCTIONAL TESTS	NON DUST EXPOSED 39 workers	ALL PNEUMOCONIOTIC 39 workers	PNEUMOCONIOTIC Profusion 2-3 Large opacities 6 workers
VC	103	96.8*	87*
RV	97.6	87.8	81
TLC	100	94.2*	86*
FEV/VC	100	102.5	102
V'max 50 %	79	77	65
V'max 25 %	47.2	49.6	35
Single breath TCO	111	100*	78*

% compared with the ECS'S reference values

* significant

A bronchoalveolar lavage was performed on eight pneumoconiotic workers, all smokers : we observed an hypercellularity with a significant increase in neutrophil and eosinophil polymorphonuclear leukocytes (table III).

Table III - BRONCHOALVEOLAR LAVAGE - 8 cases

PATIENTS	Cell x 10^3/ml	% alive cells	Macrophages %	Lymphocytes %	PMN %	PME %
1	300	100	77	13	9	1
2	220	95	95	3	2	0
3	880	70	75	5	13	7
4	530	66	89	6	4	1
5	544	80	76	13	5	6
6	74	-	90	6	2	2
7	450	-	87	11	2	0
8	1200	80	94	1	5	0
Mean value	531 ± 366	-	85 ± 8 %	7 ± 5 %	5 ± 4 %	3 ± 2 %

PMN = Polymorphonuclear neutrophils, PME = Polymorphoeosinophils

Numerous talc particles were found in all lavages fluid : uncoated plate like particles (0.5 to 40 microns), talc bodies and atypical ferruginous bodies.

A surgical pulmonary biopsy was performed on three patients. In two cases, a macrophagic alveolitis with mineral overload and a slight interstitial sclerosis could be seen, but the biopsy had been performed after a pulmonary infectious disease which had altered the pneumoconiotic radiologic signs. A third patient was suffering from a chronic obstructive pulmonary disease, his lung biopsy showed mineral overload with perilobular and peribronchial densification and emphysema lesions.

The evolution of pneumoconiosis can be evaluated from the observation of some clinical cases and the data of the mortality study. In most cases, clinical and radiological signs were stable when dust exposure were over. A worsening was observed in two cases after an acute pulmonary infection. In one case, a pneumococcal pneumonia recovered with a pseudotumoral sequela and in the other case, the patient showed an acute respiratory distress syndrom due to influenza with persistent opacities after recovery : there was an intense macrophagic alveolitis with mineral overload which cleared after corticosteroid therapy. In some cases, the evolution towards worsening was gradual after interruption of dust exposure, with signs of progressive massive fibrosis and emphysema. According to the data of the retrospective mortality study, the life expectancy of dust exposed workers was not significantly different than in the local and national population. The mortality due to cancer, especially to lung and digestive cancers was not significantly different than in the local and national population. By contrast, the mortality due to chronic respiratory diseases in a cohort of 97 workers deceased between 1970 and 1981, matched to 97 subjects dead at the same age and having lived in the same area showed a mortality ratio of 2.4. A follow-up study until 1988 confirmed these results.

As a conclusion, the French talc pneumoconiosis is frequent among talc workers and related to dust overexposure. In most cases, it is an asymptomatic micronodular pneumoconiosis. In few cases with radiological profusion, a moderate restrictive insufficiency appears ; in rare cases a progressive massive fibrosis can be observed which can be emphasized by an acute lung infection. The basic lesion is a macrophagic alveolitis with mineral overload and a slight interstitial sclerosis. There is an increased mortality rate due to chronic respiratory diseases but no carcinogenic risks and no mesothelioma.

References :

FORTUNE JP (1971) Contribution à l'étude minéralogique génétique des talcs pyrénéeens. Thèse doctorat sciences Université Paul Sabatier Toulouse

LEOPHONTE P, FABRE J, POUS J, ALBAREDE JL, DELAUDE A (1976) Les pneumoconioses par le talc. Arch. Mal. Prof. 37 : 513-532

LEOPHONTE P, FERNET P, PINCEMIN J, MOUSSET JL, RELLIER R, DELAUDE A (1980) La pathologie respiratoire chronique des travailleurs du talc. Rev. Fr. Mal. Resp. 8 : 43-45

LEOPHONTE P, BASSET MF, PINCEMIN J, LOUIS A, FERNET P, DELAUDE A (1983) Mortalité des travailleurs du talc en France. Etude épidémiologique rétrospective. Rev. Fr. Mal. Resp. 11 : 489-490

LEOPHONTE P, DE VUYST P, DUMORTIER P, ROUQUET RM, DIDIER A (1987) Données cytomorphologiques et minéralogiques du lavage broncho-alvéolaire au cours de la pneumoconiose par le talc français. Silicose et pneumoconioses à poussières mixtes. LE BOUFFANT L éd. Colloque INSERM 155 : 93-98

DE VUYST P, DUMORTIER P, LEOPHONTE P, VAN DE WEYER R, YERNAULT JC (1987) Mineralogical analysis of bronchoalveolar lavage in talc pneumonoconiosis. Eur. J. Respir. Dis. 70 : 150-156

TALC PNEUMOCONIOSIS IN ITALY

G. Chiappino, G. Rivolta, A. Todaro, E. Trillo+ and A. Forni
Research Center on the Biological Effects
of Inhaled Dusts
Institute of Occupational Health
University of Milan
Via S. Barnaba 8
20122 Milan
Italy
+National Insurance Institute for Occupational Diseases (INAIL)

Introduction

Since the reports published in the United States on contamination of commercial talc with asbestiform fibres (Cralley et al, 1968; Dement et al, 1972; Snider et al, 1972; Rohl et al, 1976), much attention has been focused on the possible risks for exposed subjects. A recent survey by the National Health Institute in Italy (Istituto Superiore di Sanità, Rome) has reported that a large number of cosmetic talc products contaminated with chrysotile or, more often, with amphiboles are still on the Italian market; most of these contaminating fibres are ultramicroscopic in size (Paoletti et al, 1984). It has been demonstrated that contaminating fibres are also present in industrial talc (Marconi et al, 1986).

Epidemiological data

Epidemiological investigations on workers employed in talc mining and milling and in talc application in the rubber industry reported cases of nodular pneumoconiosis almost exclusively among miners. The results of these studies suggested that minerals present in the rocks, and not talc *per se*, were probably

NATO ASI Series, Vol. G 21
Health Related Effects of Phyllosilicates
Edited by J. Bignon

responsible (Rubino et al, 1963). No epidemiological evidence of carcinogenic risk was ascertained in workers exposed to pure talc (Rubino et al, 1976).

The frequency of talc pneumoconioses in Italy registered by the Italian National Insurance Institute for Occupational Diseases (INAIL) in the last few years cannot be reliably assessed since these diseases in the statistic files of the Institute come under the more general category of lung diseases due to silicates and limestone. However, as a rough idea of the extent of this broader category of occupational lung diseases, the annual number of cases suspected and confirmed, together with the total number of registered occupational diseases, are reported in Table 1.

Table 1. Occupational lung diseases due to silicates and limestone compared with total occupational diseases in Italy (INAIL statistics)

Year	Occupational lung diseases due to silicates and limestone		Total occupational diseases	
	suspected	compensated	suspected	compensated
1984	2902	1655	48.544	35.111
1985	2877	1936	47.836	32.615
1986	3408	2117	47.912	28.956
1987	3361	2074	47.988	28.389

Clinical observations on selected talc exposed workers

Six workers with long and heavy exposure to industrial talc were recently studied at the Occupational Health Clinic of the University of Milan, using a broad diagnostic protocol with the aim of defining the type and severity of the lung disease and establish its etiology. Table 2 reports the type and duration of talc exposure in the patients studied.

Table 2. Occupational exposure to talc in the 6 patients investigated

Industry	Patient No.	Age	Sex	Job	Time
Rubber	1	50	M	Rubber dusting	1968-1984
"	2	31	F	"	1967-1979
"	3	50	M	"	1963-1965
Salami production	4	39	F	Salami dusting	1960-1988
"	5	44	F	"	1968-1988
"	6	39	F	"	1960-1988

All patients underwent chest X-rays, lung function tests, bronchoalveolar lavage (BAL) for mineralogic and cytologic investigations. In one case (patient No. 2), who showed at CT multiple pleural thickenings with an unusual hemispheric shape, exploratory thoracotomy was performed, with pleuro-pulmonary biopsy for histological and mineralogical examination by scanning electron microscopy (SEM). In another case (patient No. 6), thoracotomy with lung biopsy and histological examination were performed.

The radiological findings and the results of lung function tests are summarised in Table 3 A and B respectively.

Table 3. Chest X-ray findings and lung function tests in talc exposed workers

(A) Radiographic pattern	Frequency
- Pleural thickenings	2/6
- Nodular opacities	0/6
- Irregular opacities	3/6
- Normal	2/6

(B) Type of lung function alteration	Frequency
- Restrictive ventilatory insufficiency	4/6
- Impairment of gas transfer	5/6

Optical microscope examination of BAL revealed the presence of ferruginous bodies in all cases, at very high concentrations (hundreds/ml BAL fluid) in three cases, as well as naked fibres and lamellar particles with the typical morphologic characteristics of talc.

In case No. 2 the mineral particle content of lung tissue and BAL was evaluated at SEM. An enormous number of fibres was found, that could partly be identified as fibrous calcium and iron silicate (actinolite?), in addition to particles having the morphological and mineralogic characteristics of talc.

In most cases, BAL cytology showed a macrophagic alveolitis with an increase in iron-laden macrophages, and an associated slight increase in neutrophils in one case (patient No. 2) and an increase in lymphocytes with inverted T helper/T suppressor ratio in another one (patient No. 5).

Histological examination of emispherical pleural thickenings in case No. 2 showed only a collagenous hyaline connective tissue, poorly cellulated. Histological examination of lung parenchyma in cases Nos. 2 and 6 showed that the predominant alteration was a diffuse interstitial sclerosis, mainly localised in the interlobular septa and in peribronchial and perivascular areas, a pattern well known as characteristic of "early asbestosis" (Chiappino and Bertazzi, 1980). In addition to the diffuse sclerotic alterations, multiple and quite disstinctly localised granulomata were present, which consisted of epithelioid cells, foreign-body giant cells, and macrophages containing double refractile particles of talc. Only a very moderate fibrotic reaction in concentric sheaths was present at the periphery of some granulomata. No silicotic nodules were found.

Conclusions

In the cases under study, all radiological, functional and morphological data agree in indicating that the main lung changes that can be observed in subjects heavily occupationally exposed to industrial talc are very similar to those of asbestosis. These results complete previous Italian epidemiological and clinical studies and confirm that occupational exposure to talc involves above all exposure to its contaminants. These contaminants (silica and asbestiform fibres in the mining industry, asbestiform fibres in all other industrial uses) are able to cause lung changes that can be classified as silicosis or asbestosis. Lamellar talc seems to produce foreign-body granulomata with only slight tendency to sclerosis. The definition of "talcosis" should be limited to these last manifestations.

Acknowledgment. The Authors wish to thank prof. K.H. Friedrichs (Medical Institute for Environmental Hygiene at the University of Düsseldorf, FRG) for mineralogic examination at SEM of samples of case No. 2, and prof. C. Ravetto, Head, 2nd Unit of Pathological Anatomy and Histology, Civil Hospital, Busto Arsizio (VA), Italy, for providing lung specimens of case No. 6.

References

Chiappino G, Bertazzi PA (1980) The value of nonradiological indicators in the early diagnosis of asbestosis. In: Wagner JC (ed) Biological effects of mineral fibres, vol 2. IARC Scientific Publications No. 30. IARC, Lyon, p 513

Cralley LJ, Key MM, Groth DH, Lainhart WS, Ligo RM (1968) Fibrous and mineral content of cosmetic talcum products. Am Ind Hyg Ass J 29:350-354

Dement M, Shuler PJ, Zumwalde RD (1972) Fiber exposure during use of baby powders. NIOSH Report No. IS-36-6

Marconi A, Maccione m, Rossi L (1986) Asbesto e talco: determinazione del contenuto di particelle minerali fibrose in polveri di talco commerciali mediante tecniche associate di microscopia ottica. Med Lav 77:496-510

Paoletti L, Caiazza S, Donelli G, Pocchiari F (1984) Evaluation by electron microscopy techniques of asbestos contamination in industrial, cosmetic and pharmaceutical talcs. Regulat Toxicol Pharmacol 4:222-235

Rohl AN, Langer AM, Selikoff IJ, Tordini A, Klimentidis R, Bowes DR, Skinner DL (1976) Consumer talcums and powders: mineral and chemical characterization. J Toxicol Environ Health 7:255-284

Rubino GF, Maranzana P, Pettinati L, Scansetti G (1963) Aspetti etio-patologici e clinici delle pneumoconiosi da talco. Med Lav 54:496-506

Rubino GF, Scansetti G, Piolatto G, Romano CA (1976) Mortality study of talc miners and millers. J Occup Med 18:186-193

Snider DW, Pfeiffer DE, Mancuso JJ (1972) Asbestos form impurities in commercial talcum powders. The Compass: Sigma Gamma Epsilon 49:65-67

HUMAN PATHOLOGY OF KAOLIN AND MICA PNEUMOCONIOSES

A.R. GIBBS
Consultant Pathologist
Llandough Hospital
Penarth
South Wales
GB
and
Environmental Lung Disease Research Group
Llandough Hospital.

KAOLIN

Introduction

Cornish china clay deposits which are situated in the area around St. Austell have been exploited since the late 18th century. The production of kaolin from this source has continued until the present day. Radiological surveys of workers in this industry have indicated that a pneumoconiosis can develop as a consequence of exposure to dust (Sheers 1964; Oldham 1983). However, the literature contains very little information about the pathological changes in the lungs of these workers and it was not clear whether these changes were due to free silica exposure or due to the kaolinite. Since 1968 lungs from Cornish China Clay workers have been received at the MRC Pneumoconiosis Unit and the Histopathology Department of Llandough Hospital, Penarth, in order to study the pathological lesions in these workers. The majority of the cases were referred by the Cardiff Pneumoconiosis Medical Panel. Preliminary results have been published (Wagner et al 1986). Approximately a hundred cases have now been examined pathologically and half of these have also had mineralogical analysis of the lung dust burden performed using electron microprobe and x-ray diffraction techniques. The mineralogical analyses were performed by Professor F.D. Pooley and P. Salt).

Industrial Process

The production of china clay begins by washing out the clay from an open pit using a high pressure water jet. Associated minerals such as mica and sand are removed by differential sedimentation. The suspension containing the clay is piped to a drying plant. The water is removed nowadays by filter presses followed by cylindrical drum driers. Formerly it was dried by drawing off the supernatant

NATO ASI Series, Vol. G 21
Health Related Effects of Phyllosilicates
Edited by J. Bignon

fluid from settling tanks and putting the slurry into drying kilns. The clay contains less than 10% moisture after leaving the driers. It is then bulk loaded or bagged. The processed clay contains less than 1% of quartz.

Pathology

During the early phase of the study it became apparent that there were a variety of pathological lesions. In some cases there were lesions which strongly resembled silicotic nodules. When the occupational histories were checked it was evident that these workers had been exposed to china stone as well as china clay dust. Furthermore, mineralogical analysis of these cases showed a high free silica content in addition to kaolinite. Based on the mineralogical analyses three groups could be defined (Wagner et al 1986):-

(a) China clay group. The lung dust contained:-

kaolinite >90%
quartz <1.1%
feldspar <1.0%

(b) China clay and china stone group. The lung dust contained:-

kaolinite <90%
quartz >0.9%
feldspar >1.0%

(c) Miscellaneous group

The lung dust was diverse in composition and did not fit into either group (a) or (b).

Only the pathological appearances of group (a) will be described in this paper.

Macroscopic:

Lesions were virtually confined to those workers employed in the drying and loading areas. The essential lesion was a stellate grey macule which averaged 2 to 3 mm in size. A smaller proportion of the lesions and not observed in every case were palpable grey nodules which were usually slightly larger. In a few cases massive lesions measuring several centimetres in diameter were found predominantly in the upper zones of the lungs. These were pale grey in colour and had a putty-like consistency; some showed cavitation and resembled tuberculous lesions. In a small number of cases there were diffuse linear fibrotic lesions which had related fine honeycombing.

Microscopic: The earliest lesions were located around respiratory bronchioles and consisted of macrophages containing finely particulate brown material. Admixed with these collections there were fine reticulin fibrils but little collagen. Expansion of these lesions resulted in extension of the dust-containing macrophages within the interstitium along the alveolar walls (Fig.1). In some cases these lesions linked up with others from other respiratory bronchioles to give a diffuse linear pattern (Fig.2) but it was only a minority that manifested conspicuous fibrosis. The majority showed fine fibrosis. Microscopic examination of the palpable lesions showed randomly arranged collagen at the centre of the collection of dust laden macrophages but the latter exhibited marked extension along the alveolar walls. The massive lesions consisted of randomly oriented collagen fibres admixed with large numbers of dust laden macrophages. Many of the dust particles exhibited birefringence when viewed under polarised light (Fig.3). Around several of the massive lesions there were small vessels which showed infiltration of their walls by dust laden macrophages associated with variable numbers of chronic inflammatory cells and fibrosis. These probably caused ischaemic necrosis and accounted for the cavitation in some of the massive lesions. Ferruginous bodies characterised by the deposition of iron on what appeared to be plate-like particulate material were observed in some of the cases but were not conspicuous.

Discussion

Previous radiological surveys of the china clay industry have shown that workers in the more heavily exposed jobs of milling, bagging and loading are most likely to develop pneumoconiosis (Sheers 1964; Oldham 1983). The usual form is simple but occasionally in those with very heavy exposures complicated pneumoconiosis can develop (Sheers 1964; Oldham 1983). The findings in the Cornish china clay industry are similar to those reported in studies of the North American kaolin industries (Morgan et al 1988). Ventilatory impairment appears to be related to complicated pneumoconiosis. Few pathological descriptions of the pulmonary changes in workers with pneumoconiosis from the Cornish china clay industry or North American kaolin industry have been reported (Hale et al 1956; Lapenas et al 1948; Wagner et al 1986) and apart from the publication of Wagner et al they describe few cases. When those workers from the Cornish china clay industry who have been exposed to quartz in the form of china stone are excluded the pathological changes in the Cornish group appear similar to those of the North American cases and thus the changes appear to be caused primarily by the kaolin component of the dust.

Figure 1

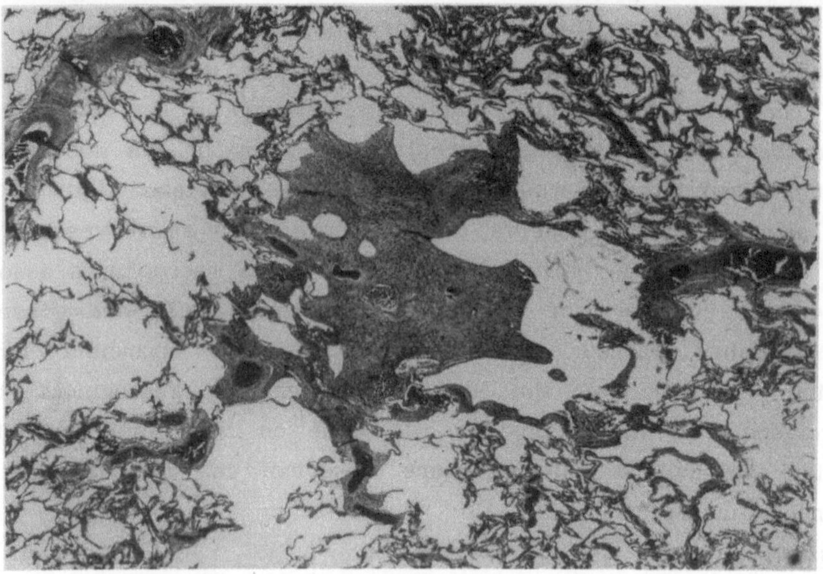

A histological section from the lung of a china clay worker showing collections of dust laden macrophages admixed with fine fibrosis located in the respiratory bronchiole and extending along the alveolar walls.

Figure 2

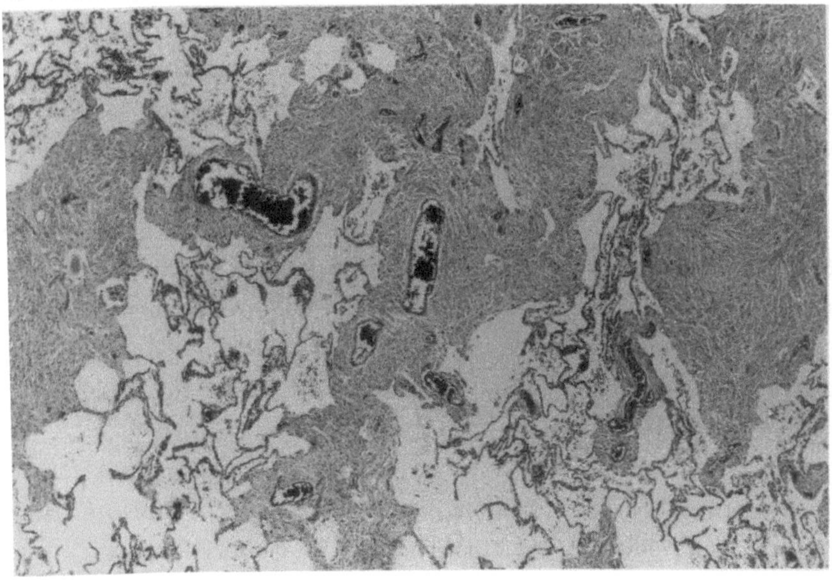

More widespread interstitial collections of dust laden macrophages with fine fibrosis from a china clay worker.

Figure 3

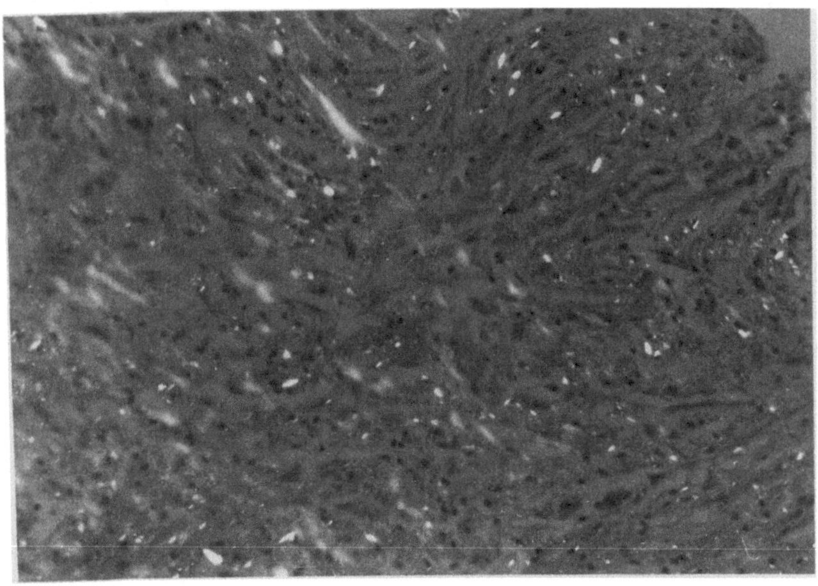

The same case as fig.2 viewed under polarised light showing numerous birefringent particles of china clay dust mainly lying within macrophages.

MICA

Mica is used as a filler in paints, cement and asphalts, in cable insulation and in drilling muds for the oil industry. It is also used in the electrical industry in vacuum tubes, condensators and insulation. Exposure to mica without the concomitant presence of other silicates or quartz is rare and some have doubted that mica pneumoconiosis exists. Skulberg et al (1985) reviewed the literature and concluded that of the 66 cases of mica pneumoconiosis reported in the literature only 26 were caused by exposure to mica alone. Few of these cases had adequate pathological documentation. Pure exposures have mainly occurred in the packing and grinding of mica. I have seen in consultation a few cases of pneumoconiosis due to relatively pure exposures to mica.

Pathology

Macroscopically, the lesions are grey, soft stellate and vary between two and several millimetres in diameter (Fig.4). Massive soft grey lesions may occur. Grey fine interstitial fibrosis may be present. The lesions are similar to those observed in kaolin pneumoconiosis. Microscopically, the earliest lesion consists of collections of dust-laden macrophages situated around the walls of the respiratory bronchiole. With increases in size they extend along the alveolar ducts and the interstitium of the adjacent alveoli (Fig.5). Lesions from adjacent secondary lobules may link up to give diffuse interstitial dust lesions associated with fine fibrosis. The dust particles are intensely doubly refractile when viewed under polarised light. In the cases I have seen prominent focal collections of foreign body giant cells have been observed in relation to the dust particles (Fig.6). Ferruginous bodies formed on the plate-like particles have also been conspicuous. The massive lesions consist of large quantities of dust admixed with randomly orientated collagen fibres.

Figure 4

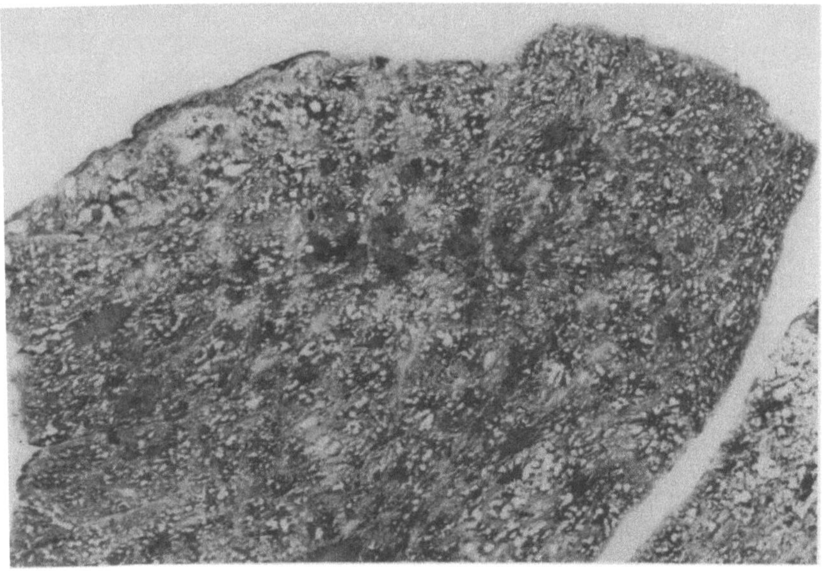

Part of the whole lung section taken from a male who had been grinding and packing mica for at least twenty years. It shows numerous grey stellate lesions, some of which are joining together.

Figure 5

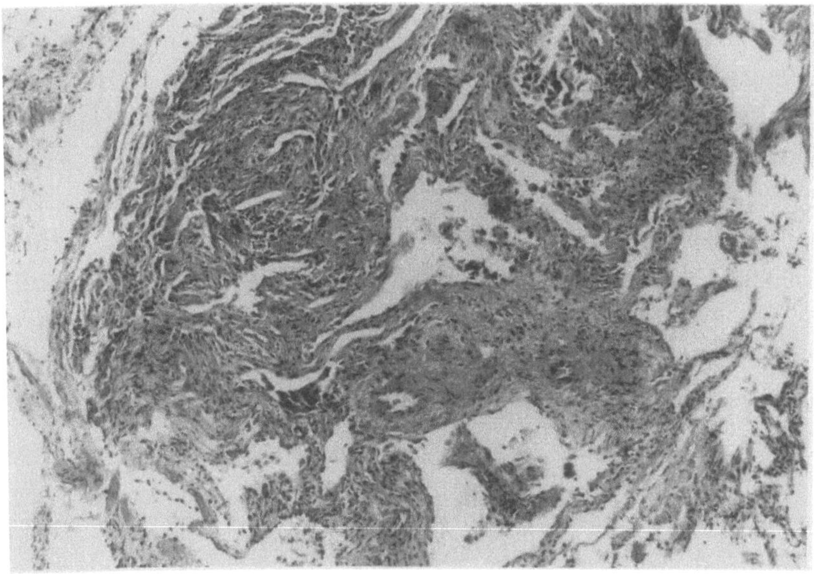

A transbronchial biopsy from a male who had packed mica for many years. It shows a fine interstitial fibrosis admixed with numerous dust laden macrophages. (Dr. M. Stephens, City Hospital, Nottingham, kindly referred this case.)

Figure 6

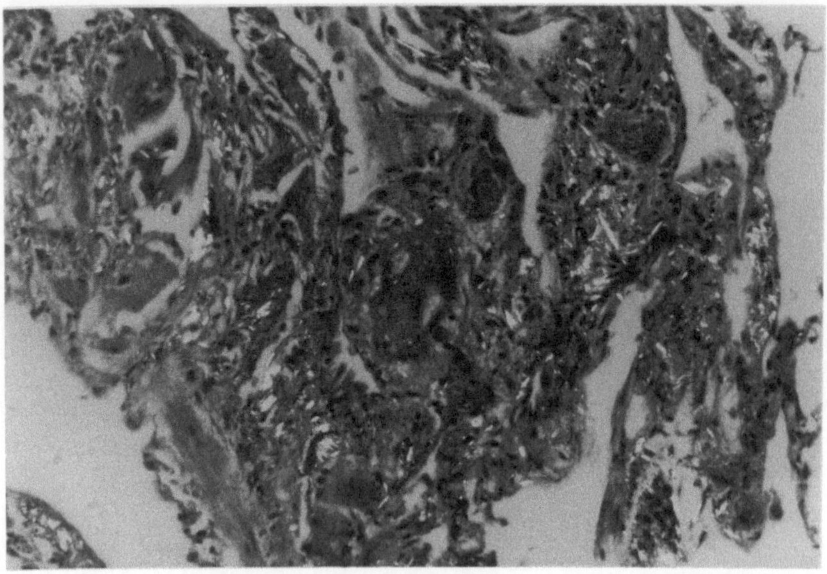

The same case as fig.5 viewed under polar sed light showing birefringent crystals within giant cells and macrophages.

Discussion

The pathological changes I have seen in cases of mica pneumoconiosis appear similar to those described by Davies and Cotton (1983). From the descriptions above it will be realised that the pathology of kaolin pneumoconiosis is similar to that of mica except that collections of foreign body giant cells and ferruginous bodies are more prominent in the latter. Although rare, I believe that pneumoconiosis can result from the inhalation of large quantities of mica in the absence of other silicates or quartz. In the cases reported by Davies and Cotton (1983) the disease was characterised by dyspnoea, a restrictive impairment of ventilation, a reduced transfer factor and hypoxaemia.

In conclusion there are few well-studied cases of pneumoconiosis due to relatively pure exposures for kaolin or mica with good pathological and mineralogical data. This is also true for other pure silicate exposures. More studies are needed to assess the human pulmonary responses to these minerals.

REFERENCES

Davies D, Cotton R (1983) Mica pneumoconiosis. Br J Ind Med 40:22-27

Hale LW, Gough J, King EJ, Nagelschmidt G (1956). Pneumoconiosis of kaolin workers. Br J Ind Med 40:131-137

Kennedy T, Rawlings W Jr., Baser M, Tockman M (1983) Pneumoconiosis in Georgia kaolin workers. Am Rev Respir Dis 127:215-220

Lapenas D, Gale P, Kennedy T, Rawlings W Jr., Dietrich P (1984) Kaolin pneumoconiosis. Radiologic Pathologic and Mineralogic Findings. Am Rev Respir Dis 130:282-288

Morgan WKC, Donner A, Higgins ITT, Pearson MG, Rawlings W Jr. With the technical assistance of MJ Bracken (1988) The effects of kaolin on the lung. Am Rev Respir Dis 138:813-820

Oldham PD (1983) Pneumoconiosis in Cornish china clay workers. Br J Ind Med 40:131-137

Sheers G (1964) Prevalence of pneumoconiosis in Cornish kaolin workers. Br J Ind Med 21:218-225

Skulberg KR, Gylseth B, Skaug MD (1985) Mica pneumoconiosis - a literature review. Scand J Work Environ Health 11:65-74

Wagner JC, Pooley FD, Gibbs A, Lyons J, Sheers G, Moncrieff CB (1986) Inhalation of china stone and china clay dusts: relationship between the mineralogy of dust retained in the lungs and pathological changes. Thorax No. 3, 41:190-196

PALABORA VERMICULITE

G.K. SLUIS-CREMER P.A. HESSEL
Epidemiology Research Unit
P O Box 4584 Johannesburg R.S.A.

At Phalaborwa some 300 km north-east of Johannesburg there is a remarkable geological phenomenon - the Loolekop volcanic pipe - of some 12 Km^2 host to two large open pit mines. One a low grade but large copper producer and the other the vermiculite mine we are concerned with. They are only 1 km apart. Additionally large quantities of foskorite are exploited by Foskor to produce phospate fertiliser.

The vermiculite is a hydrophlogopite (Libby vermiculite is hydrobiotite) the deposit is 2 km^2 and some 50 meters deep. It is underlain by phlogopite partly converted to vermiculite and then pure phlogopite to great but undetermined depth.

THE VERMICULITE MINE AND THE ORE

The mine is open pit, the mill is on site.

Production was in 1963 8000 tons a month but is now 20,000 tons.

The ore body is a large one with a high vermiculite content, Serpentine, pyroxenite, diopside, biotite and phlogopite are the associated minerals.

Moatemed et al found amphibole fibre in Palabora vermiculite (0.4% by Xray diffraction in the unexpanded ore). A similar study on mill dust in South Africa was unable to detect amphibole.

NATO ASI Series, Vol. G 21
Health Related Effects of Phyllosilicates
Edited by J. Bignon

Very little amphibole fibre was found by Moatemed in unexpanded ore by scanning electron microscopy; these were thought on analytical studies to be actinolite and anthophylite.

Electron microscopy studies by Viljoen in South Africa by qualitative and semiquantitative microprobe analysis) rendered atomic proportions very similar to tremolite (Na .27 K .23 Mg 4.75 Ca 1.68 Fe .41 Al .03 Si 8.36 O 24)

Fibre counts in the pit are very low (0.02 to 0.06 fibres/ml by RTM1 as measured in 1988).

Dust counts in the pit by konimeter after heat and acid treatment were high in some places in 1988 viz 238 p/ml during drilling rod extraction, 413 p/ml during loading.

THE MILL

In table I an attempt has been made to compare the estimated fibre counts at Libby and Palabora.

All the counts done in 1987/88 on membrane filter are below 1 fibre/ml and recent counts are very low. The extrapolations from the thermal precipitation counts on the basis of 1 fibre per million particles per cubic foot are probably gross exaggerations.

TABLE I

DUST CONDITIONS IN THE MILL FIBRES/ML

Libby (Amandus et al)	PALABORA
YEAR	
50-64 ± 168)estimated	
65 38)from midget	Dry process
68/69 27)impinger counts	No counts
67/68 65)membrane	Very dusty
74 22)filter	72 Dust extraction in mill
	78 1.6) Estimates
	80 8.3) from TP*
80/82 0.6 New Mill	
	84 2) from TP*
	87 0.27 - 0.80 RTM 1
	0.16 - 0.55 RTM 2
	88 0.008 to 0.04 RTM 1
	Despite counts 1500p/cc (tmtc) + at some places. i.e crusher, screens and conveyer (Konimeter after heat and acid).

* Thermal Precipitator.

\+ To much to count.

THE SURVEY (Hessel and Sluis-Cremer)

The Palabora mining company owns the copper mine and the vermiculite mine. The mining company recruits black labourers for both mines on the same standards of general fitness. There are no differential standards for allocation of labour to either of the mines. This point is made because in this study the workers on the copper mine were used as controls for the study on the vermiculite miners.
Black workers only were studied as the number of white workers at the vermiculite facility was small and the two ethnic groups would have to be treated separately for the purpose of comparing lung function with controls.

Of 670 potential vermiculite miners and controls, three refused the examination 82 were excluded because of a history of occupational exposure to asbestos, 59 were excluded as they had worked in more than one workplace at the Palabora mine. Accordingly 526 men were studied of whom 172 were pure vermiculite miners and 354 were control copper miners.

There were no differences in age, height, weight, between the two groups. Average length of employment of the vermiculite miners was 15.3 years, 31% had more than 20 years employment and had therefore been exposed to very dusty conditions. The copper pit mining controls had slightly less average length of exposure ($\pm$ 11 years). Also there were no differences in percent of current smokers, ever smokers or average pack - years.

A respiratory symptom (modified British MRC) questionnaire, with a smoking section and a very detailed working and residential history was administered. The point of major interest being environmental amphibole exposure because of the presence of many nearby (± 100km) asbestos containing tailings dumps from closed and active crocidolite and amosite asbestos mines.

Lung function was measured with a Vitalograph with a microprocessor (in 13 men no satisfactory curves were obtained).

Large P A films were done on all except one man and all were given a sputum bottle to try and produce a sputum specimen for detection of asbestos bodies.

RESULTS

There was no excess of any symptom among the Vermiculite miners compared to controls. This was true for morning cough, morning phlegm, ever wheeze, period of increased cough and phlegm, dyspnoea when walking on level or slight hill and chest illness within the last three years.

Lung function was compared to internally derived equations which were determined using the entire study group (except the 13 who did not produce acceptable curves). The linear regression model included age, height, weight, and pack -years.

Table II summarises the radiological findings, Table III the lung function findings and Table IV the prevalence of asbestos bodies in the two groups.

TABLE II

XRAY FINDINGS %

	Vermiculite	Controls
>0/1 Profusion S.O.*	1.2	0.7
Pleural change	3.5	9
Pleural calcif.	1.7	1.9
Oblit C.P. angle	5.2	8

* Small opacities

TABLE III

LUNG FUNCTION % PREDICTED OF VERMICULITE MINERS

	Average S.D.
VC	100.3 (13.1)
FVC	99.8 (13)
FEV_1	100.2 (14.8)

TABLE IV

ASBESTOS BODIES

Vermiculite	2/161 (1 Asbestos mining)
Controls	5/282 (3 Asbestos mining 2 unexplained)

The Pension fund of the Palabora mining company is a young one (since 1976).

Perusal of the 135 death certificates they could furnish revealed 4 deaths ascribed to bronchial cancer of whom one was a vermiculite miner. (there were only 14 death certificates for vermiculite miners).
One death certificate of a vermiculite miner was reported as "An infiltrative lesion round the heart and right kidney." Attempts to trace the person or his doctor failed.

For three deceased Vermiculite miners the lungs were examined. Two contained asbestos bodies without fibrosis. These on electron microscopy appeared to be crocidolite - testimony to the confounding effect of the presence of Transvaal crocidolite mines near Phalaborwa.

CONCLUSION

This study has revealed no evidence that the Palabora vermiculite ore has any deleterious effects on the respiratory health of the miners either due to vermiculite or the associated minerals or the small content of tremolite.

A formal mortality study is however indicated at some time in the future as the possibility of mesothelioma can not be excluded.

We thank Mr. J.M.O. van Sittert of the South African Government Mining Engineer's Department for access to the dust counts carried out at the Palabora Mine.

REFERENCES

Amandus H.E., Wheeler R., Jankovic J., Tucker J.
The morbidity and mortality of Vermiculite miners and millers exposed to Tremolite - Actinolite. Part 1 Exposure estimates.
American Journal of Industrial Medicine 1987:11:1-14.

Hessel P.A., Sluis-Cremer G.K.
Xray findings, lung function, and Respiratory Symptoms in Black South African Vermiculite Workers.
American Journal of Industrial Medicine 1989:15:21-29.

Moatamed F., Lockey J.E., Parry W.T.
Fiber contamination of Vermiculite: A Potential occupational and environmental health hazard.
Environmental Research 1986:41:207-218.

Viljoen E.A. MINTEK Mineralogy Division. 1988. Personal communication.

PART III

HEALTH RELATED EFFECTS AFTER NON OCCUPATIONAL EXPOSURE

TALC, ITS COMESTIC, ADDICTIVE AND THERAPEUTIC USES AND THEIR EFFECTS ON HEALTH

C. BOUTIN, J.R. VIALLAT
Pulmonary Disease Department
Hôpital de la Conception
147, Bd Baille
13385 Marseille Cedex 5
France

It is paradoxal but stimulating to include in the same paper the various harmful effects of talc along with its so-called "therapeutic" uses. Occupational exposure to talc and the toxic effects of talc containing chemicals such as hexachlorophene will not be discussed here. Our paper will be limited to the accidental or intentional contacts of the organism with talc. It is almost exclusively based upon a review of the literature as we have no experience in the effects of talc after inhalation, ingestion, peritoneal contamination or intravenous injection. However, we have performed pleural poudrage on thousands of patients with pleural effusion or spontaneous pneumothorax since 1965.
Among the various effects of talc, many are historical and have been known for years while others are anecdotal. Two are more interesting and should be developed: intravenous injection of talc and therapeutic talc poudrage.

I. Broncho-pulmonary talcosis from cosmetic talcum powder

Talc particle size varies depending on the process used to make the powder: common cosmetic talc has particles ranging from 0.3 μ to 50 μ in size. A small portion by weight consists of particles of respirable size but their number is nevertheless considerable.
The normal uses of cosmetic talc have not been reported to be harmful. However, accidental aspiration of excessive amounts in infants has been reported to cause serious but reversible acute respiratory disease in some instances and death in isolated cases. A review of the literature (Hildick-Smith) from 1954 to 1976 reveals 27 cases with 6 deaths. Six additional cases with 3 fatalities in infants were published between 1976 and 1981 (Butenandt, Motomatsu).
Talc can cause severe bronchiolar obstruction. Tracheal intubation and extensive broncho-alveolar lavage give good therapeutic results in association with mechanical ventilation, corticosteroids and bronchodilators.
Pulmonary fibrosis has been described in one patient by Naim and Gacey: the patient died of peritonitis which was not associated with talc fibrosis. Necropsy revealed diffuse interstitial macrophage infiltration, giant cells and numerous needle-like birefringent crystals - this patient had been using a great amount of talc at least twice a day for 4 years.
Talc is obviously safe for the health when it is normally used

NATO ASI Series, Vol. G 21
Health Related Effects of Phyllosilicates
Edited by J. Bignon

for cosmetic purposes. According to Hildick-Smith, a consumer who uses talc for 10 seconds to 1 minute per day is exposed to approximately 0.120 million talc particles per cubic foot a week, or 7,800 times less than the exposure of a miner.
Cramer et al. reported cases of ovarian cancer and talc in a case-control study of 215 women with epithilial ovarian cancer: 42.8% of the patients had regularly used talc as a dusting powder on the perineum, condom or diaphragm. The relative risk of ovarian cancer was 3.28 compared to the control group. The authors did not mention if the talc was contaminated with asbestos.

II. Surgical glove powder containing talc

There have been a great deal of reports and controversies on this talc. According to Rigdon, who wrote an extensive review on this subject, Autopol, in 1933, observed the first patient presenting with a talc granuloma 3 months after surgical drainage of a tubercular cervical lymph node.
There have been frequent cases of peritoneal granulomas containing talc crystals introduced by surgeons' gloves during laparotomy. Fibrous abdominal adhesions have often been associated. These lesions could be observed from 2 weeks to many years after surgery - the longest period published was 36 years with the talc remaining unchanged in the local area. Numerous animal experiments have confirmed these talc-induced lesions. In the late 1940's, corn starch powder was suggested as a replacement for talc. In 1955, Sneirson and Woo reported the first 2 cases of starch granuloma. The number of case reports prompted the manufacturer to replace corn starch with rice starch powder. Some cases of rice starch granulomas were then published but starch is usually subject to enzymatic digestion and does not cause dangerous reactions. Until now, glove powder has been made up of rice starch mixed with calcium carbonate or magnesium carbonate. Indomethacine or steroids are recommended for treating the problem and today it is not necessary to perform a second laparotomy.

III. Intravenous injection of talc

Krainer et al. apparently observed the first case of foreign body granulomas in the lungs of a young woman who was a meperidine drug addict. Wendt et al. described a case of fatal hypertension in a woman addicted to "blue velvet" (a mixture of pyribenzamine and paregoric).
Various drugs can be held responsible: morphine (Spain), methylphenidate (Hahn), "red devil" methadone (Murphy, Sienniewicz, Pare), Pentazocine (Farber), Heroin (Crouch and Churg).
Many pills or tablets can be the cause of such problems, some with up to 75% of their weight in talc (Farber). The total amount of talc injected several times a day for as long as 19 years (in the Crouch and Churg case) can be really enormous. For these authors, the extent of the lesions is related to particle size and cumulative particle load: they found 540 x

10^6 particles/g of dried tissue in mass lesions. Particle size ranged from less than 0.5 μ to more than 50 μ with a mean of 6.4 μ. About 10% of the particles were greater than 15 μ.
The pulmonary vascular bed acts as a filter and retains crystalline particles. Therefore, in a majority of cases, the pathologic lesions - a typical foreign body granuloma - caused by talc are limited to the lung. The lesions are irreversible and lead to chronic pulmonary hypertension which is sometimes lethal except in certain cases where corticosteroids have been reported to be active.
Pulmonary talc granulomatosis must be considered in the differential diagnosis of pulmonary infiltrates of HIV-positive drug addicts (Lewis).

Clinical Symptoms

Chronic dyspnea, associated or not with coughing, is the prominent complaint. Sometimes the patient may be symptom-free (Hopkins), with diagnosis after routine chest X-ray. Hemoptysis may be related to a complication (such as mycotic aneurysm - Hopkins and Taylor). Temperature and chills may be present, especially in the minutes following the injection. Pulse rate is increased, blood pressure is decreased. Lung crepitations may be found. Electrocardiogram may demonstrate right ventricle hypertrophy.
Radiological findings vary from a normal X-ray to a diffuse reticulo-nodular pattern which is more prominent at the base of the lung, sometimes resembling a tuberculous miliary (Mariani). A coin lesion or a lobar infiltrate may be seen.
An unusal case of progressive massive fibrosis with large and bilateral densities which resembled occupational talcosis was described by Crouch and Churg in a patient with a 19-year history of drug addiction.

Pulmonary Function

Farber et al. studied 6 young men with presumed talc granulomatosis: lung volumes and spirometry were abnormal in all 6 patients. DLco S.B. averaged 21.3 ± 2 ml/mn/mmHg (53% to 76% of predicted values). There was an increased alveolar-arterial- oxygen gradient (25 ± 6 mmHg) at rest. PaO2 was decreased.
In such cases, pulmonary arterial pressure is increased and is the prominent feature: according to Farber's experiments in dogs, a rapid elevation in mean pulmonary arterial pressure follows the injection of crushed, suspended tablets of pentazocine.
Pulmonary vascular resistance increases to approximately twice baseline values with a gradual reduction to normal in 30 to 45 minutes. This transient hypertension is due to mechanical blockage of the pulmonary vessels by talc particles. The chronic hypertension may be due to granulomas on the pulmonary vasculature (Bowen).

Biological Findings

Serum angiotensine converting enzyme values were greater than 1 SD above normal in 5 out of 6 of Farber's patients. 67 GA uptake was significantly increased, even in the patients with normal chest X-rays.
B.A.L.: lymphocyte counts ranged from 18% to 32%. Polarized microscopy disclosed free and intracellular birefringent crystals.
Lung specimens may be obtained by transbronchial lung biopsy (Ben-Haim), fine needle aspiration (Liang) or thoracotomy (Hopkins): non-caseating granuloma, intravascular, perivascular and parenchymal were found in all cases. Polarized microscopy revealed diffuse birefringent crystals consistent with talc or microcrystalline cellulose particles.
The whole pattern is thus very close to sarcoidosis and may be added to the already long list of causes of foreign body granulomas.
Extra-pulmonary talcosis may be found and is due to talc dissemination through primitive or acquired arterio-venous pulmonary shunts. Liver biopsies can reveal talc granulomas (Molos and Kyung).
Talc retinopathy may be seen (Tse, Bluth, Carman). It was present in 9 patients following as many as 9,000 crushed-tablet injections (Murphy).
Postmortem examination can reveal talc lesions in the lymph nodes, spleen, bone marrow, kidneys, etc...
Clinical evolution is most often irreversible, leading to chronic pulmonary hypertension, right cardiac failure, respiratory distress syndrome and death. Clinical and radiological improvement is seldom obtained therapeutically (Smith); Brown described improvement in clinical symptoms, chest X-ray, 67 Ga uptake, PaO2 and DLCO. Moskowitz reported the case of a patient cured by steroids.

Animal experiments

Talc inhalation experiments will not be reviewed here. Following the observation of peritoneal talc granulomatosis due to surgical glove powder, the first animal experiments were conducted as early as 1933 and have been extensively reviewed by Rigdon. Two different patterns of findings resulted from intraperitoneal and intrapleural injections. Intraperitoneally, talc reproduced approximately the same lesions found in humans after laparotomy: nodular, granulomatous lesions with giant cells, lymphocytes, and macrophages with few or no adhesions between the peritoneum and the abdominal viscera. Nodules appeared as early as the fourth week. The almost complete absence of polymorphonucular leucocytes, fibrin and fluid was a prominent feature which was not consistent with the only findings of Lee and Lehman.
Experiments with starch injected intraperitoneally (Bluemel) gave the same results as talc: granulomas were present. However, while talc remained in the tissue for an undetermined period, starch progressively disappeared and was usually completely absorbed within 60 days.

Conversely, pericardial adhesions occurred in dogs as early as 18 hours after injection of 1 g of talc (Thompson). Finally Bethune, after injecting 0.5 ml of talc into the pleura of dogs or cats obtained a fibroblastic reaction without granulomatous formations (Wagner, however, observed some pleural granulomas). Many crystals were in the fibrous zone, reticulocytes and plasma cells were numerous, fibrous adhesions between the pleura and lung were present but there was no fluid due to the close contact of the visceral and parietal layers of the pleura.
Following these experiments, Bethune began his work on pleural talc poudrage in humans in 1933.

V. Therapeutic talc poudrage

Pleurodesis can be achieved in different ways including surgery or chemical drugs. There have been some excellent general reviews on all of the known pleurodesis procedures (Hausheer, Chahinian) carried out on chronic pleural effusions or spontaneous pneumothorax. Most of the reports on pleurodesis cannot be evaluated due to either the limited number of patients in the studies or the absence of randomization. On examination of the principal randomized studies of the numerous products used, only three clearly stand out: talc, Tetracycline and Bleomycine. Other less often used items include biological products such as Corynebacterium parvum (Rossi). Tetracycline is as active as Quinacrine but the latter was not well-tolerated and has been withdrawn from use. Bleomycine is at least as active as Tetracycline but has possible side effects. Hausheer considers Bleomycine to be the most active. Talc is more efficient than Tetracycline and both have the same side effects.
The efficacy of talc poudrage varies according to the series with the percentage of cured patients ranging from 60% and 93% (Boniface). Three randomized studies compare talc and Tetracycline (Boutin, Fentiman, Muir): out of a total of 111 patients, talc cured the effusion in 44 out of 47 patients and Tetracycline in 33 out of 46 patients. Tetracycline results depended on the dosage and duration of pleural drainage.
Talc pleurodesis seems to be an excellent method. In a study of 300 patients, we obtained satisfactory and permanent results in 90% of the cases of malignant pleurisy (Boutin). In a dual series (Boutin, 1986), we compared 14 patients with mesotheliomas undergoing thoracotomy with 14 patients with mesotheliomas undergoing thoracoscopic talc pleurodesis. The use of talc gave the same results as pleurectomy in terms of local response and survival. Such a result had already been demonstrated in Frankel's experiments on animals, and is confirmed in most reports. We agree with Hausheer who wrote that "it seems likely that talc represents the most effective sclerosing agent available, even after other agents have failed to control effusion."
The duration of pleural drainage is important for the comfort of patients and lasts from 3 to 5 days after talcage. Following Tetracycline instillation, drainage lasts from 24 to 48 hours, up to a mean of 11 days (Muir) or even 16 days (Reshad).

Although the **duration of symphysis** is an important criterion in evaluating the technique, it is not reported with enough detail in some publications. Sometimes a complete response is reported when the duration of symphysis may be limited, which for us is unacceptable as the pleurodesis should be permanent. In our study (Boutin, 1985), Tetracycline gave good initial results in 90% of the cases, but 50% of the patients had relapsed after 6 months. This was also the case for Ostowski with 69% response after 1 month and 54% after 3 months; Gravelyn with 15.6% complete success, 43.8% moderate relapses, 40.6% failure requiring further drainage; Dunkel with 28 out of 60 having a late relapse; Sherman with 94% response but with only 49% of the patients asymptomatic after 3 months. As Tetracycline and Bleomycine are soluble, it is possible that they gradually disappear from the pleural cavity. Wooten noted that 4 hours after a pleural injection of 20 mg/kg, the blood concentration of Tetracycline was 3.6 ± 0.9 µg/ml. On the other hand, talc is insoluable and will remain in place indefinitely, which generally brings about permanent pleurodesis (Shedbalkar). In the Boniface series, only 10% of the patients initially cured relapsed and underwent a second poudrage.

The side effects of talc poudrage have been described in many reports. Some suggest that talc pleurodesis might be dangerous (Gaensler, Hausheer). Short-term deaths have been reported (Pearson) but seem to be due to choosing patients who were already in the late stages of cancer and only had a short life expectancy in any case (Boutin letter). Three cases of acute respiratory failure (Rinaldo) and one case of acute pneumonia (Bouchama) were reported after talc poudrage for chronic effusions. Rinaldo's patients received 10 g of talc suspended in 250 ml of saline which is a great amount, much greater than the quantity of talc pulverized during thoracoscopy.

Todd observed 7 cases of respiratory failure and/or pneumonia in 146 patients but no precise details were given. Such complications appear to be rare. In 300 cases where we used talc, we only once observed a bilateral edema as described above. However, this occurred after draining the pleura too quickly in a patient at a very advanced stage of cancer. Other fatalities have been reported (Boniface) and were related to the poor general condition of patients with cancer at stage IV, meaning a status which is depressed by denutrition and repeated pleural evacuation.

Pain is comparable following talc poudrage or Tetracycline and is usually suppressed by using intrapleural xylocaine and common analgesics. Neuroleptic analgesics can be useful for avoiding pain during thoracoscopy.

In the long term, we have never encountered extensive pleural fibrosis (Gaensler), chronic residual pain or other late complications. This seems to be the result of using only moderate amounts of talc (6 to 9 ml or 3 to 4.5 g) along with careful verification of drainage which must always completely evacuate the pleural fluid. In patients with spontaneous pneumothorax, we never instillate more than 2 ml of talc for the entire pleural cavity.

The use of talc in young patients can raise a certain number of questions on long-term tolerance.

1. **Can talc cause cancer of the pleura ?** The confusion probably

comes from the fact that certain talcs contain tremolite and may be carcinogenic. Pure talc containing no asbestos, such as the one used by Guerin, Vanderschueren and us is not carcinogenic. A retrospective study published by the Medical Research Council (1979) showed that none of the subjects treated with talc over a period of 18 years showed any signs of mesothelioma. It has been suggested that a longer period, 20 to 30 years, should be studied (Chastre). In fact, talc was first used for pneumothorax by Bethune from 1933 and onwards, published in 1935 and not one case of mesothelioma caused by talc poudrage has been published since then. On the other hand, when asbestos was put into the pericardium (Broffman) in order to revascularize the myocardium after an infarct, tumors could appear in a much shorter time. Only one case of a tumor, probably glandular, which developed after the use of talc has been reported by Jackson who admitted that talc was almost certainly not the cause but that the tumor was able to seed along the drain channel. The I.A.R.C. in Lyons recently recognized that talc was not carcinogenic in an exhaustive monograph. Leophonte reached the same conclusion after studying a group of talc miners. It should be added that the international scientific community is so anxious to discover the possible causes of mesothelioma that if talc were found to be one of the causes, the case would be published immediately.

2. Is talc fibrogenic ? Here, there has been confusion with the fibrotic action of "surgical talc", which was actually starch, previously used with surgical gloves and was the cause of progressive granuloma in the peritoneum. Gaensler has rejected the use of talc pulverized during thoracotomy because it induces fibrosis. However, he did not give any details or cite any references. In fact, not one case of evolutive pleural fibrosis has been reported in the literature. There is possibily a relationship between the dosage and effect, and the small amounts used during thoracoscopy (2ml) which are much lower than the quantity used during thoracotomy and do not lead to any discernable lesions in patients.

In our series, talc did not induce any radiological sequellae in 90% of the patients. It is indeed impossible to tell which side was treated with talc: 5% of the cases develop a fine radiological border, 1 to 3 mm thick, principally around the apex and in the remaining 5% of cases, the costo-pleural angle fills up. The radiological side effects of surgery are usually more apparent.

The functional side effects of talc are usually minimal. Vanderschueren reported almost no spirometric side effects. In a recent study, we measured pulmonary clearance with technetium D.T.P.A.: it is greatly disturbed before thoracoscopy because of the pneumothorax but improves rapidly after talc is used and is back to normal within 14 days. This lack of a measurable increase in the permeability of the alveolar membrane correlates with the negative search for talc under electronic microscope after broncho-alveolar lavage carried out in 3 patients 1 week after using talc.

Autopol W (1933) Lycopodium granuloma. Its clinical and pathologic significance, together with a note on granuloma produced by talc. Arch Pathol, 24:36.
Ben-Haim SA, Ben-Ami H, Edoute Y, Goldstein N, Barzilai D (1988) Talcosis presenting as pulmonary infiltrates in an HIV-positive heroin addict. Chest, 94:656-658.
Bethune N (1935) Pleural poudrage. A new technic for the deliberate production of pleural adhesions as a preliminary to lobectomy. J Thorac Cardiovasc Surg, 4:251-261.
Boniface E, Guerin JC (1988) Intérêt du talcage par thoracoscopie dans le traitement symptomatique des pleurésies récidivantes. A propos de 302 cas. Rev Mal Resp, In press.
Bouchama A, Chastre J, Gaudichet A, Soler P, Gibert C (1984) Acute pneumonitis with bilateral pleural effusion after talc pleurodesis. Chest, 8:795-797.
Boutin C, Rey F, Viallat JR (1985) Etude randomisée de l'efficacité du talcage thoracoscopique et de l'instillation de tétracycline dans le traitement des pleurésies cancéreuses récidivantes. Rev Mal Resp, 2:374.
Boutin C, Viallat JR, Rey F (1986) Thoracoscopy in diagnosis, prognosis and treatment of mesothelioma. In : Asbestos Related Neoplasms, K Antman and J Aisner Ed, Grune & Stratton Inc, New York, pp 301-321.
Boutin C, Viallat JR (1989) Practical thoracoscopy. Springer Verlag Ed, In press.
Boutin C, Viallat JR, Rey F (1986) A propos du talcage pleural. Presse Med, 15:123.
Bluemel G, Piza F, Zischka-Konorsa W (1962) Animal experiments on tissue reaction to the intraperitoneal use of starch and talcum powder. Wiener Klinische Wochenschrift, 74:12-13.
Bluth LL, Hanscom TA (1981) Retinal detachment and vitreous hemorrhage due to talc emboli. JAMA, 246, 9:980-981.
Bowen JH, Woodard BH, Barton TX, Ingram T, Shelburne JD (1981) Infantile pulmonary hypertension associated with foreign body vasculitis. Am J Clin Pathol, 75, 4:609-614.
Broffman BL (1958) Medical evaluation of the Beck operation for coronary artery disease. JAMA, 162:1603-1606.
Brown DG, Aguirre A, Weaver A (1980) (67) gallium scanning in talc induced pulmonary granulomatosis. Chest, 77, 4:561-565.
Butenandt I, Daeumling s, Hager C (1981) Akzidentelle badypuder-aspiration. Monatsschr Kinderheilkd, 129, 10:605.
Carman CR (1985) Talc retinopathy. J Amer Optometric Associa, 56, 2:129-130.
Chahinian PA (1985) Management of pleura tumors and malignant pleural effusions. In : The pleura in Health and Disease. Marcel Dekker Ed, New York, 30, pp. 571-584.
Cramer DW, Welch WR, Scully RE, Wojciechowski CA (1982) Ovarian cancer and talc - A case-control study. Cancer, 50:372-376.
Crouch ED, Churg A (1983) Progressive massive fibrosis of the lung secondary to intravenous injection of talc. A pathologic and mineralogic analysis. Am J Clin Pathol, 80:520-526.

Dunkel TB (1986) Intrapleural Tetracycline in the treatment of malignant pleural effusions. Minnesota Med, 69:717-725.

Farber HW, Mathers JAL, Glauser FL (1980) Gallium scans and serum angiotensin converting enzyme levels in talc granulomatosis and lymphocytic interstitial pneumonitis. South Med J, 73, 12:1663-1667.

Farber HW, Falls R, Glauser FL (1981) Transient pulmonary hypertension from the intravenous injection of crushed, suspended pentazocine tablets. Chest, 80:2.

Farber HW et al. (1982) Talc granulomatosis : laboratory findings similar to sarcoidosis. Am Rev Respir Dis, 125:258-261.

Fentiman IS, Rubens RD, Hayward JL (1986) A comparison of intracavitary talc and Tetracycline for the control of pleural effusions secondary to breast cancer. Eur J Cancer Clin Oncol, 22:1079-1081.

Frankel A, Krasna I, Baronofsky ID (1961) An experimental study of pleural symphysis. J Thorac Cardiovasc Surg, 42:43-51.

Gaensler EA, Moister MVB, Hamm J (1964) Open-lung biopsy in diffuse pulmonary disease. N Eng J Med, 270:1319-1331.

Gravelyn TR, Michelson MK, Gross BH, Sitrin RG (1987) Tetracycline pleurodesis for malignant pleural effusions. A 10-year retrospective study. Cancer, 59:1973-1977.

Hausheer FH, Yarbro JW (1985) Diagnosis and treatment of malignant pleural effusion. Seminars in Oncology, 12:54-75.

Hildick-Smith GY (1976) The biology of talc. Br J Industrial Med, 33:217.

Hopkins GB, Taylor DG (1970) Pulmonary talc granulomatosis. A complication of drug abuse. Am Rev Respir Dis, 101:101-104.

Hopkins GB (1972) Pulmonary angiothrombotic granulomatosis.J Amer Med Ass, 221, 8:909-911.

IARC (1987) Monographs on the evaluation of the carcinogenic risk of chemicals to humans - Silica and some silicates, volume 42.

Jackson JW, Bennet MH (1969) Chest wall tumour following iodized talc pleurodesis. Thorax, 28:788-793.

Krainer L, Berman E, Wishnick SD (1962) Parenteral talcum granulomatosis : a complication of narcotic addiction (Abstract). Lab Invest, 11:671.

Lee C, Lehman EP (1946) Experiments with nonirritating glove powder. Surg Gynecol Obst, 83:531.

Leophonte P, Basset MF, Pincemin J, Louis A, Fernet P, Delaude A (1983) Mortalité des travailleurs du talc en France. Rev Fr Mal Respir, 11:489.

Lewis JH, Sundeen JT, Simon GL, Schulof RS, Wand GS, Gelfand RL, Miller H, Garrett CT, Jannotta FS, Orenstein JM (1985) Disseminated tale granulomatosis. An unusual finding in a patient with acquired immunodeficiency syndrome and fatal cytomegalovirus infection. Arch of Pathol and Laboratory Med, 109:147-150.

Liang Che Tao, Morgan RC, Donat EE (1984) Cytologic diagnosis of intravenous "talc" granulomatosis by fine needle aspiration biopsy : a case report. Acta cytologica, 28:737-739.

Mariani Costantini R, Jannotta FS, Johnson FB (1982) Systemic visceral "talc" granulomatosis associated with miliary tuberculosis in a drug addict. Am J Clin Pathol, 78:785-789.

Molos MA, Litton N, Schubert TT (1987) "Talc" liver. J Clin Gastro, 9:198-203.

Moskowitz RL (1970) Talc Pneumoconiosis : a treated case. Chest, 58:37-41.

Motomatsu K, Adachi H, Uno T (1979) Two infant deaths after inhaling baby powder. Chest, 75:448-450.

Muir JF, Cerisel F, Defouilloy C, Broussier PM, Hermant A, Aubry P, Noarurinze S, Botto JM, Arlati S (1987) Pleural drainage with talc vs Doxycycline in the control of malignant pleural effusions. Am Rev Respir Dis, 135:A244.

Murphy SB, Jacson WB, Pare JAP (1978) "Talc" retinopathy. Canad J Ophthalmol, 13:152-156.

Nam K, Gracey DR (1972) Pulmonary talcosis from cosmetic talcum powder. JAMA, 221:492-493.

Ostrowski MJ (1986) An assessment of the long-term results of controlling the reaccumulation of malignant effusions using intracavitary bleomycin. Cancer, 57:721-727.

Owen M. (1937) Peritoneal response to glove powder. Tex State J Med, 32:482.

Pare JAP, Fraser RG, Hogg JC, Howlett JG, Murphy SB (1979) Pulmonary "mainline" granulomatosis. Medicine, 58:229-239.

Pearson FG, Mac Gregor DC (1966) Talc poudrage for malignant pleural effusion. J Thorac Cardiovasc Surg, 51:732-738.

Research Committee of the British Thoracic Association and the Medical Research Council Pneumoconiosis Unit (1979) A survey of the long-term effects of talc and kaolin pleurodesis. Br J Dis Chest, 73:285-288.

Reshad K, Inui K, Takeuchi Y, Takahashi Y, Hitomi S (1985) Treatment of malignant pleural effusion. Chest, 88, 3:393-397.

Rigdon RH (1975) Tissue reaction to foreign materials Critical Reviews in Food Science and Nutrition, 3:435-476.

Rinaldo JE, Owens GR, Rogers RM (1983) Adult respiratory distress syndrome following intrapleural instillation of talc. J Thorac Cardiovasc Surg, 85:523-526.

Rossi GA, Felletti R, Balbi B, Sacco O, Cosulich E, Risso A, Melioli G, Ravazzoni C (1987) Symptomatic treatment of recurrent malignant pleural effusions with intrapleurally administered Corynebacterium parvum. Am Rev Respir Dis, 135:885-890.

Shedbalkar AR, Head JM, Head LR, Murphy DJ, Mason JH (1971) Evaluation of talc pleural symphysis in management of malignant pleural effusion. J Thorac Cardiovasc Surg, 61:492-497.

Sherman S, Grady KJ, Seidman JC (1987) Clinical experience with Tetracycline pleurodesis of malignant pleural effusions. Southern Medical J, 80:716-719.

Sieniewicz DJ, Nidecker AC (1980) Conglomerate pulmonary disease : a form of talcosis in intavenous methadone abusers. Am J Roentgenol, 135, 4:697-702.

Smith RH, Graf MS, Silverman JF (1978) Successful management of drug-induced talc granulomatosis with corticosteroids. Chest, 73, 4:552-554.

Sneirson H, Woo ZP (1955) Starch granuloma. Ann Surg, 142, 1045.

Thompson SA (1939) Development of cardiopericardial adhesions following use of talc. Proc Soc Exp Biol Med, 40:260.

Todd TRJ, Delarue NC, Ilves R, Pearson FG, Cooper JD (1980) Talc poudrage for malignant pleural effusion. Chest, 78:542-543.

Tse DT, Ober RR (1980) Talc retinopathy. Am J Ophtalmol, 90, 5:624-640.

Vanderschueren R (1981) Le talcage pleural dans le pneumothorax spontané. Poumon et coeur

Wagner JC, Berry G, Cooke TJ, Hill RJ, Pooley FD, Skidmore JW (Eds)(1975) Animal experiments with talc. Symposium Edinburg. Pergamon Press.

Wooten SA, Strange C, Barbarash R, Sahn SA (1987) Pharmacokinetics of Tetracycline and Lidocaine following intrapleural instillation. Am Rev Respir Dis, 135:A245.

THERAPEUTIC AND DOMESTIC USES OF ATTAPULGITE AND SEPIOLITE

JC. Pairon*, MC. Jaurand*, A. Gaudichet**[1], P. Brochard***[2], J. Bignon*
* INSERM U. 139
CHU Henri Mondor
94010 Créteil Cédex
France

INTRODUCTION

Attapulgite and sepiolite are two clay minerals exhibiting fibrous shape and widely used over the world especially because of their remarkable sorptive and colloidal properties. Although these materials were considered as safe the biological effects observed with asbestos fibres have focused attention of scientists on the possible fibrogenic or carcinogenic potential of other fibrous minerals.

The purpose of this paper is to evaluate the health related risks resulting from non-occupational exposures to attapulgite and sepiolite.
In this connection, we will successively present :

1 - some data concerning the production and the main uses of attapulgite and sepiolite in the world ;

2 - a summary of the experimental and human occupational data evaluating their fibrogenic or carcinogenic potential ;

3 - the available data dealing with the non-occupational exposures of humans to these clay minerals.

1** Laboratoire de Microscopie Analytique Appliquée aux Sciences de la Terre, Université Paris XII, 94010 Créteil Cédex, France
2*** Laboratoire d'Etude des Particules Inhalées, 44 rue Charles Moureu, 75013 Paris, France

NATO ASI Series, Vol. G 21
Health Related Effects of Phyllosilicates
Edited by J. Bignon

PRODUCTION AND USES OF ATTAPULGITE AND SEPIOLITE

The annual world productions are summarized in the table 1. The leader producers for attapulgite and sepiolite are the USA and Spain, respectively. The total annual production of both minerals is about two millions tonnes.

Table 1. Annual world production of attapulgite and sepiolite

	Attapulgite		*Sepiolite*	
Countries	USA	84 %	Spain	90 %
	Senegal Spain Australia South Africa	9 % 4 % -	USA Turkey	
	France India Turkey USSR		France India Iran Kenya Somalia Tanzania	Madagascar Korea Japan China USSR ?
Total Production/year	≃ 1 100 000 tonnes		≃ 700 000 tonnes	

(From Clarke 1985 and personal communication)

Table 2 indicates the repartition of the different uses of attapulgite in the USA. These uses were considered as the representative reflect of world uses (WHO, 1987). Oil and grease absorbents and pet litter absorbents account for more than 70 % of the total use. When comparing data of 1983 and 1986, it can be observed that the use has decreased mainly in drilling muds, while it has increased in adhesives and paints.

Table 3 lists the uses of sepiolite in the world. Because of similar properties, sepiolite has almost the same uses as attapulgite.

Table 2. The uses of attapulgite in the USA (in short tonnes)

	1983	1984	1985	1986	(%)	Evolution
Adhesives	1,865	1,798	2,740	6,874	0.8	↗
Drilling mud	81,406	105,715	106,980	34,720	4.1	↘
Fertilizers	47,257	52 ,291	46 ,899	36,247	4.2	↘
Clarifying, filtering decolorizing mineral oils and greases	21,185	18,017	15,321	13,579	1.6	↘
Medical, pharmaceutical and cosmetic	111	105	94	379	0.04	↗
Oil and grease absorbents	269,688	209,831	291,612	264,673	30.9	→
Paints	8,238	6,606	3,244	19,885	2.3	↗
Pesticides and related products	69,731	96,200	92,266	71,314	8.3	→
Pet waste absorbents	320,587	291,943	214,800	348,710	40.7	→
Other	22,971	25,872	22,782	13,132	1.5	
Miscellaneous	66,945	28,710	17,417	47,387	5.5	
Total	909,984	849,228	814,155	856,900		

(From Ampian, 1984 and 1986)

Table 3 . The uses of sepiolite in the world

Mainly in	- Absorbents (especially pet litters)	85 %
	- Animal feeds	7.5 %
	- Pesticide carriers	4 %
Others :	decolorizer, bleaching agent (vegetable and mineral oils), drilling muds, industrial catalysts, anticaking agents, cosmetics, cigarette filters, detergents, pharmaceutical products, pipes and cigarette holders (meerschaum)	

(From Alvarez 1984, Clarke 1985, WHO 1987)

THE EVALUATION OF THE FIBROGENIC AND/OR CARCINOGENIC POTENTIAL OF ATTAPULGITE AND SEPIOLITE

Since the results and interpretation of short-term studies are presented elsewhere in this workshop (see Jaurand and Pairon), we will focuse on experimental and human occupational data.

Animal data

Experimental studies can be divided into two main classes according to the route of administration : a first class of "realistic" experiments which reflect the human exposure, that means inhalation and oral administration; and a second class of "sensitive" experiments which are the intraperitoneal injection and intrapleural inoculation. While less physiological, these experiments are probably more efficient in detecting carcinogenic properties.

Results of experimental studies are summarized in tables 4 and 5. There are few experiments performed using the same sample and the same route of administration. The distinction between palygorskite and attapulgite

Table 4. Results of experimental studies (oral - inhalation - intraperitoneal)

Route	Authors	Species Dose administered	Treatment (origin)	Results: Fibrosis	Results: Cancer	Results: N tumor / N animals	Characteristics of dusts: N/mg of dusts All L ($X10^9$)	N/mg of dusts L≥ 6 µm① ($X 10^6$)	N/mg of dusts (%)	Median length (µm)	Median diameter (µm)
Oral	Brune and Deutsch Wenzel 1983	NMRI Mice (120/group) in diet	**Attap** 1 % (Fr)		-		?	0	(0 %)	≤ 1	
			Attap 3 % (Fr)		-		?	0	(0 %)	≤ 1	
Inhalation	Wagner et al 1984, 1987	Fischer 344 Rat (40/group) 10 mg/m^3 x 6 Hours/day x 5 days/7	**Attap** Lebrija (Sp)	-	-	(2/40)	?	0	(0 %)		
			Sepiol Madrid (Sp)	-	-	(2/40)	115	0	(0 %)		
			Palyg Leicester (UK)	+	+	(4/40)	19	3650	(19 %)		
			Control (unexposed)			(0/40)					
Intra-peritoneal	Pott et al 1976	Female Wistar Rat (40/group) 3 x 25 mg	**Palyg** (Origin ?)	+	+	(26/34)	?	?	(30 %)		
			Control (saline)			(0/72)					
	Pott et al 1987	Female Wistar Rat 60 mg x 5	**Attap** Georgia (USA)		-	(4/112)	29	0.61	(0.002 %)	0.8	0.04 ②
			Attap Lebrija (Sp)		-	(4/115)	1.4	0.34	(0.02 %)	0.5	0.07 ②
			Attap Mormoiron (Fr).		-	(4/114)	0.9	0.06	(0.007 %)	0.7	0.07 ②
			Control (saline)			(5/204)					
		2 mg + 4 mg + 4 mg	**Attap** Torrejon (Sp)		+	(12/30)	8.1	240	(3 %)	1.3	0.07 ②
			Control (saline)			(2/32)					

① ≥ 5 µm for intraperitoneal injection

② From Rodelsperger et al. 1987

Attap : attapulgite ; Sepiol : sepiolite ; Palyg : palygorskite

The column "results" express the statistical comparison of treated versus control animals in term of fibrogenicity or carcinogenicity

Table 5 . Results of experimental studies (intrapleural)

Route	Authors	Species Dose administered	Treatment (origin)	Results: Fibrosis	Results: Cancer	Results: N tumor / N animals	Characteristics of dusts: N/mg of dusts: All L (X10 ^{9})	L≥ 4 µm (1)	L≥ 8 µm	Median length (µm)	Median diameter (µm)
Intrapleural	Stanton et al 1981	Female Osborne Mendel Rat (30 to 50/group) 40 mg	**Attap** n° 1 Georgia (USA)		-	(2/29)	26.5	0.5 %	0 %		
			Attap. n° 2 Georgia (USA)		-	(2/29)	20.7	0 %	0 %		
			Control (non carcinogenic pleura implant)			(17/598)					
	Jaurand et al 1987	Male Sprague Dawley Rat (36/group) 20 mg	**Attap** Mormoiron(Fr.)	-	-	(0/30)	2.6	0 %	0 %	0.77	0.06
			Control (saline)			(1/32)					
	Wagner et al 1987	SPF Fischer 344 Rat (40/group)	**Attap.** Torrejon (Sp.)		+	(14/40)	?	5.5 %	0 %		
			Attap Lebrija (Sp)		-	(2/40)	?	0 %	0 %		
			Sepiol Madrid (Sp)				(9,6)	(1 %)		(1.9)	(0.04) (2)
			* crude		-	(0/40)	?	10.5 %	0 %		
			* crude + US		-	(1/40)					
			* commercial		-	(1/40)	?	38.9 %	9 %		
			Palyg Leicester (UK)		+	(30/32)	(20)	(0.9 %)		(1.2)	(0.05) (2)
			Control (saline)			(1/40)					

(1) L ≥ 5 µm for Rodelsperger et al.

(2) From Rodelsperger et al. 1987

The column "results" express the statistical comparison of treated versus control animals in term of fibrogenicity or carcinogenicity (which mineralogically refer to the same product) has been made according to the denomination given by the authors of the experimental studies.

Sepiolite from Madrid (Spain) has been tested only once by inhalation and once by intrapleural inoculation. Results were negative for both.

Results seem to depend on the origin of the sample for attapulgite because they were negative with samples from Georgia (USA), Mormoiron (France), and Lebrija (Spain), while they were positive for attapulgite from Torrejon (Spain). Palygorskite samples gave positive results in all three experiments where they were tested. The origin of the sample was unknown in one experiment (Pott et al., 1976).

Some of the size characteristics of the particles are listed in tables 4 and 5, while figure 1 summarizes the experimental results according to the route of administration and to the fibre length. The value of the maximum length indicated for each sample represents the length under which there was more than 99 % of the fibres of the sample according to the author measurements.

It must be noted that comparison of dusts characteristics between different experiments is very difficult for two main reasons. First, the methodology used for dust analysis is not always the same in the different experiments. This might lead to dissimilar results for the same sample. Second, the expression of the results of the measurements depends on the authors : the size categories are not uniform ; in some studies, the percentage of fibres present in each size category is given but the absolute number of fibres per mg is not. In these conditions it is impossible to compare fibre concentrations in different samples of attapulgite as it was noted by the experts of the IARC (WHO, 1987).

All the experiments performed with samples containing short fibres which length was less than 4 µm were negative. In contrast samples containing fibres longer than 8 µm lead to positive results. All these results seem to be consistent with Stanton hypothesis concerning size-carcinogenicity relationship in pleural sarcoma (Stanton et al., 1981). However the negative result obtained by Wagner et al (1987) with the sepiolite sample from Madrid (containing fibres longer than 4 µm) in an inhalation experiment stresses the importance of other parameters : in most of experimental studies, fibre size and especially fibre length and diameter were generally the only dust parameters which were studied. One can arise the question of the role of other dusts characteristics in the induction of carcinogenesis, especially chemistry, physico-chemistry, durability...

Figure 1. Summary of experimental results according to the route of administration and to the fibres length

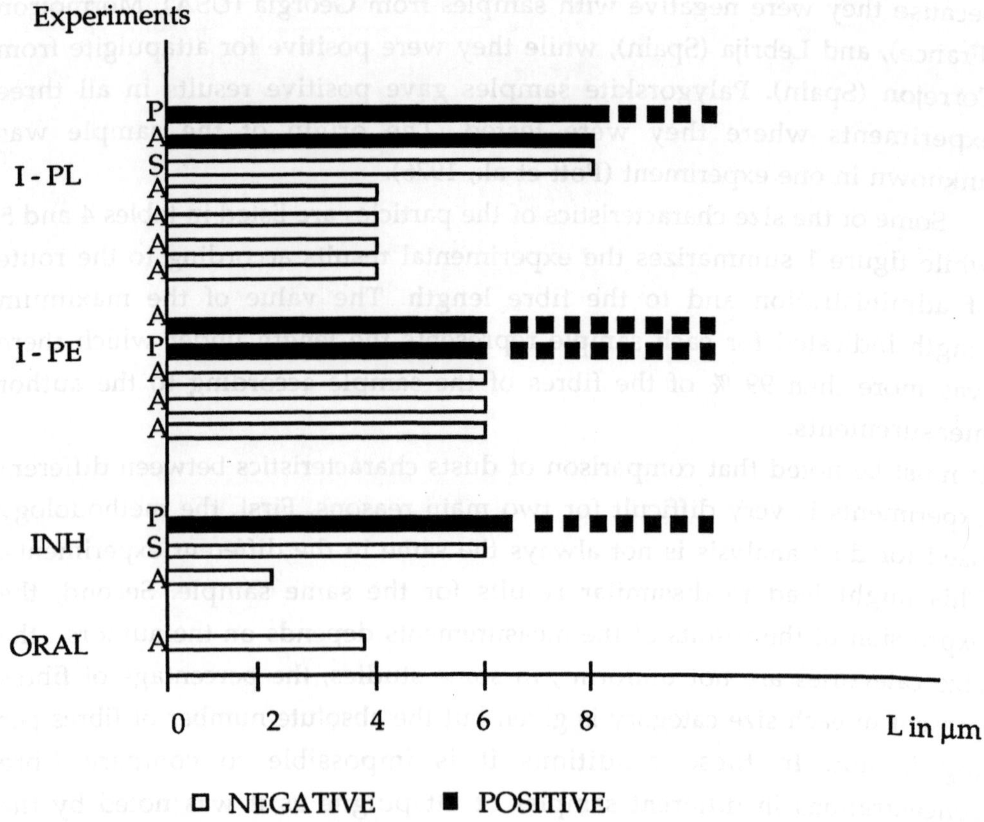

I - PL = Intrapleural inoculation
I - PE = Intraperitoneal injection
INH = Inhalation
ORAL = Oral administration

(Dotted line indicates that there was no information about the number of fibres longer than that indicated by the full line, but that their presence was highly probable)

Human occupational data

Few human occupational data have been published (Table 6).

Table 6. Human occupational data

	Authors	Population	Results
Case Report	Sors et al 1979	1 **attap.** worker (France)	Diffuse lung fibrosis
Morbidity	Baris et al 1980	63 **sepiol.** workers (Turkey)	?
	Gamble et al 1985	705 .**attap** workers (Georgia - USA)	- (?)
	McConnochie et al 1988	218 **sepiol.** workers (Spain)	- (?)
Mortality	Waxweiler et al 1988	2302 **attap.** miners and millers (USA)	+/- (?)

Sors et al. (1979) reported one case of radiological lung fibrosis in a 40-years-old mining engineer moderately exposed to French attapulgite during processing operations. A heavy concentration of attapulgite fibres (42.000/ml) was found in the alveolar lavage fluid. The hypothesis of a relationship between the occupational exposure and lung fibrosis was strongly evocated but not confirmed by further studies.

Ten cases of radiological pulmonary fibrosis were observed by Baris et al. (1980) in a clinical and radiological study in 63 sepiolite workers in Turkey. However this result was difficult to interpret because all fibrotic patients were smokers and three of them had worked in other dusty jobs. In addition more than half came from rural regions rich in asbestos and zeolite deposits.

Gamble et al. (1988) reported the results of the survey of 705 attapulgite workers from two companies (A et B) in Georgia and Florida (USA). There was no significant association between attapulgite exposure and respiratory symptoms. They observed a statistically non significant decrease of pulmonary function (FVC, FEV1, FEV%) with increasing exposure. The overall prevalence of pneumoconiosis (> 1/0 on chest radiography) was 6.4% and increased with increasing age and exposure : 11.9 % and 25.3 % in companies A and B respectively among workers exposed for more than 15 years. However this increase was not significant and the authors underlined that it could be due to other dusty preemployement. In addition this result should be interpreted with caution : median ages were 38.1 and 38.8 years in companies A and B, while median exposure durations were only 10.9 and 11.5 years respectively. Because pulmonary fibrosis has been shown to increase significantly in one company with increasing age, it seems important to follow up these workers further. In addition, it must be noted that the wording of the respiratory questionnaire was different for companies A and B, as was collection of pulmonary function parameters. The proportions of white and black workers were also dissimilar : 49%-51% and 26%-74% for A and B respectively.

In a cross-sectional study of 218 spanish sepiolite workers, Mcconnochie et al. (1988) did not observe any relationship between duration of exposure and chest x-ray abnormalities. There was a relationship between radiological small opacities and types of occupation but lower than the one observed with age. These authors reported no excess mortality from lung cancer among these sepiolite workers.

Waxweiler et al. (1988) conducted a retrospective cohort mortality study among 2302 male workers who had been employed for at least one month between 1940 and 1975 in one of the major American attapulgite mining and milling companies. A statistically significant excess of lung cancer deaths was observed among whites (standardized mortality ratio (SMR) = 193 ; 90 % confidence interval (CI) : 121-293) but not among non-whites (SMR = 53, 90% CI: 21-112). A dose effect relationship was noted since the SMR was of 309 (90% CI 122-649) among both whites and non-whites exposed to high dust levels for more than five years. Analysis of the results among employees exposed to intermittent or high levels of dusts exhibited an excess of lung cancer death among whites when latency was more than 20 years (SMR = 300, 90 % CI 131-592). Before taking definite conclusions among this cohort study, some limitations must be noted that were underlined by the authors. The more important are : the small number of employees still deceased (315); shortness of the follow-up of the workers when taking into account the importance of induction-latency period in carcinogenesis studies (the person years at risk average was of 17 years per member among the cohort) ; missing death certificates ; lack of information on smoking habits ; use of general US population rates as reference.

Overall evaluation

From available animal data and human occupational data it is not possible to establish definite conclusions on the fibrogenic or carcinogenic potential of attapulgite and sepiolite. This was underlined by the experts of the IARC (WHO, 1987) who classified both clay minerals in group number 3 (table 7).

Table 7 IARC Conclusions from IARC monograph vol. 42 (1987) and updating suppl. 7 (1987)

Attapulgite : group 3

- Evidence for carcinogenicity to experimental animals : limited
- Evidence for carcinogenicity to humans : inadequate

Sepiolite : group 3

- Evidence for carcinogenicity to experimental animals : inadequate
- Evidence for carcinogenicity to humans : no data

THE HUMAN RISKS RESULTING FROM NON-OCCUPATIONAL EXPOSURES TO ATTAPULGITE AND SEPIOLITE

Besides the potential human risks studied in attapulgite and sepiolite workers, one can ask about the harmlessness of these materials in the general population since they are found in some widely used products. Non occupational exposure may result from inhalation of pet litters fibres and from ingestion of some drugs used to cure gastrointestinal diseases.

Inhalation

The large use of attapulgite and sepiolite in pet litter absorbents arises the question of the risk of respirable dusts in many household. The air dust concentration of fibres is likely lower than that observed in working places and cannot lead to pulmonary fibrosis. However no data is presently available. The carcinogenic risk remains questionable. In addition there are no veterinary reports concerning the related effects in domestic animals resulting from the use of pet litters. It seems therefore difficult to evaluate the health related risks resulting from the inhalation by humans of pet litter absorbent fibres. However, it must be noted here that the attapulgite sample from Spain which was carcinogenic in animal experiments is not used in pet litter absorbents but only in drilling muds. This does not rule out the possibility that other carcinogenic samples would be used for these purposes. We do not know exactly the use of U.K. palygorskite and palygorskite of unknown origin which also gave positive results in animals.

Ingestion

Some drugs used for the treatment of gastrointestinal diseases may contain attapulgite.

In France, four of these drugs are available (Actapulgite®, Gastropulgite®, Mucipulgite®, Norgagil®) but no one with sepiolite. Some granulometric and chemical parameters of the fibres of the three first drugs have already been published (Bignon et al., 1980). It was shown that the median length and maximum length of fibres of these three drugs were 0.8, 0.8, 0.4 µm and 3.6, 3.2, 2.0 µm respectively. It is noteworthy that the maximum length of fibres is relatively short and always below 4 µm. Attapulgite used in these drugs comes from the deposit of Mormoiron (France). One case report was published that suggested an intestinal absorption of attapulgite (Bignon et al., 1980). It was the case of a 60 years-old woman treated with 9 g/day of attapulgite for 6 months. 300.000 attapulgite fibres per milliliter were found in the urine of the patient. No further study has been performed to validate this result in human healthy volunteers nor humans treated for gastrointestinal diseases. One experimental study was performed (Masse et al., 1980) that failed to show passage of attapulgite fibres through the intestinal mucosa in the rat. These authors collected the lymph fluid after gavage with asbestos or attapulgite fibres. Asbestos fibres were observed in the lymph, especially long fibres, but not attapulgite.

CONCLUSION

The evaluation of available animal and human occupational data do not allow to conclude on the fibrogenic or carcinogenic potential of attapulgite and sepiolite. From the results of animal experiments, fibre length seems to be the most important toxicological parameter but the role of other dust characteristics, especially chemistry and physico-chemistry should be evaluated. A follow-up of the epidemiological studies still in progress is necessary. Because they are used in some drugs and in pet litters, these fibres could present health hazards in non occupationally exposed populations. Lack of information about indoor air dust concentration measurements or on animals studies in households with pets underline the need of further studies. The evaluation of the intestinal absorption of attapulgite in patients treated for gastrointestinal diseases with this kind of drug should also be performed.

SUMMARY

Attapulgite and sepiolite are two clay minerals widely used over the world because of their remarkable sorptive and colloidal properties. The annual world consumption is probably a few under 2 millions tons.
Analysis of attapulgite and sepiolite fibres according to their granulometry have shown important discrepancies, depending mainly on the geological sources. Samples containing only short fibres ($L < 4\ \mu m$) failed to produce tumors in animals (by inhalation, intrapleural and intraperitoneal inoculation and ingestion). In contrast samples with long fibres resulted in malignant tumor formation in some experiments (by intrapleural or intraperitoneal inoculation and by inhalation). Present epidemiological data from occupational exposures by inhalation do not allow to consider attapulgite or sepiolite definitely devoided of fibrogenic or carcinogenic potential for humans, but this requires further follow up.

Human can be non-occupationally exposed mainly through two applications : pet litters and drugs. There is some concern about the potential health related risks resulting from inhalation or ingestion of attapulgite or sepiolite. Indeed few studies have been performed to evaluate these risks. No indoor air dust concentration measurements and no study in animals have been performed in households with pets. An intestinal absorption has been suggested in humans treated with drugs containing attapulgite but so far not proven.

From the available data it appears that fibres length seems to be an important toxicological parameter. However other dust characteristics (chemistry, physico-chemistry) need further studies before taking conclusion.

REFERENCES

Alvarez A (1984). Sepiolite : properties and uses. In : Singer A, Galan E (eds) Palygorskite - Sepiolite : Occurrences, Genesis and Uses, Elsevier, New York, pp. 253-287

Ampian SG (1984).Clays. In : Minerals Yearbook 1984, 1, Metals and Minerals, US Government Printing Office, Washington DC, pp. 235-268

Ampian SG (1986). Clays. In : Minerals Yearbook 1986, 1, Metals and Minerals, US Government Printing Office, Washington DC, pp. 245-279

Bignon J, Sebastien P, Gaudichet A, Jaurand MC (1980). Biological effects of attapulgite. In : Wagner JC (ed). Biological effects of mineral fibres, IARC Scientif Publ. n° 30, Lyon, 1, pp. 163-181

Baris YI, Sahin AA, Erkan ML (1980). Clinical and radiological study in sepiolite workers. Arch. Environ. Health, 35 : 343-346

Brune H, Deutsch-Wenzel R (1983). Life span feeding study with attapulgite in mice. Unpublished report

Clarke GM (1985). Special clays. Ind. Miner 216 : 25-51

Gamble J, Sieber WK, Wheeler RW, Reger R, Hall B (1988). A cross-sectional study of US attapulgite workers. Ann. Occup. Hyg. 32 (suppl. 1) : 475-481.

Jaurand MC, Fleury J, Monchaux G, Nebut M, Bignon J (1987). Pleural carcinogenic potency of mineral fibers (Asbestos, Attapulgite) and their cytotoxicity on cultured cells. J. Natl. Cancer Inst. 79 : 797-804

Mcconnochie K, Lyons JP, Bevan C, Wagner JC (1988). A study of spanish sepiolite workers. VIIth International Conference on Pneumoconiosis, 23-26 August 1988, Pittsburgh, Pennsylvania, USA. Abstracts of communications pp.96

Masse R, Sebastien P, Monchaux G, Bignon J (1980). Experimental demonstration of the penetration of asbestos fibres into the gastrointestinal tract. In : Wagner JC (ed). Biological effects of mineral fibres, IARC Scientif. Publ. n° 30, Lyon, 1, pp. 321-328

Pott F, Dolgner R, Friedrichs KH, Huth F (1976). L'effet oncogène des poussières fibreuses. L'expérimentation animale et ses relations avec la carcinogénèse humaine (in French). Ann. Anat. Pathol., 21 : 237-246

Pott F, Ziem U, Reiffer FJ, Huth F, Ernst H, Mohr U (1987). Carcinogenicity studies on fibres, metal compounds, and some other dusts in rats. Exp. Pathol. 32: 129-152

Rödelsperger K, Brückel B, Manke J, Woitowitz HJ, Pott F (1987). Potential health risks from the use of fibrous mineral absorption granulates. Br. J. Ind. Med. 44 : 337-343

Sors H, Gaudichet A, Sebastien P, Bignon J, Even P (1979). Lung fibrosis after inhalation of fibrous attapulgite. Thorax 34 : 695

Stanton MF, Layard M, Tegeris A et al. (1981). Relation of particle dimension to carcinogenicity in amphibole asbestoses and other fibrous minerals. J. Natl. Cancer Inst. 67 : 965-975

Wagner JC (1984). Health hazards of substitutes. In : Asbestos, health and society. Proceedings of the World Symposium on Asbestos. Montreal, 25-27 May 1982, Montreal Canadian Asbestos Information Centre, 1982, pp. 244-266

Wagner JC, Griffiths DM, Munday DE (1987). Experimental studies with palygorkite dusts. Br.J. Ind. Med. 44 : 749-763

Waxweiler RJ, Zumwalde RD, Ness GO, Brown DP (1988). A retrospective cohort mortality study of males mining and milling attapulgite clay. Am. J. Ind. Med. 13 : 305-315

WHO (1987) IARC Monographs on the Evaluation of the Carcinogenic Risk of Chemicals to Man. Vol. 42, Silica and some silicates. Lyon, France.

LUNG RETAINED MINERALS CORRELATED WITH SMOKING, EMPHYSEMA AND LUNG CANCER

P-L. Kalliomäki*, O. Taikina-aho**, P. Pääkkö**, S. J. Sivonen**, K. Kalliomäki**

KEY WORDS: human lungs / mineral particles / fibres / smoking / pulmonary emphysema

INTRODUCTION

The human lung is an effective filter for collecting airborne particles. For over 20 years it has been possible to analyze these particles, the main intrest having been focused on fibers. Mineral particles, metal dust and fume particles constitute an important fraction of the air pollutants to which everyone is exposed, but little information is available concerning the burden of such particles in the lungs (Churg and Wiggs 1985, Churg and Wiggs 1987, Paoletti et al. 1987). It is necessary to know the basic pulmonary level of inhaled particles in general population, when estimating lung particle burden from occupational exposure.

Cigarette smoking may alter the deposition and retention of inhaled particles and is related to diseases such as chr.bronchitis, emphysema, pre-cancerous bronchial epithelial changes and lung cancer.

The present paper describes pulmonary mineral content in a series of occupationally non-exposed subjects from Northern-Finland and correlate it with smoking, the grade of emphysema and the presence of lung cancer.

---

* Institute of Occupational Health, Helsinki, Finland.
** Institute of Electron Optics. Department of Pathology and Electronic Measurements Laboratory, University of Oulu, Oulu, Finland.

NATO ASI Series, Vol. G 21
Health Related Effects of Phyllosilicates
Edited by J. Bignon

MATERIAL

The total material in this study was obtained from 90 male who had died of non-malignant diseases or been operated on for lung cancer. Smoking habits, occupational history were ascertained from the patient or the next of kin.

For the present preliminary study we selected 14 autopsy lungs of the subjects without any malignant disease and the lungs of 13 lung cancer patients. The selection criteria was that the subject have no evidence of occupational exposure to inorganic dust. Age, smoking habits and the grade of emphysema for the subjects are presented in Table 1.

Table 1 Age, smoking habits and grade of emphysema of the subjects without cancer (autopsy lungs) and of the lung cancer patients.

	Autopsy lungs			Lung cancer patients	
	Non-smokers (N=4)	Smokers N=6)	Ex-smokers (N=4)	Smokers (N=7)	Ex-smokers (N=6)
Age (yrs)	76	68	64	52	62
Smoking time (yrs)	0	41	22	39	33
Time since stopping smoking (yrs)	0	0	15	0	14
Grade of emphysema	1.5	2.3	3.0	2.6	2.7

METHODS

The lungs and lung lobes obtained from the autopsies and surgical operations were radiographed during continuous air inflation and the severity of emphysema was graded as normal, mild, moderate or severe from these radiographs on the basis of peripheral vascular changes, tissue defect translucencies and changes in the shape of the lung or lobe (Sutinen et al. 1979). The lungs were fixed transbronchially with hypermolar formalin. The grade of emphysema was also estimated from gross specimens and histological sections and these data were used if grading from the radiographs did not produce an unambiguous result. The histological type of cancer was determined according to the WHO classification of tumours.

A piece of fresh lung tissue (about 2 g) was taken from (apico) posterior and anterior segment of the upper lobe and from the superior segment and basal part of the lower lobe (no pleural surface or cancer tissue). After low temperature plasma-ashing the samples were filtered on nucleopore filter (ø=0.1 um). A TEM-sample was prepared from the filter by dissolving it with chloroform vapour on forvar and carbon coated TEM grid. The minerals were characterized using STEM with energy dispersive X-ray spectrometer and by electron diffraction. The mass of particles was calculated from the density of identified minerals and from the dimensions. 100 inorganic particles were characterized in each case.

RESULTS

Effect of lung cancer

Table 2 presents the total number and quality of particles found in autopsy lungs and in surgical specimens from lung cancer patients. The pulmonary particles, measured in terms of the total number, total mass, volume and surface area, were not affected by the presence of lung cancer. The types of minerals were not different in autopsy lung tissue or in surgical specimens. The mean number of particles found was $86 \pm 68 \times 10^6$/g dry weight, of which 70 % were K feldspar, plagioclase, quartz, mica and kaolinite.

Table 2 The number and quality of particles in lung tissue of the subjects without malignant disease and in surgical specimens from lung cancer patients.

	Autopsy lungs (N=13) Median	95 % interval	Surgical specimens (N=13) Median	95 % interval	p
Total particles ($x10^6$/g d.w.)	78	9-170	100	40-240	N.S.
Type of minerals ($x10^6$/g d.w.)					
- Fibres	1.1	0-11	1.5	0-7	N.S.
- Al	1.1	0-13	2	0-20	N.S.
- Fe	1.1	0-53	2	0-16	N.S.
- Kaolinite	6	0-17	10	2-50	N.S.
- K feldspar	14	3-34	20	6-70	Pro obs
- Mica	9	2-50	8	0-50	N.S.
- Plagioclase	14	2-26	17	4-60	Pro obs
- Quartz	11	1-30	16	4-60	N.S.
- Talc	1.2	0-20	2	0-10	N.S.
- Ti	3.4	0-15	6	0-80	Pro obs
- Miscellaneous	8	1-20	11	2-50	N.S.

Effect of smoking and emphysema

The characteristics of the minerals in the lung tissue of the all subjects grouped as a function of smoking habits are presented in Table 3. The total number of particles was somewhat lower in smokers lungs than in non-smokers lungs but the trend was not significant. The types of minerals or fibers were not affected by smoking. No significant difference was observed in pulmonary mineral particles of fibers as a function of the severity of emphysema in our material.

Table 3 Characteristics of the mineral particles in lung tissue of the subjects as a function of smoking habits.

	Non-smokers (N=4) Mean (SD)	Smokers (N=13) Mean (SD)	Ex-smokers (N=10) Mean (SD)
Total particles ($x10^6$/g d.w.)	106 (73)	84 (68)	55 (37)
Type of minerals ($x10^6$/g d.w.)			
- Fibres	0.7 (0.7)	1.3 (1.5)	0.6 (0.7)
- Al	1 (1.3)	0.9 (1)	0.1 (0)
- Fe	3 (2.7)	0.6 (0.7)	0.5 (0.8)
- Kaolinite	8 (7)	7 (8)	5 (5)
- K feldspar	18 (14)	15 (8)	17 (16)
- Mica	12 (7)	7 (4)	11 (10)
- Plagioclase	15 (8)	16 (12)	18 (16)
- Quartz	15 (15)	10 (8)	10 (14)
- Talc	2 (3)	1 (2)	0.2 (0.3)
- Ti	5 (6)	4 (5)	2 (2)
- Miscellaneous	11 (5)	8 (6)	10 (11)

CONCLUSION

In the present preliminary study, we examined the pulmonary mineral burden in the lung specimens of 14 occupationally non-exposed male without any malignant disease and 15 lung cancer patients. The mineral analyses of the rest of the subjects are still under review.

The total number of particles ($86\pm68 \times 10^6$/g dry weight) in the lungs of the non-exposed subjects (Table 2) was somewhat lower than the mean number of particles ($465\pm295 \times 10^6$/g dry weight) in the lungs of Canadian urban cigarette smokers (Churg et al. 1987), and also lower than the mean concentration of 180×10^6/g dry weight in the lungs of 10 subjects resident in the Rome area (Paoletti et al. 1987). The most of the present subjects had lived in the country, which may explain the difference in particle retention.

The number of fibres varied from 0.1 to 11 x 10^6/g dry weight. The asbestos fibres were mostly anthofyllite and crocidolite. Minor traces of crysotile, tremolite and amusite were also found. The number and quality of particles were not affected by smoking, pulmonary emphysema or the presence of lung cancer.

The total material studied up to date was obtained from 150 subjects. Most of the subjects have had occupational exposure to different inorganic dusts. We have identified 45 inorganic compounds and elements (Table 4). Less than 5 % of the particles remain unknown. Electron microscopy applied to lung specimens has proved to be a good measure both quantitatively and qualitatively to occupational and environmental exposures.

Table 4 Minerals and elements found in human lung

	Plyllosilicates	Oxides	Fe
Nesosilicates	Clayminerals	Corundum	Ni
Almandine	Kaolinite	Hematite	Cu
Zirkon	Illite	Ilmenite	Zn
Sillimanite	Mullite	Rutile	Ag
Sphene	Talc	Chromite	Ca
	Muscovite	Magnetite	Ce
Ionisilicates	Biotite		Pb
Pyroxenes	Chlorite	Sulfides, Sulfates	Mo
Hornblende	Chrysotile	Pyrite	Sr-comp
Antophyllite		Barite	Mn-comp
Amosite	Tectosilicates	Gypsum	S-comp
Crocidolite	Plagioclase		Diatoms
Tremolite	Microcline and	Miscellaneous	Fly ash
Actinolite	Orthoclase	Al	MMMF
Wollastonite	Quartz	Ti	
		Cr	

REFERENCES

Churg A and Warnock ML (1985) Asbestos fibers in the general population. Am Rev Resp Dis 122:669-679

Churg A and Wiggs B (1985) Mineral particles, mineral fibers and lung cancer. Environ Res 37:364-372

Churg A and Wiggs B (1987) Types, numbers, sizes, and distribution of mineral particles in the lungs of urban male cigarette smokers. Environ Res 42:121-129

Paoletti L, Batisti D, Caiazza S, Petrelli MG, Taggi F, Zorzi L, Dina MA and Donelli G (1987) Mineral particles in the lungs of subjectsresident in the Rome area and not occupationally exposed to mineral dusts. Environ Res 44:18-28
Sutinen S, Pääkkö P and Lahti R (1979) Post-mortem inflation, radiography, and fixation of human lungs: a method for radiological and pathological correlations and morphometric studies. Scand J Respir Dis 60:29-35
Warnock ML and Wolery G (1987) Asbestos bodies or fibers and the diagnosis of asbestosis. Eniron Res 44:29-44

MINERAL PARTICULATE IN THE LUNG PARENCHYMA OF SUBJECTS NOT OCCUPATIONALLY EXPOSED TO DUST.

L. PAOLETTI, M. FALCHI, D. BATISTI, M.P. CARRIERI, * M.G. PETRELLI, * C. CIALLELLA # and G. DONELLI

Department of Ultrastructures
* Department of Epidemiology and Biostatistics
Istituto Superiore di Sanità
Viale Regina Elena, 299
00161 Roma
Italy

INTRODUCTION. The hypothesis according to which airborne pollutants present in urban areas could be among the causes accounting for the increased number of cases of lung cancer and chronic bronchopneumopathy in populations residing in such areas has been suggested (Cuyler Hammond and Garfinkel, 1980; Doll, 1978; Vena, 1982). Indeed, a great quantity of mineral particles is present in urban aerosols and, because of their structure many of them are able of absorbing and carrying carcinogenic and biologically active substances. (Paoletti et al., 1989).

To better identify the degree of exposition to this kind of airborne pollutants, we have studied the mineral particulate contained in the lung parenchyma of subjects not occupationally exposed to dust, lived in an urban area and not affected by tumor pathology.

MATERIALS AND METHODS. 85 subjects, 69 males and 16 females, selected among people ranging from 15-70 years of age, residing

Institute of Forensic Medicine
University "La Sapienza" of Rome
Viale Regina Elena, 336
00161 Roma
Italy

NATO ASI Series, Vol. G 21
Health Related Effects of Phyllosilicates
Edited by J. Bignon

in the Rome urban area, who had died in accidents between 1987 and the first half of 1988, were studied. Subjects occupationally exposed to mineral dusts, those suffering from serious or chronic diseases and drug addicts were excluded. Detailed information on the smoking habits and occupational background of subjects was obtained by interviewing relatives. Information on legal residence was ascertained by inquiring the municipality registers of Rome. Autopsy samples of lung tissue were all obtained from the peripheral upper lobe of the right lung. Samples were mineralized in atomic oxygen plasma. The inorganic component was collected and transferred onto carbon-coated copper grids for Transmission Electron Microscopy. The mineral particulate was analyzed by a TEM equipped with an EDAX energy dispersive spectrometer for X-ray microanalysis.

RESULTS. The distribution of the studied subjects by sex, age, smoking habit and residence is shown in tab. 1.

TABLE 1

Sex	Males 81%			Females 19%
Age	15-30 9%	31-45 22%	46-59 35%	60-75 34%
Smoke	non-smokers 23%		ex-smokers 17%	smokers 60%
Years of residence in Rome		1 - 20 28%		> 20 72%

Detected particles were grouped in 3 categories according to their morphology, composition and probable origin:

a) silicates and silica, non-fibrous particles containing Si, prevalently "soil dust";

b) mineral fibers, fibrous particles with length/diameter ratio > 3;

c) metal oxides and sulfates, non-fibrous particles of various composition, originating from several sources.

The concentration of mineral particulate in lung parenchyma ranged from 0.1 to 1.7 x 10 particles /mg of dry tissue. Silicates and silica accounted for about 52 % of the total particulate and mainly consist of phyllosilicates (70 %) of total detected mineral varieties (Tab. 2).

About 70% of silicate particles showed an equivalent diameter ranging from 1 to 5 um. Asbestos fibers were found in 16% of the subjects, in concentrations ranging from 200 to 3000 fibers/mg, and representing about 0.5-1 % of their total lung burden. Detection frequency of asbestos varieties is shown in Tab. 3.

TABLE 2
FREQUENCIES OF
PHILLOSILICATE VARIETIES

	%
MICAS	54
CLAYS	36
TALC	8
CHLORITES	1.5
SERPENTINES	0.5

Over 50% of the fibers observed had length < 3 um and diameter < 0.2 um. Talc and rutile fibers were also observed. No man-made mineral fibers, such as glass wool and rock wool, were found. Metallic elements in the form of oxides, sulfates or alloys represent about 48% of the total lung burden.

TABLE 3
FREQUENCIES OF
ASBESTOS VARIETIES

	%
CHRISOTYLE	68
TREMOLITE	19
AMOSITE	8
CROCIDOLITE	5

Analytical data were statistically compared in different groups of subjects. The influence of sex and working activity on lung burden were never observed to be significative. The concentration of mineral particulate in lung parenchyma actually tends to increase with age (Fig. 1).

MEAN CONCENTRATIONS OF MINERAL PARTICLE ACCORDING TO AGE CLASSES

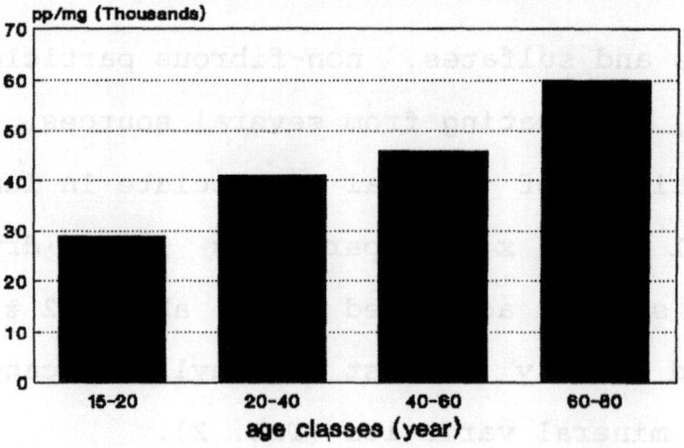

A measurable difference in the lung burden of smokers and non-smokers is shown (Fig. 2)

MEAN CONCENTRATIONS OF MINERAL PARTICLE IN SMOKERS AND NON-SMOKERS

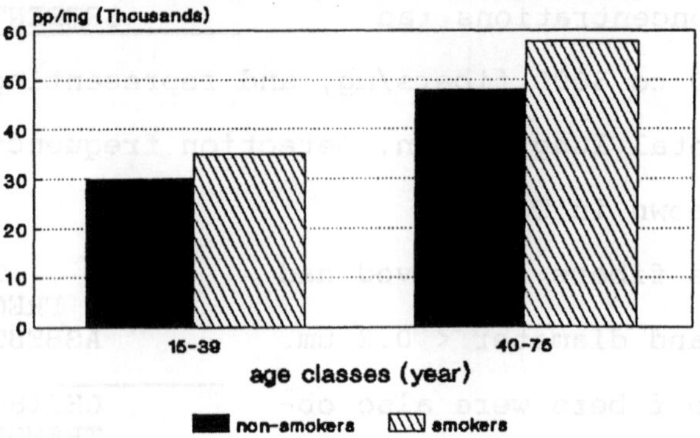

by coupling - according to age groups - the 19 non-smokers with 19 smokers. The concentration of mineral particulate in the lung tissue increases significantly in subjects who have lived in an urban environment for over 20 years ($p = 0.04$). A two-way ANOVA analysis (residence-smoke, residence-age) showed that the effect of residence is independent of age and/or smoking habits.

CONCLUSIONS. Detection frequency of mineral varieties in lung parenchyma seems to be closely related to their distribution in atmospheric aerosol in Rome. Indeed in this environment phyllosilicates account for 80 % of the minerals in the breathable particulate. Older subjects and smokers present a more severe lung burden. The relevant effect on lung burden of the time of residence in urban areas may be related to changes in pulmonary functions in response to long-term exposure to airborne irritants (Lippman, 1988).

REFERENCES

Cuyler Hammond E. and Garfinkel L. (1980) General air pollution and cancer in the United States. Prev. Med. 9: 206-211.

Doll R. (1978) Atmospheric pollution and lung cancer. Environ. Health Perspect. 22: 23-31.

Lippmann M. (1988) Health significance of pulmonary function responses to airborne irritants. JAPCA 38: 881-887.

Paoletti L., Falchi M., Viviano G., Ziemacki G., Batisti D., Pisani D. (1989) Features of airborne breathable particulate in a remote rural and in an urban area. Water, Air & Soil Pollution 42: 85-94.

Vena J.E. (1982) Air pollution as a risk factor in lung cancer. Am.. J. Epidemiol. 116: 42-56.

PART IV

FIBROGENESIS - CARCINOGENESIS

ANIMAL AND CELL RESPONSES TO PHYLLOSILICATES

HYPOTHESES ON THE MECHANISMS OF CARCINOGENESIS AND CELL TRANSFORMATION BY ASBESTOS AND OTHER MINERAL DUSTS

J. Carl Barrett,[1] Patricia W. Lamb, and Roger W. Wiseman
Laboratory of Molecular Carcinogenesis
National Institute of Environmental Health Sciences
National Institutes of Health
P. O. Box 12233
Research Triangle Park, NC 27709

NATO ASI Series, Vol. G 21
Health Related Effects of Phyllosilicates
Edited by J. Bignon

Abstract

Asbestos and other mineral fibers are carcinogenic to humans and animals but differ from many carcinogens in that they do not induce gene mutations. Evidence, however, exists that asbestos is a complete carcinogen, an initiator, and a promoter. Multiple mechanisms, therefore, must be operative to explain the diverse effects of asbestos fibers. Although asbestos is inactive as a gene mutagen, there is now clear evidence that it induces chromosomal mutations (aneuploidy and aberrations) in a wide variety of mammalian cells, including mesothelial cells. Asbestos also induces transformation of cells in culture, including mesothelial cells and fibroblasts. A mechanism for cell transformation, which is dependent on fiber dimension, has been proposed. The fibers are phagocytized by the cells and accumulate in the perinuclear region of the cells. When the cell undergoes mitosis, the physical presence of the fibers interferes with chromosome segregation and results in anaphase abnormalities. The transformed cells show aneuploidy and other chromosome abnormalities. These findings provide a mechanism at the chromosomal level by which asbestos and other mineral fibers might induce cell transformation and cancer. Identification of the critical target genes in asbestos carcinogenicity is required to understand further process, and recent progress in this area has been made. Results from several lines of investigation suggest that two distinct classes of genes, proto-oncogenes and tumor suppressor genes, are involved in the neoplastic process. In human mesothelioma, deletion of the short arm of chromosome 3 has been observed, which may result in the loss of a tumor suppressor gene on this chromosome. We have recently shown that an activated transforming oncogene exists in human mesotheliomas. Further molecular analysis of these cancers may help in understanding these neoplasms and the mechanisms of asbestos and other carcinogenic fibers.

Introduction

Asbestos and other mineral dusts are extremely fascinating chemicals to students of carcinogenesis. They are ubiquitous environmental substances, which are clearly carcinogenic to humans and to animals (IARC, 1977; National Research Council, 1984); but unlike most other carcinogens, asbestos fibers are not directly electrophilic, do not form adducts with DNA, are inactive in the Ames Salmonella test, and fail to induce mutations at specific genetic loci in mammalian cells (Chamberlain and Tarmy, 1977; Barrett, 1987; Jaurand, 1989). Some investigators have, therefore, proposed that asbestos is carcinogenic because of its tumor-promoting activity. This explanation, however, fails to explain fully the carcinogenic activity of mineral dusts (Barrett *et al.*, 1989).

Evidence exists that asbestos is a complete carcinogen, an initiator, and a promoter. The data implicating asbestos as a complete carcinogen and an initiator are summarized in Table 1. Asbestos alone appears from epidemiological studies to be a complete carcinogen in humans inducing mesotheliomas and lung cancers (IARC, 1977; National Research Council, 1984). In animals, asbestos is clearly a complete carcinogen, inducing the same types of tumors observed in humans (Wagner *et al.*, 1974; IARC, 1977; Harrington, 1981; Stanton *et al.*, 1981; Peto *et al.*, 1982). Epidemiological studies have shown that the incidence of mesotheliomas in exposed populations is independent of age at first exposure, consistent with the hypothesis that asbestos affects an early stage in the carcinogenic process (Peto *et al.*, 1982a, 1982b). On the other hand, it has been argued that asbestos must not act as an initiator, since it lacks mutagenic activity, a property of most initiators (Barrett, 1987). Although asbestos is not active as a gene mutagen in a variety of test systems, there is now clear evidence that it induces chromosomal mutations (aneuploidy and aberrations) in a wide variety of mammalian cells, including mesothelial cells (Lavappa *et al.*, 1975; Sincock and Seabright, 1975; Chamberlain and Tarmy, 1977; Babu *et al.*, 1980; Valerio *et al.*, 1980; Oshimura *et al.*, 1984;

Lechner *et al.*, 1985a and b; Linnainmaa *et al.*, 1986; Jaurand *et al.*, 1986; Barrett, 1987; Jaurand, 1989). Finally, it has been shown that asbestos can induce transformation of cells in culture, including mesothelial cells and fibroblasts (Hesterberg and Barrett, 1984; Lechner *et al.*, 1985; Patérour *et al.*, 1985a, 1985b; Hesterberg *et al.*, 1985, 1986a). As discussed below, evidence exists with these model systems that the mechanism of asbestos-induced cell transformation involves a chromosomal mutation (Oshimura *et al.*, 1984; Lechner *et al.*, 1985a; Patérour *et al.*, 1985a and b). Taken together, the evidence discussed above indicates that asbestos and other mineral fibers are complete carcinogens with tumor-initiating activity for certain cancers.

We do not wish to imply that the tumor-promoting activity of asbestos is not important in carcinogenesis. Considerable evidence exists for tumor-promoting activity associated with asbestos and other mineral fibers (Table 2). The marked synergism between asbestos exposure and smoking for the risk of lung cancer in humans (Selikoff *et al.*, 1968; Hammond *et al.*, 1979) has been cited as evidence for a tumor-promoting effect of asbestos. However, other explanations are possible for this effect; it is possible that: 1) asbestos acts as an initiator and cigarette smoke as a promoter, 2) asbestos acts as a tumor progressor for lung cancer rather than as a classical tumor promoter, or 3) asbestos and the carcinogens in cigarette smoke act as cocarcinogens. A cocarcinogenic effect of asbestos and polycyclic aromatic hydrocarbons has been demonstrated in animals (IARC, 1977). A promotion-like effect of asbestos has been demonstrated for DMBA-induced carcinomas in heterotopic tracheal transplants in rats (Topping and Nettesheim, 1980). However, in this system asbestos also has a complete carcinogenic activity. Asbestos induces changes in target tissues and some cells in culture similar to those observed with phorbol ester tumor promoters (Mossman *et al.*, 1983, 1985, 1987; Mossman and Craighead, 1980). The mechanisms by which asbestos and other mineral fibers might act as tumor promoters are no doubt quite different from the clastogenic and cell transforming effects described above. No single mechanism of action can be ascribed to asbestos and, like other potent carcinogens, these compounds exhibit multiple activities, many or all of which may contribute to their carcinogenicity.

In this review, we would like to focus on the mechanisms of mineral fiber-induced cell transformation and chromosomal mutations and the significance of these effects to the carcinogenic mechanisms of these substances.

Mechanisms of Asbestos-Induced Cell Transformation

Although asbestos is inactive as a gene mutagen in mammalian cells (Jaurand, 1989), it is able to induce heritable alterations in the growth properties and in some cases neoplastic transformation of normal cells in culture. Lechner and coworkers (Lechner *et al.*, 1985a and 1985b; Linnainmaa, 1986) reported that asbestos alters the growth properties of human mesothelial cells, and this is associated with chromosomal changes in the treated cells. Jaurand and coworkers (Jaurand *et al.*, 1986; Patérour *et al.*, 1985a, 1985b) have also shown that asbestos induces transformation and chromosomal changes in rat mesothelial cells in culture. Extensive studies from our laboratory have shown that asbestos and other mineral dusts induce neoplastic transformation of Syrian hamster embryo cells in culture. These studies have provided possible insights into the mechanism of mineral dust carcinogenicity and the role of fiber dimensions on this process. These studies will be discussed in more detail.

The Syrian hamster embryo cell transformation assay has a number of advantages for studies of chemical carcinogenesis (Barrett *et al.*, 1984, 1985). This model system uses early passage, diploid cells derived from day 13 fetuses. The cells in culture exhibit a low frequency of spontaneous transformation, but when treated with a wide variety of known carcinogens, the cells exhibit morphological and growth alterations (Fig. 1). These alterations (termed morphological transformation) can be used as a quantitative assay for carcinogens. This assay detects a wide variety of chemical carcinogens but not structurally-related noncarcinogens. Of particular interest is the observation that

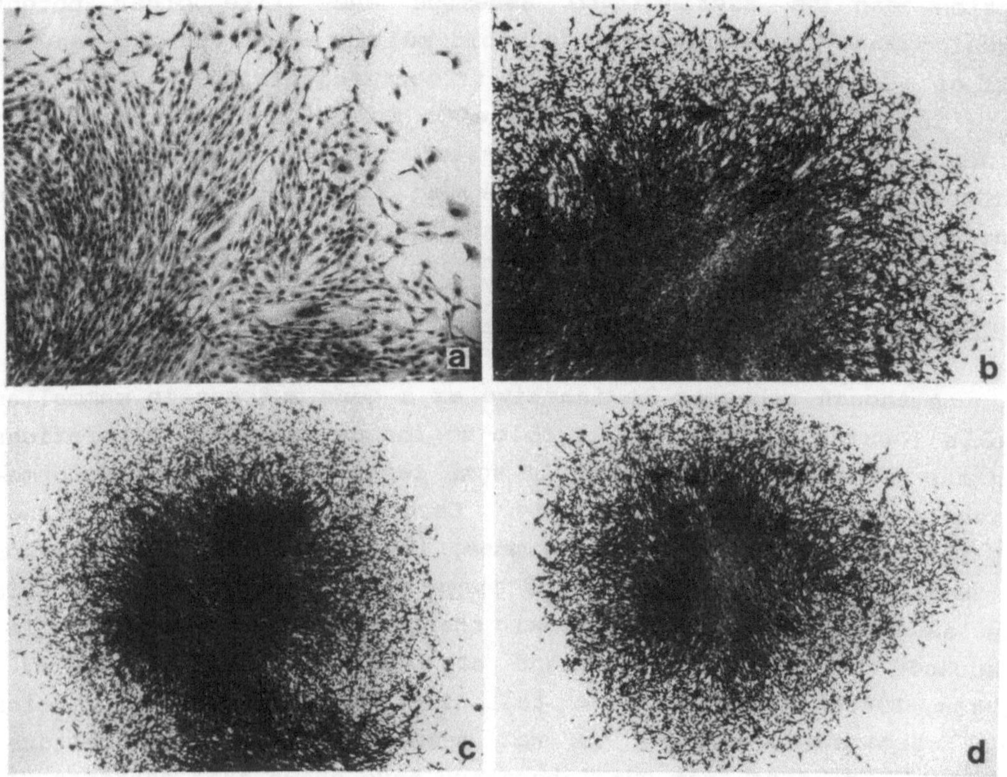

FIGURE 1. Morphologies of colonies of Syrian hamster embryo cells. Light microscope photographs show (A) a normal colony at 50X, (B) a chrysotile asbestos-transformed colony at 42X, (C) a crocidolite asbestos-transformed colony at 27X, and (D) a code 100 glass fiber-transformed colony at 27X after fixation in absolute methanol and staining in 10% Giemsa 7 days after treatment.

this assay detects a number of known carcinogens not detected in other short-term assays (e.g., asbestos, hormones, arsenic, benzene, diethylhexylphthalate). Carcinogen-induced neoplastic transformation of these cells in culture is a multistep process, and this model can be used to study the intermediate steps in the carcinogenic process.

We observed that chrysotile and crocidolite asbestos induces morphological transformation (Fig. 1) of Syrian hamster embryo cells that is indistinguishable from cell transformation induced by other carcinogens such as benzo(a)pyrene (Hesterberg and Barrett, 1984). The morphologically transformed cells progress and become tumorigenic after multiple passages, again similar to the findings with other carcinogens (Barrett and Tsó, 1978). The frequency of morphologically transformed cells increases linearly with increasing doses of asbestos (Fig. 2). The asbestos fibers are also toxic to the cells (Fig. 2). Three other laboratories also have shown that asbestos fibers transform Syrian hamster embryo cells in culture (DiPaolo *et al.*, 1983; Mikalsen *et al.*, 1988; Tu, A. personal communication). Asbestos fibers can also transform Balb/c 3T3 in culture (Lu *et al.*, 1988).

One of the more important (and intriguing) factors that influence mineral dust carcinogenicity is fiber dimension. Stanton *et al.* (1981) have shown in classical experiments that the mesothelioma-inducing activity of mineral dusts is related to fiber size, i.e., long, thin fibers are more active than short, thick fibers. Similar to the induction of mesotheliomas *in vivo*, cell transformation by mineral fibers is dependent on fiber size (Hesterberg and Barrett, 1984). Milling of chrysotile asbestos destroys its transforming activity, but this treatment reduces fiber diameter as well as fiber length. To study fiber length directly, the transformation by fiberglass was examined. Code 100 fiberglass (mean diameter = 0.1µ) is more active than thicker, code 110 fiberglass (mean diameter = 0.8µ). Code 100 fiberglass (average length = 9.5µ) is as equally potent as crocidolite asbestos of a similar length (Hesterberg and Barrett, 1984). Milling of the code 100 fiberglass reduces the fiber length without affecting the fiber diameter. Reduction of fiber length

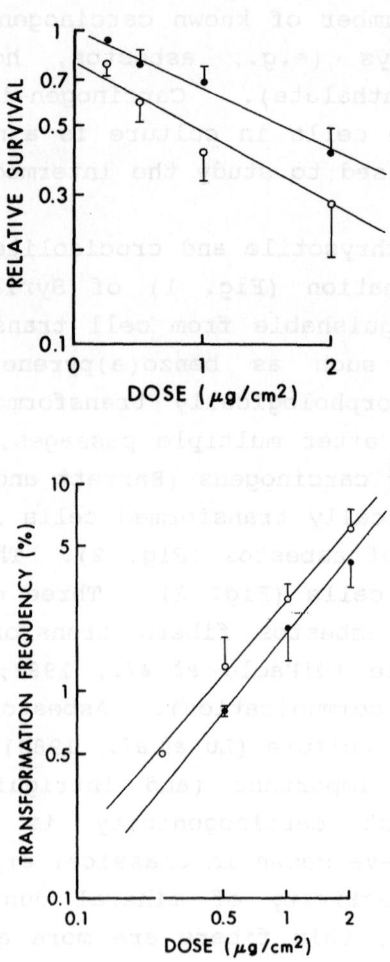

FIGURE 2. Effects of different doses of chrysotile (o) and crocidolite (•) asbestos on the relative survival (A) and the morphological transformation frequency of Syrian hamster embryo cells in culture (bars, S.E.). Reproduced from Hesterberg and Barrett (1984) with permission from *Cancer Research*.

to 2.2 μm reduces the transforming activity by 10- to 20-fold, and further reduction to <1 μm in length eliminates the transforming activity at the doses tested (Fig. 3). Thus, the fiber length dependence of cell transformation parallels the results of mesothelioma induction *in vivo*.

The effects of other mineral dusts were also examined in this

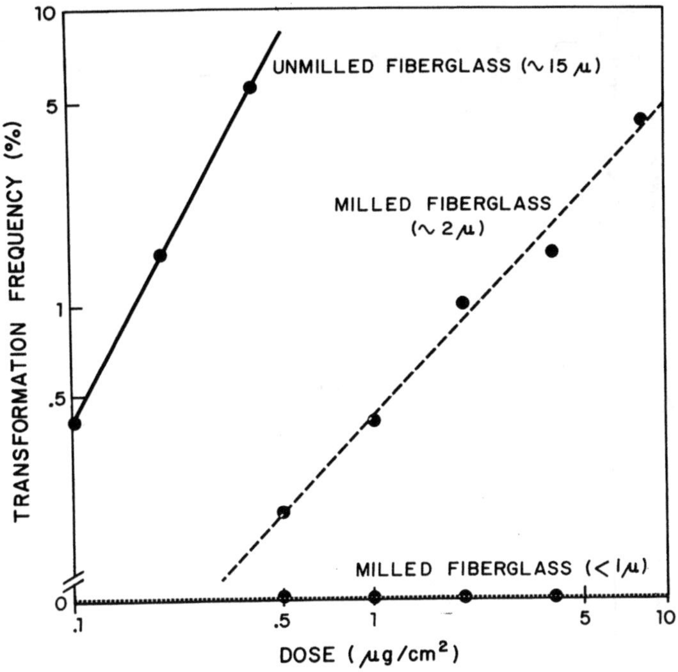

FIGURE 3. Effects of different doses of unmilled code 100 glass fibers (-) or milled code 100 (--) on transformation frequency of SHE cells in culture. Reproduced from Hesterberg and Barrett (1984) with permission from *Cancer Research*.

transformation model. Min-U-Sil, a microcrystalline silica, is active, but only at doses significantly greater than asbestos or fiberglass fibers (Fig. 4). Alpha-quartz, another crystalline silica sample, is also active, but only at higher doses. Interestingly, neither sample is very toxic as measured by reduction in the colony forming efficiencies of the treated cells (Fig. 4).

In order to explore the mechanism of the asbestos-induced cell transformation, we examined the mutagenic effects of asbestos fibers in the same cell system (Oshimura *et al.*, 1984). Consistent with other studies with mammalian cells, asbestos

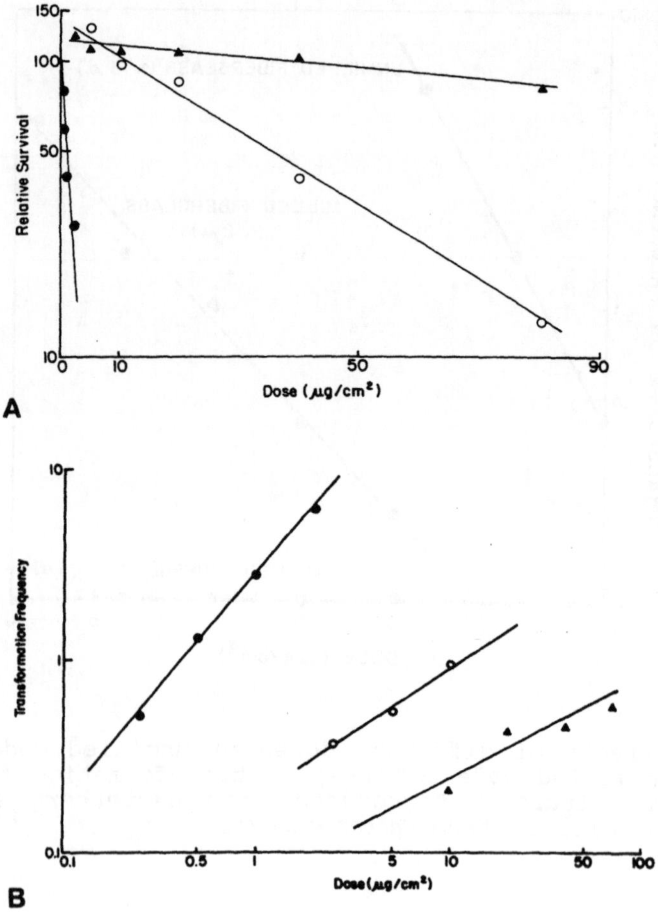

FIGURE 4. Effects of different doses of chrysotile asbestos (•), Min-U-Sil (o) α-quartz (Δ) on the relative survival (A), and the transformation frequency (B) of SHE cells in culture. Reproduced with permission from *Cancer Research*.

fibers did not induce gene mutations in the cells under conditions that induce cell transformation (Table 3). In contrast, asbestos fibers did induce chromosomal abnormalities in the treated cells, and the production of chromosomal mutations correlated with the transforming activity of the different fiber types (Table 4). Chromosome mutations induced by the fiberglass are fiber-length dependent. The mineral fibers induce numerical chromosome changes, both aneuploid cells in the near-diploid range and polyploid cells in the tetraploid range. Structural chromosome observations were also induced, but only at low levels. The induction of aneuploidy appears to be the most significant chromosome change in the transformation process. This conclusion is based on the finding of a nonrandom chromosome change, an extra copy of chromosome 11, in the transformed cells (Oshimura *et al.*, 1986). Trisomy 11 is the sole karyotypic change found in some of the asbestos-induced cell lines (Fig. 5).

We have proposed a mechanism for asbestos-induced aneuploidy, i.e., losses and gains of individual chromosomes (Hesterberg *et al.*, 1985, 1986b). In collaboration with Dr. Arnold Brody, we showed that crocidolite asbestos fibers are taken up by the cells within 24 hr after treatment by phagocytosis (Hesterberg *et al.*,

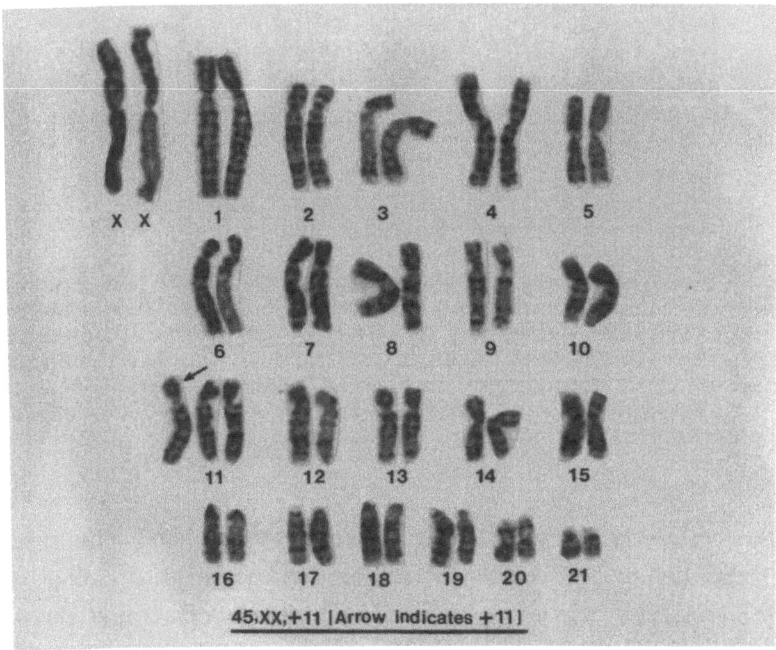

FIGURE 5. G-banded karyotype showing trisomy of chromosome 11 in the asbestos-induced Syrian hamster embryo cell line (10W). Arrow indicates the extra chromosome.

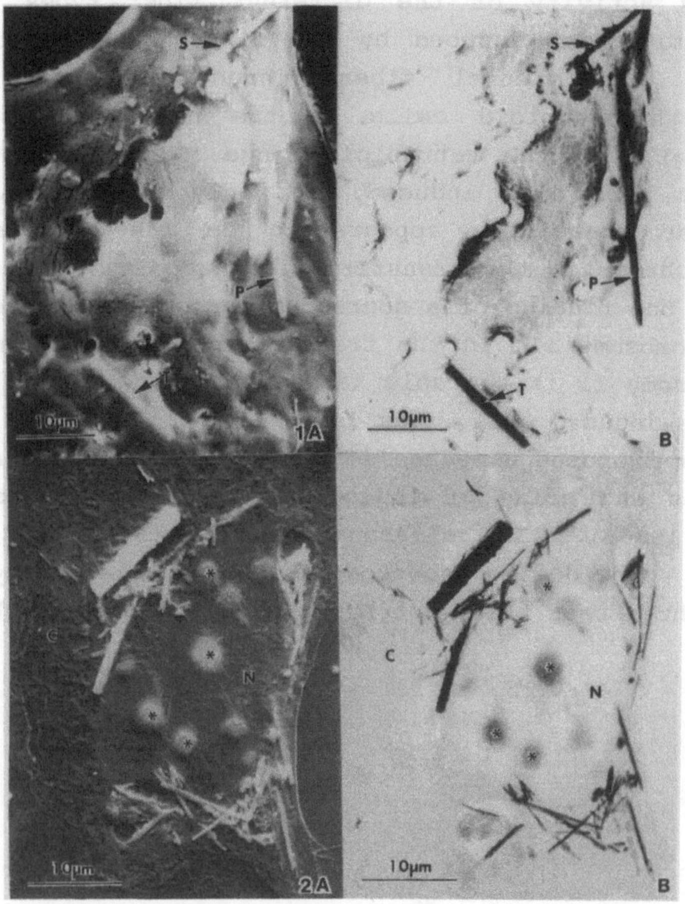

FIGURE 6. Scanning electron micrograph (A) and the corresponding backscatter electron image (B), showing the perinuclear accumulation of crocidolite asbestos fibers by a Syrian hamster embryo cell 24 hr after treatment with 1 µg/cm², N, nuclear region, which is demarcated by prominent nucleoli (asterisks); c, cytoplasm X 1,780. Reproduced from Hesterberg *et al.* (1986b) with permission from *Cancer Research*.

1986b); the intracellular fibers accumulate around the perinuclear region of the cells 24 to 48 hr after exposure (Fig. 6). When the cells undergo mitosis, the physical presence of the fibers results in interference with chromosome segregation. Analysis of crocidolite-exposed cells in anaphase (Hesterberg and Barrett,

1985) reveals a large increase in the number of cells with anaphase abnormalities including lagging chromosomes, bridges, and sticky chromosomes (Fig. 7). Asbestos fibers are observed in the mitotic cells and appear, in some cases, to interact directly with the chromosomes. Using ultrastructural analysis, Wang *et al.* (1987) have observed asbestos fibers apparently interacting with metaphase chromosomes after treatment of rat mesothelial cells in

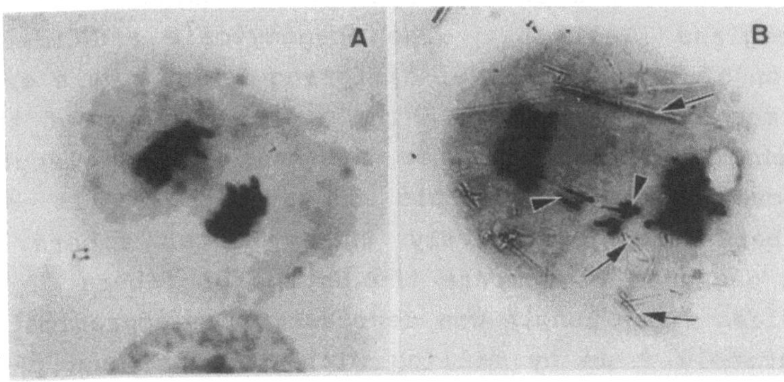

FIGURE 7. A normal (A) and an abnormal (B) anaphase from crocidolite asbestos-treated Syrian hamster embryo cells. Note the asbestos fibers (arrows), some of which appear to be associated with displaced chromosomes (arrowheads) in the abnormal anaphase. Reproduced from Hesterberg and Barrett (1985) with permission from *Carcinogenesis*.

culture. From these studies, we propose that the physical interaction of the asbestos fibers with the chromosomes or structural proteins of the spindle apparatus causes mis-segregation of chromosomes during mitosis, resulting in aneuploidy. These findings provide a mechanism, at the chromosomal level, by which asbestos and other mineral fibers might induce cell transformation and cancer. Malignant human and rat mesotheliomas are highly aneuploid (Stenman *et al.*, 1986; Gibas *et al.*, 1986; Lippus *et al.*, 1988). It is possible that asbestos exposure leads to the generation of these aneuploid cells *in vivo* as well as *in vitro*.

Nonfibrous particles such as silica are also phagocytized

by the cells and accumulate on the perinuclear region (Hesterberg *et al.*, 1986a). The presence of these particles at mitosis also results in aneuploidy induction but at higher doses than required for fibrous minerals. The cells can accumulate large numbers of silica particles without cytotoxicity, transformation, or aneuploidy induction.

Two factors appear to contribute to the fiber-size dependence of asbestos-induced cell transformation and chromosome damage. Hesterberg *et al.* (1986b) quantitated the number of intracellular fibers in Syrian hamster embryo cells exposed to long and short fiberglass (Table 5). The phagocytosis and intracellular distribution of glass fibers of differing lengths were examined at various times after treatment. The average length of the fibers that are phagocytized is greater than the average length of the fibers exposed to the cells. This indicates that cells phagocytize longer fibers more effectively than shorter fibers. In an experiment designed to compare the uptake of fibers of different lengths, glass fiber length was decreased from approximately 15 µm to approximately 2 µm by milling with a mortar and pestle, and cells were treated with an equal dose (1 µg/cm^2) of milled or unmilled fibers. Since fiber length was reduced 7-fold in the milled sample, the cells were exposed to 7-fold more fibers with the milled sample. However, cells exposed to the milled fiberglass had a similar number of fibers per cell as those exposed to unmilled fibers. This again indicates that the shorter fibers are less readily phagocytized. Thus, fiber length affects intracellular fiber accumulation. Fiber length, however, did not appear to affect the migration of intracellular fibers to the perinuclear region of the cytoplasm. Even though cells treated with milled glass fibers contained a similar number of fibers as those treated with unmilled glass fibers, the resulting cytotoxicity, transformation frequency, and frequency of micronuclei were greatly reduced in the cultures treated with milled glass fibers. Thus, fiber length appears to affect the phagocytosis of fibers as well as the ability of intracellular fibers to induce cytogenetic damage and the resultant transformation.

Two Classes of Genes Involved in Carcinogenesis

There is increasing evidence for the involvement of two classes of genes in carcinogenesis, oncogenes and tumor suppressor genes (Table 6). Oncogenes are the activated forms of a family of normal cellular genes, proto-oncogenes. These genes when activated by a variety of mutational events, which include point mutations, chromosome translocation, and gene amplifications, result in a positive, proliferative signal for the cells. In contrast, tumor suppressor genes negatively regulate tumor cell growth and are frequently inactivated or lost in tumors. These genes are inactivated by chromosome loss, chromosome deletion, and recombination, as well as by point mutations (Barrett and Wiseman, 1987). Asbestos fibers fail to induce point mutations and therefore would not be expected to activate proto-oncogenes or to inactivate tumor suppressor genes by this mechanism. Asbestos-induced chromosome breaks may result in activation of proto-oncogenes. Asbestos-induced chromosome loss or deletion could result in inactivation of tumor suppressor genes. A further understanding of the mechanism of action of asbestos should focus on addressing whether either of these types of genes is altered by fibers.

Molecular Alterations in Human Mesotheliomas

One approach to elucidating the mechanisms of asbestos carcinogenicity is to define the critical molecular alterations in asbestos-induced cancers. Since mesotheliomas are induced predominately, if not exclusively, as the consequence of asbestos exposure, these are ideal cancers for this analysis. Recently, two significant alterations in mesotheliomas have been reported, which may shed light on the mechanisms of fiber carcinogenesis. Karyotypic analyses of human mesotheliomas (Gibas *et al.*, 1986; Popescu *et al.*, 1988) reveal multiple chromosome changes. Chromosomes 1, 2, 3, 6, 11, 17, and 22 are most frequently involved in either numerical or structural rearrangements. Chromosome mutations caused by asbestos may be involved in the induction of some of these changes. Among these chromosomal changes, deletions, inversions, or translocations of chromosome 3 are most frequent. The deletion or break points in different tumors are varied between 3p13-p21. Interestingly, chromosome deletions and loss of heterozygosity of this chromosomal region frequently are observed in many lung cancers (Kok *et al.*, 1987; Whang Peng *et al.*, 1982). It is possible that loss of a tumor suppressor gene in this chromosomal region is a critical step in the carcinogenic process. We have recently established a highly tumorigenic mesothelioma cell line (Lamb and Barrett, unpublished), and attempts to suppress this tumor by introduction of a normal human chromosome 3 are in progress.

In collaboration with Drs. Y. Suzuki and P. Chahinian at Mt. Sinai Hospital, we have also examined human mesotheliomas for activated transforming genes by the ability of DNA from these tumors to neoplastically transform NIH3T3 cells. DNAs from three human mesotheliomas, diagnosed by histochemical and ultrastructural criteria, were isolated and cotransfected into NIH3T3 cells by the calcium phosphate precipitation method along with a cosmid vector DNA encoding resistance to the antibiotic G418. Drug-resistant cells were selected, and when sufficient cells were obtained (2 weeks), the cultures were injected into nude mice. Cultures cotransfected with the cosmid vector and normal DNA were nontumorigenic for up to 16 weeks after injection, while multiple cultures treated with the mesothelioma DNAs formed tumors between 7 and 14 weeks. DNAs were isolated from nude mouse tumors and retransfected without additional cosmid vector. Secondary transfectants from two primary tumors were highly positive in the tumorigenicity assay with latency periods of 5 weeks. DNAs from the NIH3T3 tumors were screened by Southern analyses for the presence of newly-acquired sequences homologous to the H-*ras*, K-*ras*, or N-*ras* oncogenes and hybridization was observed only for the endogenous mouse genes. However, the primary and secondary transfectants did contain human sequences detectable by hybridization to an *Alu* repetitive DNA probe. These results indicate that human mesotheliomas contain activated transforming genes that are not members of the *ras* gene family. The cosegregation of G418 resistance and tumorigenicity during secondary transfection suggests that the transforming genes and the cosmid vector are closely linked in at least two NIH3T3 tumors. Further experiments to characterize and clone these genes by cosmid rescue are in progress. Identification of the molecular basis for the activation of this transforming gene or genes may yield new insights into cancer and asbestos carcinogenicity.

Conclusion

Asbestos and other mineral fibers induce multiple types of cancers and are likely to act by multiple mechanisms. Although asbestos fibers do not induce gene mutations, they are effective inducers of chromosomal alterations, which may result in either activation of proto-oncogenes or inactivation of tumor suppressor genes. The identification in human mesothelioma of activated transforming genes and the loss of a region on the short arm of chromosome 3 are indicative that proto-oncogenes and tumor suppressor genes are altered in these cancers, and it will be important in the future to understand the role of asbestos fibers in these alterations.

Table 1. Evidence for complete carcinogenic and initiating activity of asbestos

(1)	Asbestos is a known human carcinogen.
(2)	Asbestos and other mineral fibers are complete carcinogens in animals by multiple routes of exposure including intrapleural and intraperitoneal injection, inhalation, and tracheal transplants.
(3)	Incidence of mesotheliomas in exposed populations is independent of age of first exposure, which is consistent with the hypothesis that asbestos affects an early stage in the carcinogenic process.
(4)	Asbestos is inactive as a gene mutagen but induces chromosomal mutations (aneuploidy and aberrations) in a wide variety of mammalian cells.
(5)	Asbestos induces transformation of human and rat mesothelial cells and hamster fibroblast cells in culture.

Table 2. Evidence for tumor promoting activity of asbestos and other mineral fibers

(1)	Marked synergism exists between asbestos exposure and smoking for risk of lung cancer in humans.
(2)	Rodents exposed by instillation to a combination of asbestos and chemical carcinogens such as polycyclic aromatic hydrocarbons have synergistic incidence of lung cancers (cocarcinogenic effect).
(3)	Asbestos enhances dimethybenzanthracene-induced carcinomas in heterotopic tracheal transplants in rats (asbestos also has complete carcinogenic activity in this model).
(4)	Asbestos induces changes in target tissues similar to those observed with tumor promoters such as hyperplasia, metaplasia, induction of DNA synthesis, induction of ornithine decarboxylase, and stimulation of production of oxygen-free radicals.

Table 3. Specific locus mutation and toxicity of Syrian hamster embryo cells after treatment with transforming doses of asbestos and benzo(a)pyrene

Treatment	Dose (µg/cm²)	Relative survival[a] (%)	Mutation frequency[b] Oua^R	TG^r
Control	0	100	$<10^{-6}$	$<10^{-6}$
Chrysotile	1.0	41	$<10^{-6}$	$<10^{-6}$
Chrysotile	2.0	28	$<10^{-6}$	$<10^{-6}$
Crocidolite	1.0	69	$<10^{-6}$	$<10^{-6}$
Crocidolite	2.0	41	$<10^{-6}$	$<10^{-6}$
Benzo(a)pyrene	1 µg/ml	69	$1.5 (\pm 0.7) \times 10^{-4}$	$1.0 (\pm 0.4) \times 10^{-4}$

[a]Measured immediately after treatment.

[b]Mutation frequencies were corrected for relative survival measured after mutant expression period. Resistance to ouabain (Oua^R) measures mutations at the Na^+/H^+ATPase locus and resistance to 6-thioguanine (TG^r) measures mutations at the hypoxanthine phosphoribosyl transferase locus.

These data were reproduced from Oshimura *et al.* (1984) with permission from *Cancer Research*.

Table 4. Cytogenetic effects of Syrian hamster embryo cells *in vitro* with various mineral dusts (2 μg/cm²)

	Transformation frequency[a] (%)	Aneuploid cells[b] (%)	Cells with chromosome aberrations[c] (%)	Cells with micronuclei[d] (%)	Tetraploid cells[d,e] (%)	Binucleated cells (%)
Control	0	1.7	1	0.3	5	0.3
Chrysotile	6.2*	12.5*	5	2.6*	33*	25.0*
Crocidolite	4.6*	9*	4	1.1*	14*	11.2*
Fiberglass[f]	3.0*	7*	4	3.0*	20*	18.4*
Milled fiberglass	0	2	1	0.5	6	0.4
Alpha-quartz	0	3	1	0.5	5	0.3

[a]Cited from Hesterberg and Barrett (1984). The transformation frequency was calculated by dividing the number of morphologically transformed colonies by the total number colonies examined x 100%.

[b]This represents the percentage of metaphases that are aneuploid with near-diploid number of chromosomes.

[c]Percentage of metaphases containing the following aberrations: chromatid breaks, isochromatid breaks, chromosome fragments, chromatid exchanges, or dicentric chromosomes.

[d]1000 cells were scored per treatment group.

[e]Cells with a tetraploid (4N = 88) or near tetraploid (70 - 100) number of chromosomes.

[f]This fiberglass was obtained from Johns Manville (Code 100) and processed as described previously (Hesterberg and Barrett, 1984).

*Statistically significant from the control ($p < 0.05$, Fisher Exact Test).

These data reproduced from Oshimura *et al.* (1984) with permission from *Cancer Research*.

Table 5. Relationship between average fiber length of unmilled and milled glass fibers and the biological effects induced by treatment with 1 μg/cm²

Glass fibers	Av. no. of phagocytized fibers/cell[a]	Fiber length in stock suspension	Intracellular fiber length[b] (μm)	Relative survival[c] (%)	Transformation frequency[d] (%)	Frequency of micronuclei[e] (%)
Unmilled	13.1 ± 0.3	15.5 ± 1.5	25.5 ± 1.3	14	12	5.0 ± 1.0
Milled	15.0 ± 0.4	2.2 ± 0.2	7.1 ± 0.8	84	0.3	0.7 ± 0.05

[a] Fibers that were partially or totally phagocytized were counted (± SE).

[b] Length was determined using electron backscatter imaging at x6000 (± SE).

[c] $$\text{Relative survival} = \frac{\text{no. of colonies in 10 glass fiber-treated cultures}}{\text{no. of colonies in 10 control cultures}} \times 100\%$$

[d] $$\text{Transformation frequency} = \frac{\text{no. of morphologically transformed colonies in 10 cultures}}{\text{total no. of surviving colonies in 10 cultures}} \times 100\%$$

[e] There were 1000 cells examined per treatment group. A frequency of 5.0 ± 1.0% is significantly different but a frequency of 0.7 ± 0.05% is not significantly different from control (0.6 ± 0.05)(Fisher's exact test). Values are followed by ± SE.

Table 6. Two classes of genes involved in carcinogenesis

Proto-oncogenes	Tumor Suppressor Genes
1. Involved in cellular growth and differentiation	1. Function unknown but possibly involved in cellular growth and differentiation (negative regulators of cell growth?)
2. Family of genes exists	2. Family of genes exists
3. Must be activated in cancers (quantitatively or qualitatively)	3. Must be inactivated, lost, or mutated in cancers
4. Mutational activation by point mutation, chromosome translocation or gene amplification	4. Mutational inactivation by chromosome loss, chromosome deletion, point mutation, somatic recombination, or gene conversion
5. Little evidence for involvement in hereditary cancers	5. Clear evidence for involvement in hereditary and nonhereditary cancers

References

Babu KA, Lakkad BC, Nigam SK, Bhatt DK, Karnik AB, Thakore KN, Kashyap SK, Chatterjee SK (1980) *In vitro* cytological and cytogenetic effects of an Indian variety of chrysotile asbestos. Environ Res 21:416-422

Barrett JC, Ts'o POP (1978) Evidence for the progressive nature of neoplastic transformation *in vitro*. Proc Natl Acad Sci USA 75:3761-3765

Barrett JC, Hesterberg TW, Thomassen DG (1984) Use of cell transformation systems for carcinogenicity testing and mechanistic studies of carcinogenesis. Pharmacological Reviews 36:53S-70S

Barrett JC, Hesterberg TW, Oshimura M, Tsutsui T (1985) Role of chemically induced mutagenic events in neoplastic transformation of Syrian hamster embryo cells. In Barrett JC and Tennant RW (eds) *Carcinogenesis - A Comprehensive Survey: Mammalian Cell Transformation: Mechanisms of Carcinogenesis and Assays for Carcinogens*. New York Raven Press, Vol. 9, pp 123-137

Barrett JC, Wiseman RW (1987) Cellular and molecular mechanisms of multistep carcinogenesis: relevance to carcinogen risk assessment. Environ Health Perspec 76:65-70

Barrett JC (1987) Relationship between mutagenesis and carcinogenesis. In Barrett JC (ed) *Mechanisms of Environmental Carcinogenesis: Role of Genetic and Epigenetic Changes*. Boca Raton CRC Press, Vol. I, pp 129-142

Barrett JC, Lamb PW, Wiseman RW (1989) Multiple mechanisms for the carcinogenic effects of asbestos and other mineral fibers. Environ Health Perspec. In press.

Chamberlain M, Tarmy EM (1977) Asbestos and glass fibers in bacterial mutation tests. Mutat Res 43:159-164

DiPaolo JA, DeMarinis AJ, Doniger J (1983) Asbestos and benzo(a)pyrene synergism in the transformation of Syrian hamster embryo cells. Pharmacology 27:65-73

Gibas Z, Li FP, Antman KH, Bernai S, Stahel R, Sandberg AA (1986) Chromosome changes in malignant mesothelioma. Cancer Genet Cytogenet 20:191-201

Hammond EC, Selikoff IJ, Seidman H (1979) Asbestos exposure, cigarette smoking and death rates. Ann NY Acad Sci 330:473-490.

Harrington JS (1981) Fiber carcinogenesis: Epidemiologic observations and the Stanton hypothesis. J Natl Cancer Inst 67:977-988

Hesterberg TW, Barrett JC (1984) Dependence of asbestos- and mineral dust-induced transformation of mammalian cells in culture on fiber dimension. Cancer Res 44:2170-2180

Hesterberg TW, Barrett JC (1985) Induction by asbestos fibers of anaphase abnormalities: Mechanism for aneuploidy induction and possibly carcinogenesis. Carcinogenesis 6:473-475

Hesterberg TW, Oshimura M, Barrett JC (1985) Transformation of mammalian cells in culture by asbestos and other mineral dusts: A mechanism involving chromosomal mutation. In: *In vitro* Effects of Mineral Dusts. Third International Workshop, NATO ASI Series, Vol. G3 (Beck EG, Bignon J, eds), Springer-Verlag Berlin, pp 185-196

Hesterberg TW, Brody AR, Oshimura M, Barrett JC (1986a) Asbestos and silica induce morphological transformation of mammalian cells in culture: A possible mechanism. In: Silica, Silicosis, and Cancer (Goldsmith DF, Winn DM, Shy CM, eds), Praeger Press New York, pp 177-190

Hesterberg TW, Butterick CJ, Oshimura M, Brody AR, Barrett JC (1986b) Role of phagocytosis in Syrian hamster cell transformation and cytogenetic effects induced by asbestos and short and long glass fibers. Cancer Res 46:5795-5802

IARC Monographs on the Evaluation of Carcinogenic Risk of Chemicals to Man (1977) Vol. 14, Asbestos. International Agency for Research on Cancer, Lyon France

Jaurand MC (1989) A particulate state carcinogenesis: Recent data on the mechanisms of action of fibers. IARC Publications 90:54-73

Jaurand MC, Kheuang L, Magne L, Bignon J (1986) Chromosomal changes induced by chrysotile fibers or benzo-3,4-pyrene in rat pleural mesothelial cells. Mutation Res 169:141-148

Kok K, Osinga J, Carritt B, Davis MB, van der Hout AH, van der Veen AY, Landsvater RM, de Leij LFMH, Berendsen HH, Postmus PE, Poppema S, Buys CMCM (1987) Deletion of a DNA sequence at the chromosomal region 3p21 in all major types of lung cancer. Nature 330:578-581

Lavappa KS, Fu MM, Epstein SS (1975) Cytogenetic studies on chrysotile asbestos. Environ Res 10:165-173

Lechner JF, Tokiwa T, LaVeck M, Benedict WF, Banks-Schlegel S, Yeager Jr H, Banerjee A, Harris CC (1985a) Asbestos-associated chromosomal changes in human mesothelial cells. Proc Natl Acad Sci USA 82:3884-3888

Lechner JF, Tokiwa T, Yeager Jr H, Harris CC (1985b) Asbestos-associated chromosomal changes in human mesothelial cells. In: *In vitro* Effects of Mineral Dusts. Third International Workshop, NATO ASI Series, Vol. G3 (Beck EG, Bignon J, eds), Springer-Verlag Berlin, pp 197-202

Linnainmaa K, Gerwin B, Pelin K, Jantunen K, LaVeck M, Lechner JF, Harris CC (1986) Asbestos-induced mesothelioma and chromosomal abnormalities in human mesothelial cells *in vitro*. NIOSH Symposium Extended Abstract

Lu Y-P, Lasne C, Lowy R, Chouroulinkov I (1988) Use of the orthogonal design method to study the synergistic effects of asbestos fibers and 12-*O*-tetradecanoylphorbol-13-acetate (TPA) in the BALB/3T3 cell transformation system. Mutagenesis 3:355-372

Mikalsen SO, Rivedal E, Sanner T (1988) Morphological transformation of Syrian hamster embryo cells induced by mineral fibres and the alleged enhancement of benzo(a)pyrene. Carcinogenesis 9:891-899

Mossman BT, Craighead JE (1980) Asbestos-induced epithelial changes in organ cultures of hamster trachea: Inhibition by retinyl methyl ether. Science 207:311-313

Mossman BT, Light WG, Wei ET (1983) Asbestos: mechanisms of toxicity and carcinogenicity in the respiratory tract. Ann Rev Pharmacol Toxicol 23:595-602

Mossman BT, March JP (1985) Mechanisms of toxic injury by asbestos fibers: Role of oxygen free radicals. In: *In vitro* Effects of Mineral Dusts. Third International Workshop, NATO ASI Series, Vol. G3 (Beck EG, Bignon J, eds), Springer-Verlag Berlin pp 66-74

Mossman BT, Marsh JP, Doherty J, Gilbert R, Hill S (1987) Implication of active oxygen species as second messengers of asbestos toxicity. Drug Chem Toxicol 10:157-180

National Research Council (1984) Asbestiform Fibers: Nonoccupational Health Risks. National Academy Press, Washington, DC

Oshimura M, Hesterberg TW, Tsutsui T, Barrett JC (1984) Correlation of asbestos-induced cytogenetic effects with cell transformation of Syrian hamster embryo cells in culture. Cancer Res 44:5017-5022

Oshimura M, Hesterberg TW, Barrett JC (1986) An early, nonrandom karyotypic change in immortal Syrian hamster cell lines transformed by asbestos: trisomy of chromosome 11. Cancer Genet Cytogenet 22:225-237

Patérour MJ, Bignon J, Jaurand MC (1985) *In vitro* transformation of rat pleural mesothelial cells by chrysotile and/or benzo-a-pyrene. Carcinogenesis 6:523-529

Patérour MJ, Renier A, Bignon J, Jaurand MC (1985) Induction of transformation in cultured rat pleural mesothelial cells by chrysotile fibers. In: *In vitro* Effects of Mineral Dusts. Third International Workshop, NATO ASI Series, Vol. G3 (Beck EG, Bignon J, eds), Springer-Verlag Berlin, pp 203-207

Peto J, Henderson BE, Pike MC (1982a) Trends in mesothelioma incidence in the United States and the forecast epidemic due to asbestos exposure during World War II. Quantification of Occupation Cancer 9:51-69

Peto J, Seidman H, Selikoff IJ (1982b) Mesothelioma mortality in asbestos workers: Implications for models of carcinogenesis and risk assessment. Br J Cancer 45:124-135

Popescu NC, Chahinian AP, DiPaolo JA (1988) Nonrandom chromosome alterations in human malignant mesothelioma. Cancer Res 48:142-147

Renier A, Fleury J, Monchaux G, Nebut M, Bignon J, Jaurand MC (1989) Toxicity of an attapulgite sample studied *in vivo* and *in vitro*. In Bignon J, Peto J, and Saracci R (eds) *Non-occupational Exposure to Mineral Fibres*. ARC Scientific Publications No. 90, 180-184

Selikoff IJ, Hammond EC, Churg J (1968) Asbestos exposure, smoking, and neoplasia. J Amer Medical Assoc 204:101-110

Sincock AM, Seabright M (1975) Induction of chromosome changes in Chinese hamster cells by exposure to asbestos fibers. Nature (Lond) 257:56-58

Stanton MF, Layard M, Tegeris A, Miller E, May M, Morgan E, Smith A (1981) Relation of particle dimension to carcinogenicity in amphibole asbestoses and other fibrous minerals. J Natl Cancer Inst 67:965-975

Stenman C, Olofsson K, Mansson T, Hagmar B, Mark J (1986) Chromosomes and chromosomal evolution in human mesotheliomas as reflected in sequential analyses of two cases. Herreditas 105:233-239

Topping DC, Nettesheim P (1980) Two-stage carcinogenesis studies with asbestos in Fischer 344 rats. J Natl Cancer Inst 3:627-630

Valerio F, DeFerran M, Ottaggio L, Repetto E, Santi L (1980) Cytogenetic effects of Rhodesian chrysotile on human lymphocytes *in vitro*. In: Biological Effects of Mineral Fibers, Vol. I (Wagner JC, ed), International Agency for Research on Cancer, Lyon France, pp 485-489

Wagner JC, Berry G, Skidmore JW, Timbrell V (1974) The effects of the inhalation of asbestos in rats. Br J Cancer 29:252-269

Wang NS, Jaurand MC, Magne L, Kheuang L, Pinchon MC, Bignon J (1987) The interactions between asbestos fibers and metaphase chromosomes of rat pleural mesothelial cells in culture. Am J Pathol 126:343-349

Whang-Peng J, Kao-Shan CS, Lee EC, Bunn PA, Carney DN, Gazdar AT, Minna JD (1982) Specific chromosome defect associated with human small-cell lung cancer: deletion 3p14-23. Science (Wash DC) 215:181-182

REVIEW ON PULMONARY EFFECTS OF PHYLLOSILICATES AFTER INHALATION

J.C. WAGNER
Recently retired MRC External Staff Team
on Occupational Lung Diseases.
Now at "Foxstones", 59 Coombe Valley Road,
Preston, Weymouth, Dorset. DT3 6NL

I had hoped that the abstract that I wrote for this conference would be sufficient and I would not have to attempt a paper on this subject. The reason for this is the limited amount of information that I could cull from the literature.

In the inhalation experiments with the phyllosilicates I could only find two papers on the inhalation of talc and I had thought that a single experiment that we had undertaken with kaolin was the only paper on record. However, at the end of the conference the last paper was on kaolin inhalation in rats. This provided a most interesting example of the results of very heavy exposure.

In these four experiments the animals were exposed to relatively fine dusts of talc and kaolin. It would be of interest to see how far the human experience could be repeated if animals were exposed to talc dust with significant tremolite content or to kaolin containing large amounts of free quartz as would be obtained from China stone.

At the conference there has been a tendency to include the palygorskite minerals sepiolite and attapulgite as phyllosilicates. I have not done so but our inhalation studies with dusts derived from these materials has been published (Wagner, Griffiths and Munday 1987).

Talc

(a) With hamsters

These were the experiments undertaken by Wehner (Wehner et al 1977 and 1979), in which a large number of golden hamsters were exposed to Vermont talc which was 95% pure but did contain traces of magnesite, dolomite, chlorite and rutile.

NATO ASI Series, Vol. G 21
Health Related Effects of Phyllosilicates
Edited by J. Bignon

In the first study the animals were exposed for short periods over thirty days. Two further groups were exposed for 30 or 150 minutes per day for 300 days or until they died. The total aerosol concentration was 27.4mg/m^3 average diameter 6.0μg and total aerosol concentration 27.4mg/m^3 respirable fraction 8.1mg/m^3. No lung tumours were observed in these animals. The incidence of alveolar hyperplasia was 25% in the longer exposed group. This could be compared with 10% in a control group.

(b) Using rats

Wagner et al 1979 (An animal model for inhalation exposure to talc). The material used in this study was Italian 00000 grade. It was taken from a commercial sample with an upper particle size of 70μm and a mass median particle size of 25μm. The sample was found to contain 92% talc mineral by weight together with 3% chlorite, 1% carbonate mineral. Quartz was present at the 0.5 1% level. For comparison a sample of super fine chrysotile asbestos was used. Non-exposed animals were used as a negative control. The experimental animals were barrier protected caesarian derived rats of the Wistar strain. The rats were exposed in 1.4m^3 chambers which hold 48 rats in cages of six. The dusts clouds were generated for 7½ hours a day and 5 days per week. The respirable dust concentrations were measured daily and the dosage adjusted to give the required concentration. The talc cloud was generated using the generator designed for asbestos dusts. Initially 48 rats (24 of each sex) were exposed to the dusts. After six months 24 rats were transferred to cages in ambient air and replaced by a further 24 animals. These rats were in turn removed and replaced after three months exposure and all exposure ceased after a further three months. Thus there were 96 rats exposed to each dust - 48 for 3 months, 24 for 6 months and 24 for 12 months. The mean respirable dust concentration was 10.8mg/m^3 and the cumulative doses were approximately 4100, 8200 and 16400 mg/m^3 depending on length of exposure. Rats were sacrificed after each exposure period and at the end of two years. The remaining rats were allowed to survive until they died. The tissue reaction was scored (1) normal lung, (2) evidence of dust in lung, (3) minimal tissue reaction, (4) minimal fibrosis, (5) definite fibrosis.

TABLE 1

Talc Inhalation Experiment

Pulmonary Fibrosis Occurring in Rats

Material	Time	Length of exposure 3 mths.	6 mths.	12 mths.
Italian talc	End of exposure	2.2 (8)	2.7 (6)	3.4 (6)
	1 year later	2.4 (8)	3.4 (4)	4.6 (4)
SFA chrysotile	End of exposure	2.8 (8)	3.0 (6)	3.2 (6)
	1 year later	2.2 (8)	3.2 (4)	4.2 (4)
Controls	End of exposure	1.8 (8)	1.9 (6)	1.3 (6)
	1 year later	1.6 (8)	1.5 (3)	1.9 (3)

TABLE 2

Talc Inhalation Experiment

Incidence of Bronchiolar Alveolar Hyperplasia (BAH) and Lung Tumours

Material	Exposure	No. at risk*	BAH	Number of lung tumours Adeno-Carcinomas	Mesotheliomas
Italian talc	3 mths.	39	0	0	0
	6 mths.	18	0	0	0
	12 mths.	24	2	0	0
SFA Chrysotile	3 mths.	40	0	0	1
	6 mths.	18	3	1	0
	12 mths.	22	6	2	0
Controls		71	1	0	0

* Number surviving at least 300 days from start of exposure.

Results

Both dusts produced minimal pulmonary fibrosis in the animals exposure for a year and surviving a further year (Table 1).

Two of the rats exposed to the talc developed bronchiolar alveolar hyperplasia (BAH) and a similar lesion was seen in one of the control animals. In the animals exposed to S.F.A. chrysotile there were six with BAH, two developed adeno-carcinomas, and there was one mesothelioma.

In spite of the large median diameter of the talc dust a reasonable amount was retained in the lungs. This was calculated at 2.8; 4.5 and 12.3mg per rat at the end of the 3, 6 and 12 month exposures respectively.

Kaolin

Introduction

Following discussions with The China Clay Association it was considered that it would be of value to study the effects of the inhalation of the dusts from two china clays (Kaolins) representative of current commercial production. The experiment was undertaken to establish whether either of these dusts would produce pulmonary fibrosis of either the nodular or interstitial type, and whether the lesions were progressive if no further dusting took place. It was considered that six other dusts which were being investigated at the same time could be used for comparison. In addition, it was planned to quantify the total amount of dust retained in the lungs at various stages of the study and to establish the quantity of dust retained in the lungs when the animals were removed from the dusting chambers. The remaining animals would be allowed to survive until they died of natural causes. Any pulmonary lesions which developed were recorded.

Materials

A filler grade (Grade D) and a coating grade (SPS) of china clay were used in the investigation, these were supplied by English Clays Lovering and Pochin and which is representative of present commercial production. Clouds of dust were generated from each sample for inhalation by Fischer 344 SPF rats. Twenty rats of each sex, approximately 8 weeks old, were used for each dust. As controls similar groups of rats were exposed to clouds generated from (1) crocidolite,

(2) mordenite, (3) non-fibrous zeolite, (4) long fibred zeolite, (5) sepiolite and (6) attapulgite. There was also an unexposed group.

Methods

The rats were placed in 8 cages in each dusting chamber and dust clouds were generated for 7 hours on 5 days of each week over a period of 12 months. The cloud generators were adusted to provide a respirable dust concentration of $10mg/m^3$ as measured by Casella Type 113A size selective gravimetric samplers. These samplers operated throughout each daily exposure period and the dust collected for subsequent mineralogical analysis. The size distributions of the airborne respirable dusts were determined from samples collected by the size selective long running thermal precipitator.

Dust exposure commenced when the rats were 47-67 days old. Eight rats, four of each sex, were withdrawn at the end of 3 months from each group, and sacrificed and four rats were sacrificed at 6 months, 12 months and 24 months after the first day of exposure. The lungs of these animals were studied histologically and the dust content of the lungs was determined gravimetrically. At the end of the twelfth month dust exposure ceased and the rats were removed to a clean atmosphere in a barrier protected room. With the exception of the rats sacrificed twenty-four months after the first day of exposure, these rats were allowed to live out their full lifespan and were only killed if they appeared to be distressed.

A full necropsy examination was undertaken on each rat, the thoracic organs, liver, spleen, kidney and any other significant tissue were prepared to histological examination. The lungs from sacrificed rats were air inflated and fixed in formalin by submersion. In the remaining rats the lungs were inflated by formalin instilled through the trachea. After fixation, the lungs were cut in sagittal axis and prepared for histological examination.

Analysis, by XRD of the bulk china clay samples and the samples of the respirable dust collected in the exposure chambers was made by ECLP and the results are given in Table 3. The kaolinite content of the respirable dust is higher than that of the bulk sample for both grades.

TABLE 3

Analysis of china clay bulk samples and respirable dust samples by XRD.

	Filler Grade		Coating Grade	
	Bulk %	Respirable %	Bulk %	Respirable %
Kaolinite	85	92	93	95
Mica	13	7	6	4
Quartz	0.7	0.2	0.3	0.3
Feldspar	1.3	0.8	0.7	0.7

The dust particles in the clouds were composed of aggregates of small particles of kaolinite and ancillary materials, the individual particles contained in the coating and filler grades were of similar size. The size distributions on the respirable dust particle aggregates making up the clouds was determined microscopically and the relative surface area and mass of distributions computed. These are given in Table 4, differences between the clouds are possibly attributable to differences in the performance of the cloud generators.

TABLE 4

Size distributions of respirable china clay samples with computed relative surface areas (ND^2) and mass distributions (ND^3).

Coating Grade

Size range µm	Number %	ND^2 %	ND^3 %
0.8 - 1.2	19.3	2.3	0.5
1.2 - 1.6	19.5	4.7	1.5
1.6 - 2.3	19.4	8.9	3.8
2.3 - 3.3	22.4	21.2	13.1
3.3 - 4.6	10.8	20.5	17.9
4.6 - 6.6	6.7	25.4	31.4
6.6 - 9.3	1.8	13.8	24.2
9.3 - 13.2	0.2	3.1	7.6

3000 particles (>0.8µm) ≡ $10mg/m^3$

TABLE 4 (Cont'd.)

Filler Grade

Size range µm	Number %	ND^2 %	ND^3 %
0.8 - 1.2	9.1	0.5	0.1
1.2 - 1.6	13.0	1.5	0.3
1.6 - 2.3	17.3	3.8	1.1
2.3 - 3.3	21.5	9.7	4.1
3.3 - 4.6	16.8	15.3	9.2
4.6 - 6.6	12.2	22.1	18.8
6.6 - 9.3	7.7	28.2	34.1
9.3 - 13.2	2.6	18.8	32.2

1000 particles (>0.8µm ≡ 10mg/m^3

An average concentration of 10mg/m^3 respirable dust was maintained in the chambers and the cumulatived dust dosage, concentration (mg/m^3) x time (hours), offered to the rats after 3, 6 and 12 months, were 4417, 9997 and 17002 respectively for the coating grade kaolin and 4325, 10026 and 17020 for the filler grade kaolin.

Results

In studying the lung sections the severity of the tissue reaction was recorded using internationally accepted standards which had been agreed at Copenhagen (McConnell et al, 1983).

TABLE 5

Inhalation Results

	Mean Reaction Grade			
	3 mths.	6 mths.	12 mths.	24 mths.
Kaolin Coating	1.2	3.4	2.9	3.1
Kaolin Filler	2.6	3.7	2.9	3.2
Controls Unexposed	1.1	1.0	1.1	1.0

No lung tumours were observed in these groups.

Two animals exposed to kaolin coating developed BAH (Bronchiolar Alveolar Hyperplasia).

One animal exposed to kaolin filler BAH.

There was no evidence of the nodular type of fibrosis seen in any of the sections. Exposure to each of the eight dusts elicited a minimal interstitial lesion in the animals which survived for the two year period. Neither of the two "kaolins" produced a reaction greater than grade 4, whilst in concurrent experiments long attapulgite produced two scores of 5 and non-fibrous zeolite six scores of grade 5 and one of grade 6.

Kaolin coating clay produced a slightly lower level of response than did kaolin filler clay (Table 5). The mean scores, excluding the 3 months results, were 2.46 $\pm$ 0.098 and 2.79 $\pm$ 0.133 respectively, a statistically significant ($p<0.001$) but trivial difference.

In similar experiments the long fibred zeolite produced a very high rate of mesotheliomas whilst all the dusts other than the two kaolins were associated with a few malignant tumours. Only hyperplastic dust reaction occurred in two animals with broncho-alveolar hyperplasia among those exposed to coating grade kaolin and one BAH in the kaolin filler group, indicating that the long residence of the dust in the lungs had caused a very slight irritation. The leukaemias occurring in this experiment were at the rate expected in rats of this strain.

Discussion

The presence of any mineral dust retained in the lung tissue will cause some reaction. In this experiment this was recorded following exposure to all the dust used. It was only significant as a result of two of the exposures, but not in the animals exposed to kaolins. There was no evidence of nodular tissue reaction.

Conclusions

This experiment has demonstrated that prolonged exposure of rats to a high concentration of the kaolin dusts produces a minimal fibrotic reaction. No cancers of the lung or mesotheliomas occurred in any of the animals. The difference between the dusts being trivial.

Acknowledgments

This experiment is a summary of the report on Commission 260 undertaken for the British Health & Safety Executive. I wish to thank W.B. Jepson and P.D. Salt of English Clays Lovering & Pochin for producing the material and analysing the samples in this experiment.

The other kaolin inhalation study in which Cornish china clay was used was reported on by Wastiaux and Daniel from CERCHAR during this conference. They exposed 45 animals to a concentration of 300mg/m^3 respirable Cornish kaolin dust for 6 hours a day for 3 months. Animals were killed at month 3, 6, 12, 24 and 28. These animals appeared to develop a florid dust reticulation with foreign body granulomata. There was a focal interstitial fibrosis in the animals that survived for 2 years. In addition there was evidence of squamous metaplasia and hyperplasia. Three squamous carcinomas were observed in the nine animals which survived for 28 months. No nodular lesions as has been described in human cases, were observed.

These are the results of a preliminary study with the animals exposed to a very high level of respiratory dust which has no relevance to the human situation. It is hoped that this experiment will be repeated with controls, including animals which will receive a much smaller dose.

Apart from this most recent study the talc and kaolin dust used in these inhalation studies produced evidence of minimal fibrosis and no malignant tumours were observed in the animals exposed to relatively high dosage of dust with significant retention. There will always be a tissue reaction to such dust if retained in the lung tissue for long periods. It must be remembered that the results of these studies apply only to actual dust samples used in the experiments. These results should not be used in considering the biological effects of any other material, although it may be closely related.

REFERENCES

Wagner JC, Berry G, Hill RJ, Skidmore JW, Pooley FD (1979) An animal model for inhalation exposure to talc. In: Lemen R and Dement JM (eds) Dusts and Disease, Pathotox Publishers Inc, Illinois USA, 389-392.

Wagner JC, Griffiths DM, Munday DE (1987) Experimental studies with palygorskite dusts, Brit J Industr Med, 44:749-763.

Wehner AP, Stuart BO, Sanders CL (1979). Inhalation studies with Syrian golden hamsters. Prog Exp Tumour Res 24:177-198.

Wehner AP, Zwicker GM, Cannon WC, Watson CR, Carlton WW (1977) Inhalation of talc baby powder by hamsters. Food Cosmet Toxicol 15:121-129.

INTRAPERITONEAL INJECTION STUDIES FOR THE EVALUATION OF THE CARCINOGENICITY OF FIBROUS PHYLLOSILICATES

F. Pott[1], B. Bellmann[2], H. Muhle[2], K. Rödelsperger[3], R. M. Rippe[1], M. Roller[1], M. Rosenbruch[1]

[1]Medical Institute for Environmental Hygiene at the
Heinrich Heine University of Düsseldorf
Auf'm Hennekamp 50
D-4000 Düsseldorf

Summary

A review of the inhalation studies performed with crocidolite shows that, in the case of mineral fibres, realistic exposure can lead to irrelevant (false negative) results. The intraperitoneal test with mineral dusts in rats is much more sensitive and the qualitative analogy to the carcinogenicity found in the lung is good. Evaluation of the carcinogenicity of different fibre types according to their activity after i.p. injection may be justified. New measurements of fibres tested in earlier experiments indicate that at least 0.5 to 1 x 10^9 fibres longer than 5 µm should be injected for a negative result to become valid. About 10^9 fibres of chrysotile Calidria did not show a clear carcinogenic effect in contrast to 4 x 10^7 fibres of UICC chrysotile from Canada which induced tumours in about 30 % of the rats.

Why Injection Studies with Inhalable Fibres?

The use of natural and man-made fibrous materials has grown in many industries. Most of these materials can emit respirable fibres into the air.

---

[2]Fraunhofer Institute of Toxicology and Aerosol Research, Nikolai-Fuchs-Straße 1, D-3000 Hannover 61
[3]Institute for Occupational and Social Medicine of the Justus Liebig University of Gießen, Aulweg 129, D-6300 Gießen

NATO ASI Series, Vol. G 21
Health Related Effects of Phyllosilicates
Edited by J. Bignon

At least twenty years ago, questions were raised as to whether those fibres which do not belong to the asbestos group may cause cancer. Mineral fibres having a geometric diameter of less than about 5 µm can reach the bronchi. About 95 % of all lung cancer cases develop in the ciliated airways and only 5 % or less in the terminal airways (Greenberg, 1987; Noltenius, 1987). We, therefore, need suitable experimental models for detecting fibres which are probably carcinogenic in humans or probably non-carcinogenic in humans.

In view of the high lung cancer incidence in asbestos workers and the lung tumours in rats exposed to chrysotile, the thesis of Wehner et al. (1986) seems to be reasonable:

Thesis 1: Inhalation exposure is the method of choice if the potential risk to humans warrants the effort.

On the one hand, in the inhalation experiments with fibres from glass, rock, slag and ceramic described in eight publications, it was not possible to detect a clear carcinogenic effect of these materials. On the other hand, the administration of man-made and natural mineral fibres directly at the serosa tissue (pleura or peritoneum) resulted in dose-response relationships dependent on the number, length, diameter and durability of the fibres applied (review see IARC 1987, 1988). The working groups of Wagner and Stanton used the pleura whilst Davis and our group preferred the peritoneum. The intraperitoneal test is less troublesome, more sensitive and probably less painful than intrapleural administration.

The contrast between the findings after inhalation and those from the serosa tests evoked a second thesis:

Thesis 2: Non-physiologic routes of exposure lead to irrelevant (false positive) results.

These two theses may be correct for some substances, but do they indeed hold true in the case of respirable fibres? If the negative inhalation studies with fibres in experimental animals are also valid for humans, then mineral fibres which are proven to be a very strong carcinogen in humans should be a suitable substance for a reliable positive control group in inhalation studies. Crocidolite is regarded as a very strong carcinogen in humans; however, considerable tumour rates were reported in only one of the inhalation experiments (Wagner et al., 1974). This result could not be confirmed in two further studies by the same research group. Table 1 summarizes the inhalation studies with crocidolite. Conversely, intraperitoneal injection or intrapleural administration of crocidolite showed clearly the carcinogenicity of this amphibole asbestos (Pott et al., 1976; 1987; Stanton et al., 1981; Wagner et al., 1973). The carcinogenic effect of fibres in the pleural or peritoneal cavity after injection of the dust suspensions cannot be explained by an unspecific reaction to foreign bodies, as was described by Nothdurft (1955; 1961) and Oppenheimer et al. (1948), because these authors and others observed tumours after implantation of a single foreign body, the size of a coin (diameter about 20 mm). A carcinogenic effect was not detected when the same materials (glass, metal, synthetic materials) were applied in much smaller pieces or as a powder.

The findings after inhalation of crocidolite in experimental animals show unequivocally that false negative results have to be expected when the carcinogenic potency of a fibrous dust is lower than that of crocidolite.

Table 1. Inhalation experiments with crocidolite

Concentration/m^3 [mg]	fibers x 10^6	Time of exposure h/week	months	Species	Anim. with tum. no.	(mes)	%	Anim. exam.	References (first author only)
12.5		7	0.03	rat	7	(1)	16	43	Wagner 1974
12.6		7 x 5	3	rat	15	(1)	42	36	Wagner 1974
10.7		7 x 5	6	rat	4	(0)	22	18	Wagner 1974
10.6		7 x 5	12	rat	20	(2)	77	26	Wagner 1974
10.3		7 x 5	24	rat	13	(0)	72	18	Wagner 1974
	control			rat	7	(0)	6	126	Wagner 1974
50	1105	4 x 4	24	rat	5	(0)	8	66	Reeves 1974
13.5		7 x 5	35	baboons	0	(0)	0	10	Goldst. 1983
5	430	7 x 5	12	rat	3	(1)	7	43	Davis 1978
10	860	7 x 5	12	rat	1	(0)	3	40	Davis 1978
10	1630	7 x 5	12	rat	1	(0)	4	28	Wagner 1985
10		6 x 5	12	rat	1	(0)	4	28	Wagner 1987
7		6 x 5	24	rat	3	(1)	5	57	Smith 1987
7		6 x 5	24	hamster	0	(0)	0	58	Smith 1987
2.2*	162	5 x 4	12	rat	1	(0)	2	50	Muhle 1987

* Sample contained longer fibers than UICC crocidolite; other samples: UICC crocidolite, except Reeves.

Therefore, in contrast to the theses mentioned above, two antitheses can be put forward:

Antithesis 1: In the case of mineral fibres, realistic exposure leads to irrelevant (false negative) results.

Antithesis 2: Non-physiologic exposure leads to relevant results which are suitable for the classification of fibres according to their carcinogenicity.

The second antithesis becomes plausible in consideration of the analogy which exists between limited or sufficient evidence of the carcinogenicity of some substances in the lungs of humans and carcinogenicity in the peritoneal cavity of rats. Table 2 shows this analogy which is a good one and justifies a general interpretation: if a mineral dust induces tumours after intraperitoneal injection in rats, it is probably also carcinogenic in the lung of humans. This analogy primarily concerns to the carcinogenicity, in principle, and not the dose-response relationships.

After this discussion on the validity of the i.p. test, some experiments with phyllosilicates will be described. Although most of the carcinogenicity data have already been published, the combination of these results with the number of injected fibres provides a new basis for an interpretation.

Table 2. Carcinogenicity of hardly soluble dusts in the lung and at the serosa of humans and rats

	Lung		Pleura/Peritoneum	
	Human	Rat	Human	Rat
Asbestos	+	+	+	+
Erionite	+		+	+
Glass fibers		+		+
Other min. fibers	(+)			+
Nickel	+	+		+
Cadmium	(+)	+		+
Quartz	(+)	+		+
Several other dusts				-

Materials and Methods

The origin and the preparation of most of the dusts have been described in earlier publications which are quoted together with the results in Table 3. Sepiolite "Uicaluaro" was received from Tolsa S.A., Madrid, Spain. Chrysotile "Calidria" originates from the mine Idria in California, U.S.A. It was recommended for experimental use in a report by NIOSH (1979) because sufficient amounts of UICC chrysotile are no longer available.

One of the two samples [Calidria (1) in Fig. 1] applied in the intraperitoneal test was received as a flaky dust. It was characterized by NIOSH (1979) as follows: contaminants included mica (less than 2 %), quartz (less than 2 %), carbonates (less than 3 %), other unidentified minerals (less than 1 %) and hematite. Approximately 1 % of the sample consisted of metallic fragments, probably introduced during mining and processing of the crude asbestos ore. Fibre bundles ranged from 0.5 to 68 µm in width and from 1 to 300 µm in length. This material was milled in a knife mill for ten minutes. After this treatment, the distribution of the fibre lengths was similar to that of UICC chrysotile from Canada (see Figure 1). The second sample of the Calidria chrysotile [Calidria (2) in Fig. 1] was obtained from the Asbestos Institute, Neuss, F.R.G. The material was prepared for the production of asbestos paper and was rather short (see Figure 1). However, it has to be underlined that measurement of the particle size distribution of chrysotile is very difficult; the results cannot be considered to be reliable enough.

The names of the dusts and the calculated numbers of injected fibres longer than 5 µm are listed in Table 3. The data on the attapulgite and sepiolite samples have already been published (Rödelsperger et al., 1987), the data of Bellmann on chrysotile and actinolite partly (in Pott et al., 1989). It has to be noticed that Rödelsperger counted fibres with an aspect ratio > 3 : 1 whilst Bellmann counted fibres with an aspect ratio > 5 : 1. However, in the case of attapulgite and sepiolite, there is no large difference because these fibres are nearly all very slender.

Female Wistar rats (Wistar-WU/Kißlegg) from the breeding farm S. Ivanovas (Kißlegg, Allgäu, F.R.G.) were used. The method of animal keeping,

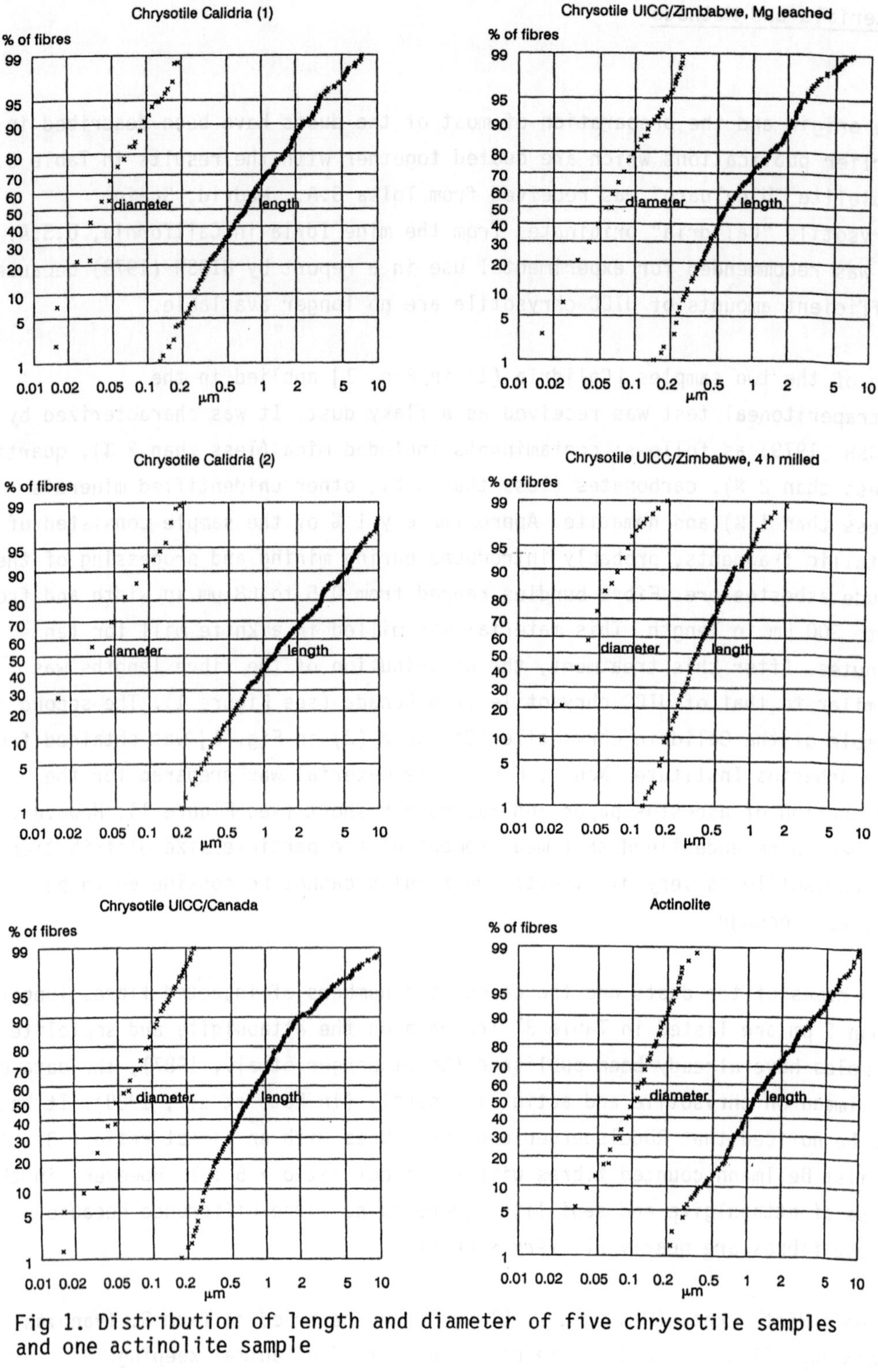

Fig 1. Distribution of length and diameter of five chrysotile samples and one actinolite sample

administration of the substances and examination are described in an earlier paper (Pott et al., 1987a).

Results

The tumour incidences observed in the abdominal cavity after i.p. injection of the used phyllosilicates and in some control groups are summarized in Table 3.

Table 3. Results of carcinogenicity studies with phyllosilicates in comparison to some other dusts (Pott et al.)

Substance	Mass i.p. [mg]	Fibre no.x 10^6 L > 5 µm per mg	i.p.	Rats with abd. tumours* [no.]	[%]	Median survival [weeks]	Pub-lished
Attapulgite Caceres	10	240	2400	12/ 30	40	109	1987
Attapulgite Georgia	60	0.61	37	4/112	3.6	108	1987
Attapulgite Lebrija	60	0.34	20	4/115	3.5	116	1987
Attapulgite Mormoiron	60	0.06	3.6	4/114	3.5	116	1987
Sepiolite Finland	10	5500	55000	24/ 36	67	62	
Sepiolite Uicaluaro	80	180	14000	2/ 32	6	112	this paper
Chrysotile Calidria (1)	1		940	3/ 34	9	126	
Chrysotile Calidria (2)	0.5		920	2/ 32	6.3	116	1987
Chrys. UICC/Zimb	6.3	(not measured)		27/ 35	77	~70	1976
Chrys. UICC/Zimb leached	6	147	880	0/ 38	0	78	1987
Chrys. UICC/Zimb milled	100	***		15/ 30	50		1972
Chrys. UICC/Zimb milled	100	***		12/ 37	32	~73	1976
Chrysotile UICC/Can	1	800	800	30/ 36	83	63	1989
Chrysotile UICC/Can	0.25**	800	200	23/ 34	67	90	1989
Chrysotile UICC/Can	0.05**	800	40	12/ 36	33	105	1989
Actinolite (amphibole)	0.25**	400	100	20/ 36	56	90	1989
Actinolite	0.05**	400	20	15/ 36	42	101	1989
Titanium dioxide	100	non-fibrous		2/ 53	4	109	1989
Titanium dioxide	90	non-fibrous		6/113	5.3	120	1987
Corundum	90	non-fibrous		4/115	3.5	112	1987
NaCl-solution	-	-	-	2/102	2	111	1989
NaCl-solution	-	-	-	4/204	2		1987

* sarcomas or mesotheliomas excluding tumours of the uterus
** very low doses, high tumour rates, no scars, only very small fibrosis
*** among 400 measured fibres no fibre was longer than 5 µm, see Fig. 1

Discussion

Figure 2 shows the carcinogenicity of some phyllosilicates in relation to the effect of many other fibrous dusts. According to these results, a tumour incidence of more than 30 % can be expected after intraperitoneal injection of 10^9 mineral fibres longer than 5 µm if the fibres are sufficiently durable. (But there exist long, thin fibres whose durability in the body is too low for them to be carcinogenic [Pott et al., 1987].)

From a practical point of view, it should be considered whether mineral fibres which do not show a clear carcinogenic effect (this means a tumour incidence of more than 10 to 15 % after i.p. injection of more than 10^9 fibres longer than 5 µm) can be exonerated from the suspicion of being carcinogenic. Such a convention would, for example, raise the question of whether chrysotile Calidria is a carcinogen. Moreover, sepiolite Uicaluaro would be evaluated as a substance unlikely to be carcinogenic in humans, but the sepiolite sample from Finland and the attapulgite sample from Caceres would be classified as carcinogens.

A further point of interest is the milled UICC chrysotile from Zimbabwe. The sample did not contain any fibres longer than 5 µm in 400 fibres counted and resulted in tumour incidences of 32 and 50 % in two experiments. However, as already discussed earlier (Pott, 1978), the number of fibres injected in 100 mg dust was extremely high because most of the fibres were very short and thin. On the basis of the recent measurement, it can be calculated that 1.5 x 10^9 fibres longer than 5 µm would have been applied if just one in 10,000 fibres was 5.5 µm in length and 0.25 µm in diameter. Nevertheless, this calculation cannot exclude a carcinogenicity of fibres which are shorter than 5 µm.

Acknowledgements

This work was supported by the Bundesanstalt für Arbeitsschutz (Federal Institute for Occupational Savety and Health), Dortmund, F.R.G.

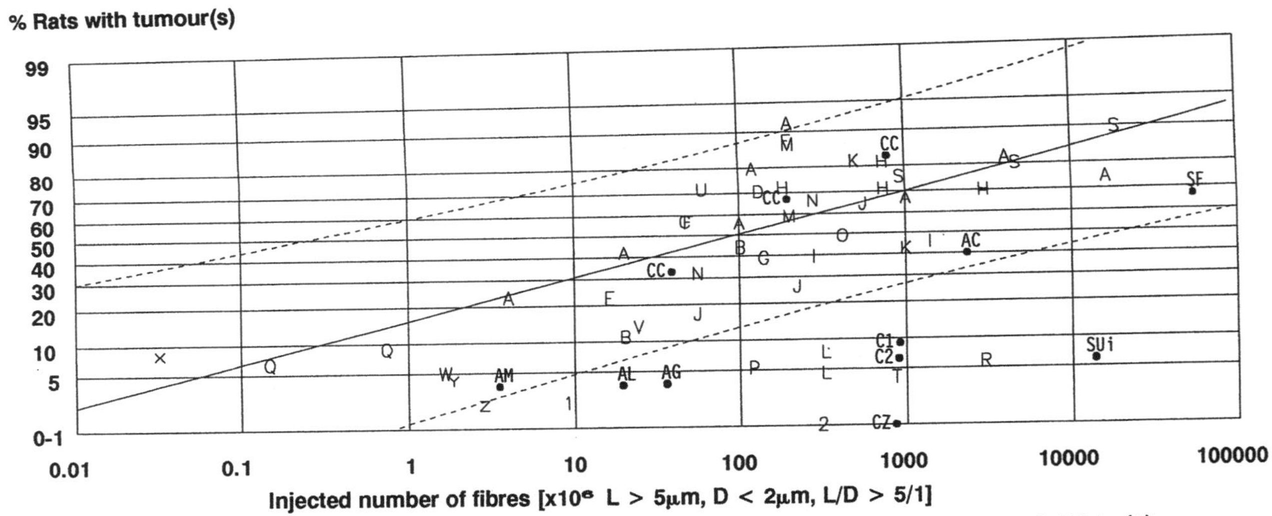

Phyllosilicates

CZ=Chrys. UICC/Zimb leached
C2=Chrys. Calidria (2)
AC=Attapulgite Caceres
AM=Attapulgite Mormoiron
CC=Chrys. UICC/Can
SF=Sepiolite Finland
AG=Attapulgite Georgia
C1=Chrys. Calidria (1)
SUi=Sepiolite Uicaluaro
AL=Attapulgite Lebrija

Other materials

A Actinolite
B Anthophyllite
C Basalt G+H
D Ceramic Fiberfrax
E Ceramic Manville
F Crocidolite SA
G Crocidolite UICC
H Erionit Turkey
I Glass 100 Pen
J Glass 104/475
K Glass 104/E1
L Glass 104/E1 HCl
M Glass 104/E1 NaOH
N Glass 104/E2
O Glass 104/E3a
P Glass 106
Q Glass filam. 3
R Kevlar
S Nemalite
T Polypropylene
U Rock S
V Rock S fine
W Slag Ko fine
X Slag Rh coarse
Y Slag Rh fine
Z Slag Zi coarse
1 Slag Zi fine
2 Wollastonite

Fig 2. Dose-response relationship of phyllosilicates in relation to the effect of many other fibrous dusts after intraperitoneal injection. (The animal experiments with the non-phyllosilicates are described by Pott et al., 1987 and 1989)

References

Davis JMG, Addison J, Bolton RE, Donaldson K, Jones AD (1978) Inhalation and injection studies in rats using dust samples from chrysotile asbestos prepared by a wet dispersion process. Br J exp Path 67:113-129

Goldstein B, Rendall REG, Webster I (1983) A comparison of the effects of exposure of baboons to crocidolite and fibrous-glass dusts. Environ Res 32:344-359

Greenberg SD (1987) Carcinomas of the peripheral airways. In: McDowell EM (ed) Lung carcinomas, Churchill Livingstone, Edingburgh London Melbourne New York, p 286

IARC (ed)(1987) IARC Monographs on the evaluation of the carcinogenic risk of chemicals to humans, Vol. 42, Silica and some silicates. World Health Organization International Agency for Research on Cancer, Lyon

IARC (ed)(1988) IARC Monographs on the evaluation of the carcinogenic risk of chemicals to humans, Vol. 43, Man-made mineral fibres and radon. World Health Organization International Agency for Research on Cancer, Lyon

Muhle H, Pott F, Bellmann B, Takenaka S, Ziem U (1987) Inhalation and injection experiments in rats to test the carcinogenicity of MMMF. Ann occup Hyg 31:755-764

NIOSH (ed)(1979) Publication No 79-139 Preparation and characterization of analytical reference minerals. US Department of Health, Education and Welfare, Cincinnati, USA

Noltenius H (1987) Tumoren der Lungen und Pleura. In: Noltenius H, Tumor Handbuch: Pathologie u. Klinik d. menschl. Tumoren, Bd 2. Urban & Schwarzenberg, München Wien Baltimore, S 691

Nothdurft H (1955) Über die Sarkomauslösung durch Fremdkörperimplantationen bei Ratten in Abhängigkeit von der Form der Implantate. Naturwissenschaften 42:106

Nothdurft H (1961) Sarkomerzeugung bei Ratten durch implantierte Fremdkörper. Therapie des Monats No 8:262-274

Oppenheimer BS, Oppenheimer ET, Stout AP (1948) Sarcomas induced in rats by implanting cellophane. Proc Soc Exper Biol & Med 67:33-34

Pott F (1978) Some aspects on the dosimetry of the carcinogenic potency of asbestos and other fibrous dusts. Staub-Reinhalt. Luft 38:486-490

Pott F (1989) The carcinogenicity of fibres in experimental animals - Data and evaluation. In: Mohr U (ed) Assessment of inhalation hazards: Integration and extrapolation using diverse data. February 19-24 1989, in Hannover, West Germany. Springer, Berlin Heidelberg New York London Paris Tokyo (in press)

Pott F, Huth F, Friedrichs KH (1972) Tumoren der Ratte nach i.p. Injektion von gemahlenem Chrysotil und Benzo(a)pyren. Zbl Bakt Hyg I Abt Orig B 155:463-469

Pott F. Friedrichs KH, Huth F (1976) Ergebnisse aus Tierversuchen zur kanzerogenen Wirkung faserförmiger Stäube und ihre Deutung im Hinblick auf die Tumorentstehung beim Menschen. Zbl Bakt Hyg I Abt Orig B 162:467-505

Pott F, Ziem U, Reiffer FJ, Huth F, Ernst H, Mohr U (1987) Carcinogenicity studies on fibres, metal compounds, and some other dusts in rats. Exp Pathol (Jena) 32:129-152

Pott F, Roller M, Ziem U, Reiffer FJ, Bellmann B, Rosenbruch M, Huth F (1989) Carcinogenicity studies on natural and man-made fibres with the intraperitoneal test in rats. In: Bignon J, Peto J, Saracci R (eds) Non-occupational exposure to mineral fibres. Intern. Agency for Research on Cancer, Lyon (IARC Sci Publ No 90, p 173)

Reeves AL, Puro HE, Smith RG (1974) Inhalation carcinogenesis from various forms of asbestos. Environ Res 8:178-202

Rödelsperger K, Brückel B, Manke J, Woitowitz HJ, Pott F (1987) Potential health risks from the use of fibrous mineral absorption granulates. Br J industr Med 44:337-343

Smith DM, Ortiz LW, Archuleta RF, Johnson NF (1987) Long-term health effects in hamsters and rats exposed chronically to man-made vitreous fibers. Ann occup Hyg 31:731-754

Stanton MF, Layard M, Tegeris A, Miller E, May M, Morgan E, Smith A (1981) Relation of particle dimension to carcinogenicity in amphibole asbestoses and other fibrous minerals. J natl Cancer Inst 67:965-975

Wagner JC, Berry LG, Timbrell V (1973) Mesotheliomata in rats after inoculation with asbestos and other materials. Br J Cancer 28:173-185

Wagner JC, Berry G, Skidmore W, Timbrell V (1974) The effects of inhaltion of asbestos in rats. Brit J Cancer 29:252-269

Wagner JC, Skidmore JW, Hill RJ, Griffiths DM (1985) Erionite exposure and mesotheliomas in rats. Br J Cancer 51:727-730

Wagner JC, Griffiths DM, Munday DE (1987) Experimental studies with palygorskite dusts. Br J Ind Med 44:749-753

Wehner AP, Dagle GE, Clark ML, Buschbom RL (1986) Lung changes in rats following inhalation exposure to volcanic ash for two years. Environ Res 40:499-517

REVIEW ON SHORT-TERM TESTS USED FOR THE TOXICOLOGICAL EVALUATION OF PHYLLOSILICATES

M.C. Jaurand and J.C. Pairon
INSERM Unité 139
CHU Henri Mondor
94010. Créteil, Cedex, France

1 - Introduction

During the last 20 years, several in vitro studies have been carried out to determine the cytotoxicity of phyllosilicates. Results are summarized in this paper to state the present knowledge on the potential toxicity of these minerals. We recorded data on kaolin, talc, bentonite and montmorillonite ; attapulgite, palygorskite and sepiolite ; finely antigorite, micas and slate dusts.

In vitro studies have been performed with different models mainly red blood cells (RBC) and macrophages. However, other cell types have been used such as, P388 D1, rat tracheal epithelial cells, rat pleural mesothelial cells or fibroblasts. In addition, liposomes, a model of membranes have been also used in some experiments.

Different end-points have been determined. Haemolysis of red blood cells has been often measured as a consequence of cell killing. Composition of red blood cell membranes has also been analyzed after treatment with particles. Macrophages either peritoneal or alveolar and stimulated or unstimulated have been also largely used. In a normal situation macrophages produce metabolites such as lactic acid ; molecules such as 2, 3, 5-triphenyltetrazolium chloride (TTC) are reduced by cell metabolism. Impairement of the macrophage metabolism has been studied following treatment with phyllosilicates ; end-points were measurement of lactic acid production or of TTC reduction. A decrease was taken as a sign of macrophage injury. Maintenance of functionallity of macrophage membrane was also studied to detect macrophage injury. An increase in membrane permeability due to membrane damage has been detected using 2 end-points : measurement of lactic dehydrogenase (LDH) in the extracellular medium and numeration cells stained in the presence of vital stain (trypan blue). P388 D1 a "macrophage-like" cell line has been used in some experiments. Other parameters such as membrane potential measurement and transport of amino-acids have been studied with these cell types.

Proliferative cells have been also tested to assess the toxicity of phyllosilicates. A large range of cell lines have been used. Several end-points have been

NATO ASI Series, Vol. G 21
Health Related Effects of Phyllosilicates
Edited by J. Bignon

determined to assess effects on cell growth : cell number measurement or amount of thymidine incorporated into DNA. Genotoxic potency has been evaluated using tests of DNA repair and of sister chromatid exchanges (SCEs). In addition, several ultrastructural studies have been performed to determine if phyllosilicates were ingested by cultured cells.
Results will be reported according to the type of minerals tested.

2 - In vitro studies carried out with kaolin

2.1 In vitro effects of kaolin using red blood cells

Numerous experiments have been carried out to study if RBC damage resulted from incubation of RBC with kaolin (Brown et al. 1980 ; Daniel and Le Bouffant, 1980 ; Depasse and Leonis 1974 ; Gormley et al. 1978 ; Le Bouffant et al. 1980 ; Manyai et al. 1969 ; Oscarson et al. 1981, 1986 ; Ottery and Gormley, 1978 ; Woodworth et al. 1982). These authors reported an haemolytic potency of kaolin. Membrane damage was confirmed using liposomes which released intravesicular chromium following treatment with kaolin (Woodworth et al. 1982).

It has been observed that pretreatment of kaolin with different molecules protected against haemolysis. Depasse and Leonis (1974) reported an inhibition of sheep erythrocytes haemolysis following pretreatment with lyzozyme. Oscarson et al (1981) showed that coating of positively charged aluminium-hydroxypolymers significantly reduced haemolysis. This indicates that the negative charges at the surface of the mineral might be important in RBC-particle interaction. Coating with lecithin resulted in a complete inhibition of kaolin hameolysis as demonstrated by Wallace et al.(1985). It is interesting to note that these authors demonstrated a restoration of haemolysis after treatment of the coated particles with phospholipase A2 (Wallace et al. 1988b). Using poly-2-vinyl pyridine-N-oxide, a polymer forming hydrogen bonds, Oscarson et al. (1981) reported that haemolysis of bovine RBC was reduced after pretreatment when compared to untreated particles. The protective effect seems likely due to adsorption of the molecules to the particle surface as demonstrated by in vitro studies using acellular systems.

2.2 Effects of kaolin using acellular systems

In vitro studies are reported on Table I. They have shown that kaolin adsorbed positively-charged molecules such as methylene blue (Timar et al. 1979). Other molecules such as NO and NO_2 have also affinity for kaolin (Robertson et al. 1982). A reduction of cytotoxicity towards P388 D1 cells after adsorption was observed ; however, the authors noted that the results do not rule out the possibility of synergistic effects with oxides of nitrogen and respirable dusts arising from desorption of nitrogen oxides. Adsorption of dipalmitoyl phosphatidyl chroline (DPPC) is of particular interest since DPPC is a component of pulmonary surfactant. Wallace et al. (1975, 1985) reported an adsorption of DPPC on kaolin. This resulted in an abolition of cytotoxicity when particles were pretreated at maximal surface coverage. In these experiments cytotoxicity was determined by haemolysis of sheep erythrocytes and enzyme release from rat alveolar macrophages. Interactions between DPPC and kaolin might impair the lung function since DPPC plays an important role in the maintenance of surface tension of the alveoli. Ma and Weber (1986) reported that sprinkling of kaolin onto the DPPC surface and onto rat lung surfactant resulted in increased minimum surface tensions and decreased hysteresis areas. This indicates that particles might prevent surface tension to reach low values. The adsorption of the surface active materials to the particles is probably responsible for these effects (Ma and Weber, 1986).

In vitro studies using acellular systems have also been performed to assess production of radical molecules by kaolin. Kennedy (1987) reported a production of OH° when the reactive medium contained cystein.

Rahman et al (1973) studied the solubility of kaolin in physiological buffer or in biological fluids. Aluminium and silicium were released in the incubation media.

2.3 Effects of kaolin on other cell systems

Numerous in vitro studies performed with cultured cells have used macrophages (Table II). Davies et al. (1984) demonstrated that kaolin was ingested by mouse peritoneal macrophages. Other studies have generally shown a cell damage (Adamis and Timar, 1976 ; Brown et al. 1980 ; Davies, 1983 ; Davies et al. 1984 ; Le Bouffant et al. 1980 ; Timar et al. 1979 ; Wallace et al. 1985). However Styles and Wilson (1973) treated suspensions of peritoneal and alveolar rat macrophages with kaolin and these authors did not observe cytotoxicity using trypan blue exclusion as an indicator of cell death. This lack of effect does not seem related to an absence of particle uptake since the authors noted that particles were associated with cells ; however it might be related to the short time of incubation of 2 hours. In other studies particles were incubated with cells for at least 18 h, except Wallace et al. (1985) who incubated rat alveolar macrophages for 2 hours. Difference might be due to other parameters such as origin of the kaolin used in these different experiments (Table II) or to the performance of the in vitro tests.

As observed with red blood cells, cytotoxicity of kaolin to macrophages seems likely related to its sorptive properties. Davies (1983) and Davies et al. (1984) have studied the activity of kaolin towards macrophages after treatment of kaolin with either poly-2-vinyl pyridine-N-oxide (PVPNO), polyacrylic acid (PAA or aluminium chloride ($AlCl_3$). These authors reported that PVPNO substantially reduced cytotoxicity of kaolin and that it was not related to an inhibition of uptake of kaolin particles. PAA pretreatment of kaolin resulted in a significant but small reduction of cytotoxicity and aluminium chloride treatment induced little modification of kaolin activity. The authors concluded that positive charges at the edges of kaolin play a minor role in cytotoxicity. The role of negative charges at the surface of the particles was difficult to assess since $AlCl_3$ did not abolish the negative net charge of the particles. It was suggested that cytotoxicity was related to the amorphous silica-rich gel coating the particles. However, results indicate that surface interactions are of importance since PVPNO had a strong inhibitory effect. In vitro damage has been also detected by Low et al. (1980) who reported that amino-acid transport was modified and amino-acid incorporation into proteins was decreased after treatment of rabbit alveolar macrophages with kaolinite.

Cytotoxicity was also observed with macrophage-like cells, P 388 D1 (Table III). Gormley and Addison (1983) concluded to a low cytotoxicity when using trypan blue exclusion and Robertson et al. (1982) reported a decreased cell viability and a release of enzymes. This cytotoxicity was reduced after particle treatment with nitrogen oxides as discussed above. Cytotoxicity on hamster tracheal cells has been also reported (Woodworth et al. 1982). In this paper comparison between activity of kaolin, montmorillonite and talc indicated that kaolin was intermediate between the other samples (Table III). In one study it was shown that neutrophils or monocytes treated with a well-ordered form of kaolinite produced chemiluminescence which is related to formation of oxygen species (Gormley et al. 1985). Comparison with the effect of other pathogenic dusts indicated that there was no correlation between cytotoxicity, as determined by trypan blue exclusion, and chemiluminescence levels. The authors suggest that production of oxygen species was not related to pathogenicity.

2.4. Summary of results on in vitro effects of kaolin

About 30 studies were reviewed. It appears that kaolin particles are ingested by macrophages and probably by other cells capable of phagocytosis. There is an interaction with cells which results in a membrane alteration and a metabolism impairement. Kaolin has affinity for several molecules covering of particles by adsorption with certain molecules (DPPC, NO, NO_2, PVPNO) decreases cell damage indicating that surface properties are important in cell-particle interactions. This coating might play a role in the in vivo cytotoxicity of kaolin.

No data on DNA damage or on effect of kaolin on cell multiplication was recorded.

Using acellular systems it has been reported that kaolin may release some constituents (Si, Al) and, under certain conditions, may produce hydroxyl radicals (see also Kennedy et al. this meeting).

3 - In vitro studies carried out with talc : Effects of talc using cells systems

A particle uptake has been reported when lung fibroblasts (Henderson et al. 1975) or rat pleural mesothelial cells (Endo-Capron et al., this meeting) were incubated with talc (Table IV).

Haemolysis was either not detectable or very low (Brown et al. 1980 ; Schnitzer and Pundsack 1970). Woodworth et al. (1982) reported that haemolysis was low and much lower than that observed with kaolinite or montmorillonite, on a per weight basis.

Styles and Wilson (1973) and Brown et al. (1980) reported an absence of effect on macrophages. Davies et al. (1983) compared the activity of 7 different talc samples for their ability to induce enzyme release from mouse peritoneal macrophages. The authors concluded to a "modest but consistent macrophage cytotoxicity" (Table IV). In those experiments there was no great differences between the different talc samples and no clear relationship between their cytotoxicity and either the size distribution or the content in chlorite and quartz.

Using replicative cells, no effect (Brown et al. 1980) or a low cytotoxicity was observed with V 79 cells (Chamberlain and Brown, 1978, Pigott and Pinto 1983) (Table V). LD 50 (dose giving 50 % reduction in cloning efficiency) was greater than 50 µg/ml, about of the same order as with quartz. When hamster tracheal epithelial cells were treated with talc, cytotoxicity determined with chromium release was low ; it was lower than that observed with montmorillonite. An effect was observed with 300 µg/ml of talc. (Woodworth et al. 1982).

In vitro experiments carried out with liposomes confirmed a low interaction between talc and membranes (Woodworth et al. 1982) (Table VI). Studies of sister chromatid exchanges (SCEs) in talc-treated rat pleural mesothelial cells failed to detect enhancement of SCEs (Table V). An inhibition of cell growth was observed but without cell death (Endo-Capron et al. this meeting).

From these results it can be suggested that interaction between cells and talc produce some low level of cell damage, when the cytotoxicity was compared to other particles. However, it must be noted that comparison are made on a per weight basis and not a particle basis.

4. In vitro studies carried out with bentonite or with montmorillonite

4.1 In vitro effects of bentonite using cell systems. Bentonite particles seem highly active on red blood cells. Haemolysis has been reported (Ottery and Gormley 1978, Gormley et al. 1978) as well as cytotoxicity on macrophages

(Table VII). This latter effect was detected using reduction of 2, 3, 5-triphenyltetrazolium chloride (TTC) (Adamis and Timar 1976, 1978), modification of membrane potential (Gormley et al. 1978) stain exclusion or decrease in ATP formation (Hatch et al. 1985). However it must be noted that in this latter experiment the amount of bentonite used was high (1 mg/ml/ 106 cells).

4.2 <u>In vitro effects of montmorillonite using red blood cells and macrophages</u>

Montmorillonite was reported to be highly active on the different cell systems used (Table VII). Montmorillonite damaged red blood cells at concentrations lower than 10 µg/ml (Woodworth et al. 1982, Oscarson et al. 1986). Oscarson et al. (1981, 1986) reported an inhibition of haemolysis following pretreatment of montmorillonite with PVPNO or with positively charged molecules. This suggests, as in the case of kaolinite, that surface interactions are predominant mode of interactions between particles and red blood cell membranes.

Using macrophages (Daniel and Le Bouffant, 1980) and macrophage-like cells (Gormley and Addison, 1983) a cytotoxicity was detected. With the samples tested in those experiments, cytotoxicity was higher than that observed with 2 samples of kaolin. Gormley and Addison (1983) studied cell viability according to surface area ; they observed a good inverse correlation between viability and surface area, as measured by N_2 adsorption. Cell viability decreased when surface area increased. Moreover, the authors noted an absence of correlation between cation exchange capacity and viability.

Uptake of montmorillonite by rabbit macrophages has been studied by Hahn et al. (1974). Montmorillonite was followed by ^{169}Yb which was incorporated into the particles by cation exchange. It was found that small particles (0.32 µm) were little phagocytosed while larger particles (0.97 to 2.02 µm in diameter) were similarly phagocytosed.

4.3 <u>In vitro effects of montmorillonite using other cell systems (Table VIII)</u>

A chemiluminiscence was reported following treatment of neutrophils or monocytes with montmorillonite (Gormley et al. 1985). Using hamster tracheal cells, Woodworth et al. (1982) found that montmorillonite was the most cytotoxic sample, as detected by chromium release, when compared to one sample of kaolin and of talc.

4.4 <u>Effects of bentonite or montmorillonite using acellular systems (Table IX).</u>

Adsorption of several molecules, methylene blue, paraquat, PVPNO, on montmorillonite has been reported (Adamis and Timar, 1976 ; Oscarson et al. 1981). This adsorption explains the protective effect of these molecules on haemolysis as described above.

Studies with liposomes confirmed that montmorillonite particles damage membranes (Woodworth et al. 1982).

Some papers will be quoted here besides they are not directly related to cytotoxicity on mammalian cells. Those however bring aditional informations on the interactions between montmorillonite and living organisms and on the physico-chemical properties of the mineral particles relevant to biological effects. These papers refer to metabolic studies of *histoplasma capsulatum*, a fongi highly pathogen to human (Lavie and Stotzley, 1986 a,b). It was found that metabolism was reduced by montmorillonite and that these reduction resulted from adsorption of iron-binding compounds produced by fungi to accumulate iron necessary for their normal metabolism.

4.5 Summary data on bentonite and montmorillonite.

From these experiments it seems that bentonite and montmorillonite are able to damage cells in culture and that surface properties are important in cell-particle interactions. Adsorption of different molecules might have antagonist effects on cell damage as observed with kaolin.

No report on effect of these minerals on cell growth or on DNA was recorded.

5 - In vitro effects of attapulgite, palygorskite or sepiolite.

5.1 In vitro effects using red blood cells and macrophages (Table X).

Attapulgite, palygorskite or sepiolite are clay fibres belonging to the same family. Therefore, their effects have been gathered.

Two electron microscopic studies have shown that attapulgite, and sepiolite can be ingested by rat cells (Denizeau et al. 1985a, Achard et al. 1987).

Studies carried out with red blood cells have shown that the samples studied were haemolytic (Harvey et al. 1984 ; Jaurand et al. 1979 ; Koshi et al. 1968 ; Nadeau et al. 1987 ; Oscarson et al. 1986 ; Perderiset et al. 1989 ; Schnitzer and Pundsack. 1970 ; Wright et al. 1980).

A protective effect was observed when these minerals were pretreated with positively charged polymers or PVPNO (Oscarson et al. 1981, 1986) and with albumin or DPPC (Perderiset et al. 1989). This indicate that the surface reactivity is important in cell-particle interactions.

Chamberlain et al. (1982) studied cytotoxicity of attapulgite and sepiolite of

6 - In vitro data obtained with other phyllosilicates : micas, illite, antigorite

6.1 In vitro studies carried out with micas One sample of biotite, phlogopite and muscovite were tested on V 79 cells (Pigott and Pinto, 1983). All samples showed a similar minimal cytotoxicity with a CD50 (dose giving 50% cell death) higher than 100 µg/ml. Cytotoxicity and haemolytic effect of one sample of muscovite were low and null respectively (Daniel and Le Bouffant, 1980). Hahon and Booth (1987) studied production of interferon by influenza virus in Rhesus monkey kidney cell monolayers. These authors reported that phyllosilicates (muscovite, pyrophyllite) showed inhibitory activity for the induction process. This effect was abolished following pretreatment of muscovite with PVPNO. Using acellular system, Rahman et al. (1973) reported that silica and aluminium were solubilized in physiological buffer and in biological media. One can wonder if this might explain the lower fibrogenicity of mixed dusts containing micas when compared to silica quartz alone. Indeed, aluminium has been shown to reduce the fibrogenic potency of quartz (Le Bouffant, 1985)

6.2 In vitro studies carried out with illite
Red blood cells were damaged by illite as reported by Le Bouffant et al. (1980) and Manyai et al.(1969). In addition, a cytotoxicity on macrophages and an adsorption of methylene blue has been shown (Timar et al. 1979).

6.3 In vitro studies carried out with antigorite
Two studies have been recorded. Koshi et al. (1969) studied the haemolytic activity of one sample of antigorite submitted to heating. Haemolytic activity was not detected when specimen was heated below 400°C and above 850°C ; a slight haemolysis was found when specimen was heated at 650 or 750°C. Using hamster tracheal organ cultures, Woodworth et al. (1983) failed to detect metaplasia after treatment with one sample of antigorite.

6.4 In vitro studies carried out with slate dusts
Several studies have been carried out with slate dusts from India (Singh et al. 1983 ; 1984 ; 1987). These particles were found to be haemolytic. There was a modification of the erythrocyte membrane composition characterized by a decrease in phospholipids and glucosamine content. A lipid peroxidation was also detected. Slate dust haemolysis was inhibited by polyvinyl pyrrolidone, bovine serum albumin, plasma and by some fractions from lung lavage. Superoxide dismutase and catalase did not modified haemolysis while inhibiting lipid peroxidation. This indicate that lipid peroxidation and haemolysis are independent processes. The authors suggest that haemolysis is related to particle solubilisation.

Table I : IN VITRO EFFECTS OF KAOLIN (Kao)

PARTICLE ORIGIN	IN VITRO SYSTEM	RESULT	REFERENCE	
India	dissolution	Si, Al	Rahman et al.	1973
~ 90% Kao	methylene blue adsorption	yes	Timar et al .	1979
Colloidal Fisher	Surface tension surfactant	increased	Ma and Weber	1986
?	NO, NO_2 adsorption	yes	Robertson et al.	1982
Georgia	DPPC adsorption	yes	Wallace et al.	1975, 85
Georgia	Cr release from liposomes	yes	Woodwroth et al.	1982
?	OH° production from H_2O_2	yes	Kennedy et al.	1987

TABLE II : IN VITRO EFFECTS OF KAOLIN (Kao) ON CELL SYSTEMS

MINERAL ORIGIN	CELL SYSTEM	END POINT	EFFECT	REFERENCE	
Cornish	macrophages	ultrastructure	uptake	Davies et al.	1984
?	macrophages	stain exclusion	no effect	Styles and Wilson	1973
?	"	enzyme release	moderate	Le Bouffant et al.	1980
Georgia	"	"	yes. Protection by DPPC	Wallace et al.	1985
?	"	"	yes	Brown et al.	1980
Cornwall 98% Kao	"	"	yes. Protection by PVPNO (±$AlCl_3$)	Davies Davies et al.	1983 1984
?	"	TTC reduction	decreased	Adamis and Timar	1976
98% Kao, 10% mica	"	"	"	Timar et al.	1979
Georgia	"	amino-acid transport " incorporation	modified decreased	Low et al.	1980

TABLE III : IN VITRO EFFECTS OF KAOLIN (Kao) ON CELL SYSTEMS

MINERAL ORIGIN	CELL SYSTEM	END POINT	EFFECT	REFERENCE	
Georgia	P 388 D1	stain exclusion	slightly modified	Gormley and Addison	1983
?	"	" enzyme release	decreased yes ↓ N0 NO_2 ads	Robertson et al.	1982
natural	neutrophils monocytes	chemiluminescence "	yes "	Gormley et al.	1985
Georgia	hamster tracheal	Cr release	yes Mont > Kao > Talc	Woodworth et al.	1982

Mont : montmorillonite

TABLE IV : IN VITRO EFFECTS OF TALC ON CELL SYSTEMS

ORIGIN OF TALC	CELL SYSTEM	END POINT	EFFECT	REFERENCE	
Italian 000 00	lung fibroblasts	ultrastructure	particle uptake	Henderson et al.	1975
European	rat pl. mesothelial	"	"	Endo-Capron et al.	1989
?	red blood cells	haemolysis	no effect	Schnitzer and Pundsack	1970
Cyprus, USA	" "		low < kao < Mont*	Woodworth et al.	1982
?	" "		no effect	Brown et al.	1980
?	macrophages	strain exclusion	5% killed cells	Styles and Wilson	1973
?	?	enzyme release	no effect	Brown et al.	1980
Indian finex	mouse peritoneal	"	"modest but	Davies et al.	1983
Italian 000 00	macrophages	"	consistent"		
micro "	"	"			
Spanish SS	"	"	"		
Australian	"	"	"		
French OXO	"	"	"		
Chinese	"	"	"		

* Mont = montmorillonite

TABLE V : IN VITRO EFFECTS OF TALC ON CELL SYSTEMS

ORIGIN OF TALC	CELL SYSTEM	END POINT	EFFECT	REFERENCE	
Italian	V 79	colony formation	LD50 > 50 µg/ml	Chamberlain and Brown	1978
Lab reagent	V 79	"	LD50 > 50 µg/ml	Pigott and Pinto	1983
?	V79-A549	"	no effect	Brown et al.	1980
Cyprus, USA	Hamster tracheal	Cr release	low <Mont	Woodworth et al.	1982
European	Rat pl. mesothelial	growth	inhibition	Endo-Capron et al.	1989
"	"	SCEs	no effect		

TABLE VI : IN VITRO STUDIES CARRIED OUT WITH TALC

ORIGIN OF TALC	IN VITRO SYSTEM	RESULT	REFERENCE	
India	Solubility in physiological media	leaching Si, Mg	Rahman et al	1973
	Adsorption of biological molecules	ads. albumin		
Cyprus, USA	liposomes, Cr release	low < Kao < Mont	Woodworth et al.	1982

TABLE VII : IN VITRO EFFECTS OF BENTONITE (Bent.), MONTMORILLONITE (Mont.) ON CELL SYSTEMS

MINERAL	CELL SYSTEM	END POINT	EFFECT	REFERENCE
Bent	red blood cells	haemolysis	Bent > Kao	Ottery and Gormley 1978
"	red blood cells	haemolysis	Yes ++	Gormley et al. 1978
"	macrophages P 388 D1	membrane potential "	decreased "	
"	macrophages	TTC reduction	decreased	Adamis and Timar 1976
Bent.	macrophages	TTC reduction	decreased	Adamis and Timar 1976,78
75 % Mont.	"	enzyme release	no effect	
Bent.	"	strain exclusion ATP content	decreased "	Hatch et al. 1985
Mont.	"	uptake (^{169}Yb)	no 0.32 um yes 0.97 to 2.02 µm	Hahn et al. 1974
"	red blood cells	haemolysis	Mont >Kao* > talc	Woodworth et al. 1982
"	"	"	Mont >Paly *> Kao inhibition by PVPNO polymers +	Oscarson et al. 1986
"	"		yes	Daniel and Le Bouffant 1980
"	macrophages	enzyme release	yes	
"	P 388 D1	strain exclusion	↓ Mont >Att *>Kao	Gormley and Addison 1983

* Att : attapulgite ; Kao : kaolin ; Paly : palygorskite

TABLE VIII : IN VITRO EFFECTS OF MONTMORILLONITE (Mont.) ON CELL SYSTEMS

MINERAL	CELL SYSTEM	END POINT	EFFECT	REFERENCE
Mont	neutrophils monocytes	chemiluminescence "	yes,not related to pathogenicity	Gormley et al. 1985
"	Hamster tracheal	Cr release	Mont >Kao>talc	Woodworth et al. 1982

TABLE IX : IN VITRO STUDIES CARRIED OUT WITH BENTONITE OR MONTMORILLONITE

ORIGIN OF TALC	IN VITRO SYSTEM		RESULT	REFERENCE
Bentonite	methylene blue paraquat	adsorption "	yes yes	Adamis and Timar 1976
Montmorillonite	PVPNO	adsorption	yes	Oscarson et al. 1981
Montmorillonite	liposomes	Cr release	yes	Woodworth et al. 1982

TABLE X : IN VITRO EFFECTS OF ATTAPULGITE (A), PALYGORSKITE (P) OR SEPIOLITE (S) ON CELL SYSTEMS

MINERAL ORIGIN	CELL SYSTEM	END POINT	EFFECT	REFERENCE	
A, S	Hepatocyte	Ultrastructure	Particle uptake	Denizeau et al.	1985a
A	rat pleural mesothelial	"	"	Achard et al.	1987
P.S.	red blood cells	haemolysis	yes	Schnitzer and Pundsack	1970
A	"	"	yes.Protection by albumin,DPPC	Perderiset et al.	1989
"	"	"	mont > P.S.> kao Inhib. PVPNO polymers+	Oscarson et al.	1981,86
A,P,S	red blood cells macrophages	haemolysis enzyme release	yes yes	Koshi et al.	1968
A, France	idem	idem	idem	Jaurand et al.	1979
A	"	"	"	Harvey et al.	1984
A Georgia	"	"	"	Nadeau et al.	1987
S, commercial	red blood cells	haemolysis	yes	Wright et al.	1980
A, S, (s) Short/long (l)	mouse per. macro A 549 V 79-4	enzyme release giant cell formation colony formation	l+, s ± l+, s- l+, s-	Chamberlain et al.	1982
A, American, French	P 388 D1	growth	no effect	Lewis and Lipkin	1985

TABLE XI : IN VITRO EFFECTS OF ATTAPULGITE (A), PALYGORSKITE (P) OR SEPIOLITE (S) ON CELL SYSTEMS

MINERAL	CELL SYSTEM	END POINT	EFFECT	REFERENCE	
A, Georgia	I - 407	colony formation	low effect	Reiss et al.	1980
A, Nevada	Hamster tracheal	metaplasia	low effect*	Woodworth et al.	1983
A, France	rat pl . mesothelial	growth	no effect	Jaurand et al.	1987
A, S,	hepatocytes	DNA repair	no effect	Denizeau et al.	1985b
A, Senegal	rat pl. mesothelial	SCEs	no effect	Achard et al.	1987
S, Spain	fibroblasts	growth on sepiolite -collagen	no effect	Olmo et al.	1988

* statistically not significant

TABLE XII : IN VITRO STUDIES CARRIED OUT WITH ATTAPULGITE

IN VITRO SYSTEM	RESULT	REFERENCE	
Adsorption of phospholipids	yes	Baillif and Touray	1988
	"	Perderiset et al.	1988
Adsorption of chemical carcinogens *B (a) P, NNN, NAAF*	yes high	Harvey et al.	1984
Adsorption of P V P N O	yes	Oscarson et al.	1981

7 - Concluding remarks

In vitro experiments carried out with phyllosilicates mainly dealt with cytotoxicity. Little data was obtained on genotoxicity or on effects using in vitro tests relevant to carcinogenicity. However results reported yet did not demonstrate genotoxicity (Table V and XI). Due to the large variety of the end points determined in those experiments as well as the vast range of particles used, it is difficult to draw a general conclusions.

Before doing a summary of reports, some important points must be raised. First, experiments reported here have been performed with different samples of the same mineral, sometimes from unknown origin and not always well characterized. Therefore it might be misleading to extrapolate results obtained with certain samples to what would be the effect of a given type of phyllosilicate. This is particularly true if one remember that different data were obtained with similar samples but of different size.

Comparison between toxicity of one type of phyllosilicate versus another type must be made with caution. Indeed the number of particles for a given weight of particles is not comparable and a similar quantitative response on a per weight basis can be different on a per number basis. This was well discussed for mineral fibres (Hesterberg and Barrett, 1984) and must be taken into consideration to compare activities of different phyllosilicates. In addition, the method of preparation of the particles for examination of their in vitro effects is also of importance. Size and number of particles can be modified by the preparation method (Griffiths and Hill, 1983).

Extrapolation of in vitro results to in vivo situation must be cautious since the phyllosilicates reactivity is modified by many components readily adsorbed at their surface.

Keeping in mind the above remarks, a summary of the in vitro effects of phyllosilicates can be made. Phyllosilicates such as kaolinite, bentonite, montmorillonite, attapulgite, palygorskite and sepiolite were toxic to RBC. Particle pretreatment with molecules able to adsord at the particle surface reduced the haemolytic effect. All phyllosilicates seem to have affinity to biological molecules or to polymers. We do not know except with kaolinite from Georgia or Cornwall (Table II) if this protective effect occurs where the particles are tested on macrophages ; however, it has been reported elsewhere that cytotoxicity might be related to sorptive properties for several particles (Jaurand et al.1981). Talc tested in the present experiments had a low effect on RBC or macrophages.

From the results reported here it appears that surface interactions are of importance to explain cytotoxicity. Negative charges seem to play a role in the destabilisation of the cell membrane and binding of polymers by hydrogen bonds prevent from damaging interactions. Positive charges seem to have a smaller role, as demonstrated with kaolinite.

From these in vitro tests, what have we learn on the toxic potential of phyllosilicates ? These tests brought some indications on the potential hazard of phyllosilicates. It appears that some phyllosilicates if they reach respiratory airways and alveolar space after inhalation can interact with pulmonary cells, macrophages and epithelial cells. This interaction can result in a cell injury as well as in a cell death. It is well known that injuried cells can release inflammatory factors or other molecules playing a role in the development of lung fibrosis. However, we also know from physico-chemical studies quoted here that phyllosilicates have physico-chemical surface properties such as sorptive properties which might modify or prevent toxicity. Stability or degradation of the protective molecules is different according to the particle type (Wallace et al. 1988a). This might explain differences in in vivo toxicity. Further experiments are strongly needed to understand the mechanisms of phyllosilicates activity on mammalian cells.

References

Achard S, Perderiset M, Jaurand MC (1987) Sister chromatid exchanges in rat pleural mesothelial cells treated with crocidolite, attapulgite, or benzo 3-4 pyrene. Br J Ind Med 44:281-283

Adamis Z, Timar M (1976) Effects of various mineral dusts on macrophages *in vitro*. Int Arch Occup Environ Hlth 37:301-307

Adamis Z, Timar M (1978) Studies on the effect of quartz, bentonite and coal dust mixtures on macrophages *in vitro*. Br J Exp Path 49:411-415

Baillif P, Touray JC (1988) Etude comparative, par spectroscopie de photo-électrons de l'adsorption de phospholipides par trois minéraux silicatés fibreux (chrysotile, crocidolite, attapulgite).CR Acad Sci, Paris, 307, Série III, 83-87

Brown RC, Chamberlain M, Davies R, Morgan DML, Pooley FD, Richards RJ (1980) A comparison of 4 in vitro systems applied to 21 dusts. In: Beck EG, Bignon J (eds) In vitro effects of mineral dusts.NATO ASI Series, Springer-Verlag Berlin, pp 47-52

Chamberlain M, Brown RC (1978) The cytotoxic effects of asbestos and other mineral dust in tissue culture cell lines. Br J Exp Path 59:183-189

Chamberlain M, Davies R, Brown RC, Griffiths DM (1982) In vitro tests for the pathogenicity of mineral dusts. Ann Occup Hyg 26:583-592

Daniel H, Le Bouffant L (1980) Study of a quantitative scale for assessing the cytotoxicity of mineral dusts. In: Beck EG, Bignon J (eds) In vitro effects of mineral dusts.NATO ASI Series, Springer-Verlag Berlin, pp 33-42

Davies R (1983) Factors involved in the cytotoxicity of kaolinite towards macrophages *in vitro*. Environ Hlth Perspect. 51:249-252

Davies R, Griffiths DM, Johnson NF, Preece AW, Livingston DC (1984) The cytotoxicity of kaolin towards macrophages *in vitro* Br J Exp Path 65:453-466

Davies R, Skidmore JW, Griffiths DM, Moncrieff CB (1983) Cytotoxicity of talc for macrophages *in vitro*. Fd Chem Toxic 21:201-207

Denizeau F, Marion M, Chevalier G, Cote MG (1985a) Ultrastructural study of mineral fiber uptake by hepatocytes *in vitro*. Toxicol Lett 26:119-126

Denizeau F, Marion M, Chevalier G, Cote MG (1985b) Absence of genotoxic effects of nonasbestos mineral fibers. Cell Biol Toxicol 1:23-32

Depasse J, Leonis J (1974) Toxicity of quartz and kaolin towards membranes : effect of lysozyme. Arch Int Physiol Bioch, 82:941-1027

Endo-Capron S, Saint-Etienne L, Renier A, Pillières F, Janson X, Fleury-Feith J, Bignon J, Jaurand MC (1989) In vivo and in vitro effects of talc samples, this meeting.

Gormley IP, Addison J (1983) The *in vitro* cytotoxicity of some standard clay mineral dusts of respirable size. Clay Minerals 18:153-163

Gormley IP, Kowolik MJ, Cullen RT (1985) The chemiluminescent response of human phagocytic cells to mineral dusts. Br J Exp Path 66:409-416

Gormley IP, Wright MO, Ottery J (1978) The effect of toxic particles on the electrophysiology of macrophage membrane. Ann Occup Hyg 21:141-149

Griffiths DM, Hill RJ (1983) The effects of dispersion on fibrous clays *in vitro* and *in vivo*. Ann Occup Hyg 27:405-414

Hahn FF, Newton GJ, Springstead KL, Karen L, Cordes JC (1974) Influence of particle size on phagocytosis by rabbit alveolar macrophages in vitro. Annual Rep. Inhalation Toxicol, Res Inst pp 257-260

Hahon N, Booth JA (1987) Silicate minerals and the interferon system. Environ Res 43:395-409

Harvey G, Page M, Dumas L (1984) Binding of environmental carcinogens to asbestos and mineral fibres. Br J Ind Med 41:396-400

Hatch GE, Boykin E, Graham JA, Lewtas J, Pott F, Loud K, Mumford JL (1985) Inhalation particles and pulmonary host defense *in vivo* and *in vitro* effects of ambient air and combustion particles. Environ Res 36:67-80

Henderson WJ, Blundell G, Richards R, Hext PM, Volcani BE, Griffiths K (1975) Ingestion of talc particles by cultured lung fibroblasts. Environ Res 9:173-178

Hesterberg TW, Barrett JC (1984) Dependence of asbestos and mineral dust-induced transformation of mammalian cells in culture on fiber dimension. Cancer Res. 44:2171-2180

Jaurand MC, Bignon J, Magne L, Renier A, Lafuma J (1979) Interaction des fibres avec les globules rouges et les macrophages alveolaires *in vitro*. Rev Fr Mal Resp 7:717-722

Jaurand MC, Magne L, Boulmier JL, Bignon J (1981) *In vitro* reactivity of alveolar macrophages and red blood cells with asbestos fibres treated with oxalic acid, sulfur dioxide and benzo 3-4 pyrene. Toxicology 21:323-342

Kennedy T, Baser M, Ghio A, Rawlings WR, Rao NV, Hoidal JR Kaolin generates OH° and causes hemolysis by acting as a Fenton reagent, this meeting.

Kennedy T, Melnicoff P, Dodson R, Rawlings W, Hoidal J (1987) Kaolin catalyzes hydroxyl radical generation from hydrogen peroxide. Chest 91:304

Koshi K, Hayashi H, Sakabe H (1968) Cell toxicity and hemolytic action of asbestos dust. Ind. health 6:69-79

Koshi K, Hayashi H, Sakabe H (1969) Biological effects of serpentine minerals in heated state. XVI Int Congress on Occupational Health. Sept 22-27, Tokyo, Japan

Lavie S, Stotzky G (1986a) Adhesion of the clay minerals Montmorillonite, kaolinite and attapulgite reduces respiration of Histoplasma capsulatum. Applied Environ Microbiol 51:65-73

Lavie S, Stotzky G (1986b) Interactions between clay minerals and siderophores affect the respiration of Histoplasma capsulatum. Applied Environ Microbiol 51:74-79

Le Bouffant L (1985) Relationships between *in vitro* toxicity, tissue effects and mineralogical composition of coal mine dusts. In: In Vitro Effects of Mineral Dusts. Beck EG and Bignon J (eds) NATO ASI Series, Springer-Verlag, Berlin, Heidelberg RFA pp 421-431

Le Bouffant L, Daniel H, Martin JC (1980) The value and limits of the relationship between cytotoxicity and fibrogenicity of various mineral dusts.In: *In vitro* Effects of Mineral Dusts, RC Brown, M Chamberlain and R Davies (eds) pp 333-338

Lewis E, Lipkin MD (1985) Failure of attapulgite to produce tumors : prediction of this result by *in vitro* cytotoxicity test. In: In Vitro Effects of Mineral Dusts. Beck EG and Bignon J (eds) NATO ASI Series, Springer-Verlag, Berlin,Heidelberg RFA pp 539-542

Low RB, Leffingwell CM, Bulman CA (1980) Effects of kaolinite on amino acid transport and incorporation into protein by rabbit pulmonary alveolar macrophages. Arch Environ Hlth, 35:217-223

Ma JYC, Weber KC (1986) Dipalmitoyl lecithin and lung surfactant adsorption at an air-liquid interface by respirable particles. Environ Res 41:120-129

Manyai S, Kabai J, Kis J, Suveges E, Timar M (1969) The *in vitro* hemolytic effect of various clay minerals. Med Lavoro 60:331-342

Nadeau D, Fouquette-Couture L, Paradis D, Khorami J, Lane D, Dunnigan J (1987) Cytotoxicity of respirable dusts from industrial minerals : comparison of two naturally occurring and two man-made silicates. Drug Chem Toxicol 10:49-86

Olmo N, Lizarbe MA, Turnay J, Müller KP, Gavilanes JG (1988) Cell morphology, proliferation and collagen synthesis of human fibroblasts cultured on sepiolite-collagen complexes. J Biomed Mat Res 22:257-270

Oscarson DW, Van Scoyoc GE, Ahlrichs JL (1981) Effect of poly-2-vinylpyridine-N-oxide and sucrose on silicate-induced hemolysis of erythrocytes. J Pharm Sciences, 70:657-659

Oscarson DW, Van Scoyoc GE, Ahlrichs JL (1986) Lysis of erythrocytes by silicate minerals. Clays and Clays minerals 34:74-80

Ottery J, Gormley IP (1978) Some factors affecting the haemolytic activity of silicate minerals. Ann Occup Hyg 21:131-139

Pairon JC, Jaurand MC, Gaudichet A, Brochard P, Bignon J (1989) Therapeutic and domestic uses of attapulgite and sepiolite. This meeting.

Perderiset M, Baillif P, Jaurand MC (1988) Chemical analysis and photoelectron spectroscopy of the adsorption of macromolecules on the surface of attapulgite. J Colloid Interface Sci 121:381-391

Perderiset M, Saint-Etienne L, Bignon J, Jaurand MC (1989) Interactions of attapulgite (fibrous clays) with human red blood cells. Toxicol Lett 47:303-309

Pigott GH, Pinto PJ (1983) Effects of non fibrous minerals in the V79-4 cytotoxicity test.Environ Hth Perspect. 51:173-179

Rahman Q, Beg MU, Viswanathan PN, Zaidi SH (1973) Solubility of kaolin, talc and mica dusts under physiological conditions. Environ Physiol Biochem 3:286-294

Reiss B, Millette JR, Williams GM (1980) The activity of environmental samples in a cell culture test for asbestos toxicity. Environ Res 22:315-321

Robertson A, Dodgson J, Gormley IP, Collings P (1982) An investigation of the adsorption of oxides of nitrogen on respirable mineral dusts and the effects on their cytotoxicity. Ann Occup Hyg 26:607-623

Schnitzer RJ, Pundsack FL (1970) Asbestos hemolysis. Environ Res 3:1-13

Singh SV, Das B, Kahn MF, Rahman Q (1984) Interaction of silicate dust with erythrocyte ghost membrane : in vitro studies. J Appl Toxicol 4:236-240

Singh SV, Das B, Rahman Q (1983) Relationship between solubility and hemolytic effects of toxic dusts. J Appl Toxicol 3:14-17

Singh SV, Rahman Q (1987) Interrelationship between hemolysis and lipid peroxidation of human erythrocytes induced by silicic acid and silicate dusts. J Appl Toxicol 7:91-96

Styles JA, Wilson J (1973) Comparison between *in vitro* toxicity of polymer and mineral dusts and their fibrogenicity. Ann Occup Hyg 16:241-250

Timar M, Adamis Z, Ungvary G (1979) Biological efects of mineral dusts. In vitro and *in vivo* studies. Arh Hig rada toksikol 30:871-874

Wallace WE, Headley LC, Weber KC (1975) Dipalmitoyl lecithin surfactant adsorption by kaolin dust *in vitro*. J Colloid Interf Sc 51:535-537

Wallace WE, Keane MJ, Mike PS, Hill CA, Vallyathan V (1988a) Mineral surface-specific differences in the adsorption and enzymatic removal of surfactant and their correlation with cytotoxicity. In : Fourth Int. Workshop : Effects of Mineral Dusts on Cells. NATO Advanced Research Workshop, sept 21-23, Quebec, Canada.

Wallace WE, Keane MJ, Vallyathan V, Hathaway P, Regad ED, Castranova V, Green FHY (1988b) Suppression of inhaled particle cytotoxicity by pulmonary surfactant and re-toxification by phospholipase : distinguishing properties of quartz and kaolin. Ann Occup Hyg 32:291-298

Wallace WE, Vallyathan V, Keane MJ, Robinson V (1985) In vitro biological toxicity of native and surfacxe-modified silica and kaolin. J Toxicol Environ Hlth 16:415-424

Woodworth CD, Mossman BT, Craighead JE (1982) Comparative effects of fibrous and nonfibrous minerals on cells and liposomes. Environ Res 27:190-205

Woodworth CD, Mossman BT, Craighead JE (1983) Induction of squamous metaplasia in organ cultures of hamster trachea by naturally occuring and synthetic fibers. Canc Res 43:4906-4912

Wright A, Gormley IP, Collings, Davis JMG (1980) The cytotoxicities of asbestos and other fibrous dusts. In : In vitro effects of Mineral Dusts (eds) NATO ASI Series, Springer-Verlag, Berlin, RFA pp 25-31

TOXICITY OF PHLOGOPITE AND MUSCOVITE *IN VITRO*

M. Holopainen[1], V. Vallyathan[2], M. Hedenborg[3], M. Klockars[3]

[1] National Public Health Institute, Department of Environmental Hygiene and Toxicology. P.O.B. 95, SF-70701 Kuopio, Finland.

Introduction

Mica minerals are used for many industrial purposes. The traditional view suggests that mica dust is not harmful to the health of exposed workers (Parkes 1982), but the issue is controversial (Skulberg et al. 1985). *In vitro* studies suggested that mica was toxic (Holopainen et al. 1986). Intratracheal instillation of mica induced mild pulmonary fibrosis in rats (Dianova et al. 1976). Furthermore, sporadic case reports of mica pneumoconiosis have been reported (Davies and Cotton 1983). Although some epidemiologic studies suggested that mica may cause pneumoconiosis (Krasnopejeva et al. 1964, Dreessen et al. 1940), in other studies such a risk was not observed (Heimann et al. 1953, Smith 1952).

This study describes the *in vitro* effects of two mica dusts, phlogopite and muscovite, using conventional toxicity bioassays. The results were compared with positive (quartz) and negative (TiO_2 and barite) controls to assess relative toxicities. One sample of phlogopite was treated with acid to extract the cationic elements. This sample was used in the tests to assess the role of metal elements in the toxicity of phlogopite.

[2] NIOSH, Division of Respiratory Disease Studies, 944 Chestnut Ridge Road, Morgantown, WV 26505-2888, U.S.A.

[3] Institute of Occupational Health, Topeliuksenkatu 41 a,A, SF-00250 Helsinki, Finland

NATO ASI Series, Vol. G 21
Health Related Effects of Phyllosilicates
Edited by J. Bignon

Materials and Methods

Dust samples

The phlogopite sample, used in the tests at the National Institute for Occupational Safety and Health, USA, (NIOSH), was selected by a geologist from the Siilinjärvi mine. The sample was initially ground by means of an Abbie pebble mill and the respirable dust was separated aerodynamically using the Bacho Micro particle classifier. The silica used in this study was Min-U-Sil (Pennsylvania Sand and Glass Corp., Pittsburgh, PA) and barite was from mines in Georgia, USA. All the dust samples used in the study were less than 7 µm in size.

The phlogopite and muscovite samples used in the tests at the National Public Health Institute, Finland (NPHI) were from the Kemira company, Finland, and they were ground by means of a jet mill. The TiO_2 was a standard sample of rutile obtained from the Institute of Occupational Health, Helsinki, whereas the silica (min-u-sil) sample was obtained from NIOSH, USA. One sample of phlogopite was treated with nitric and sulfuric acids to extract metals from the mineral and then washed several times with water. This acid treated sample, which was almost pure SiO_2, was ground by means of a jet mill into respirable size. The chemical analysis of the mica samples is shown in table 1. The particle size distributions of the jet milled phlogopite and muscovite mica samples, used in the tests in Finland, were almost identical. Over 98% of the particles were smaller than 5µm and about 40% were smaller than 1µm. One fine dust sample of phlogopite was

Table 1. Chemical composition of phlogopite and muscovite (%)

	SiO_2	Al_2O_3	K_2O	MgO	Fe	F
Phlogopite	42	11	11	23	6.4	1.1
Muscovite	44	35	10	0.4	1.2	0.4

collected by water elutriation. The typical shape of mica was better retained and the particles were larger in the elutriated sample than in the jet-milled sample. The particle size analysis of the samples used in the tests at NIOSH is shown in table 2.

Hemolysis tests

Hemolytic activity was determined in a 2% sheep erythrocyte preparation according to the method of Harington et al.(1971), with minor modifications (Collan et al. 1986).

Macrophage tests

Rat alveolar macrophages, collected by alveolar lavage, were used in the enzyme release assay for lactic dehydrogenase (LDH), ß-glucuronidase (ß-GLUC) and ß-N-acetyl-glucosaminidase (ß-NAG) according to the methods of Reeves and Fimignari (1963), Lockard and Kennedy (1976) and Sellinger et al. (1960), respectively. Bronchoalveolar lavage was carried out by repeated instillation and aspiration of 8ml of ice cold calcium-and magnesium-free Hanks balanced salt solution, until a total volume of 80ml of lavage fluid was collected. The cells were pelleted by centrifugation at 500g for 10 minutes and washed with ice cold HEPES buffer containing 149 mM NaCl, 5 mM KCl, 10 mM HEPES and 5 mM

Table 2. Particle characteristics of barite, silica and phlogopite.

Mineral	Surface area m^2/g	Particle Size			Silica %
		MCD	MMAD	AD	
barite	5.8	o.87	7.8	6.9	3.1
silica	1.7	1.24	5.2	7.8	98.0
phlogopite	1.3	o.66	6.4	10.7	1.6

MCD = median circular diameter
MMAD = mass median aerodynamic diameter
AD = aerodynamic diameter

glucose. The washed cells were pooled and resuspended in a HEPES medium. Cell counts were made using a grid hemocytometer. The purity of macrophage preparations obtained was 90-95%.

Aliquots of alveolar macrophage suspensions in HEPES medium (2×10^6 cells/ml) were incubated with mineral dusts (1mg/ml) for 2 hours at 37°C in a rotary shaker water bath. After the incubation the cells and dusts were sedimented by centrifugation and the supernatant fluid was assayed for enzyme activity. Total enzyme activities were determined in macrophages that were incubated for 2 hours without dust and subsequently lysed with triton-X-100.

The amount of H_2O_2 released by alveolar macrophages was measured as the oxidation of scopoletin in the presence of horseradish peroxidase. The samples contained alveolar macrophages (5×10^6 cell/3ml), 2.4 µM scopoletin, 6.6 units of peroxidase and 1mg of dust. The H_2O_2 released by the cells oxidized the scopoletin and decreased the fluorescence of the solution. The change in fluorescence was monitored using a Perkin-Elmer fluorescence Spectrophotometer (Model MP-3L) equipped with a temperature controlled water bath (37°C) and a magnetic stirring unit. The excitation wavelength was 350nm and emission wavelength 460nm. The H_2O_2 release (in nano moles) was calculated from a standard graph.

Chemiluminescence assay

The ability of the dusts to induce production of reactive oxygen metabolites in human polymorphonuclear leukocytes (PMN) was studied by luminometric method. PMNs were isolated from heparinized human venous blood samples taken from apparently healthy adults, as has been described earlier (Hedenborg et al. 1987), and cells were adjusted to concentrations of 5×10^6 cells/ml. Chemiluminescence (CL) was measured for 40 min at 37°C using a microcomputer-controlled luminometer (LKB-Wallac 1251). The reaction mixture consisted of 50 µl of cell suspension, 700 µl of 1×10^{-4} M luminol in PBS, and 25-200 µl of dust.

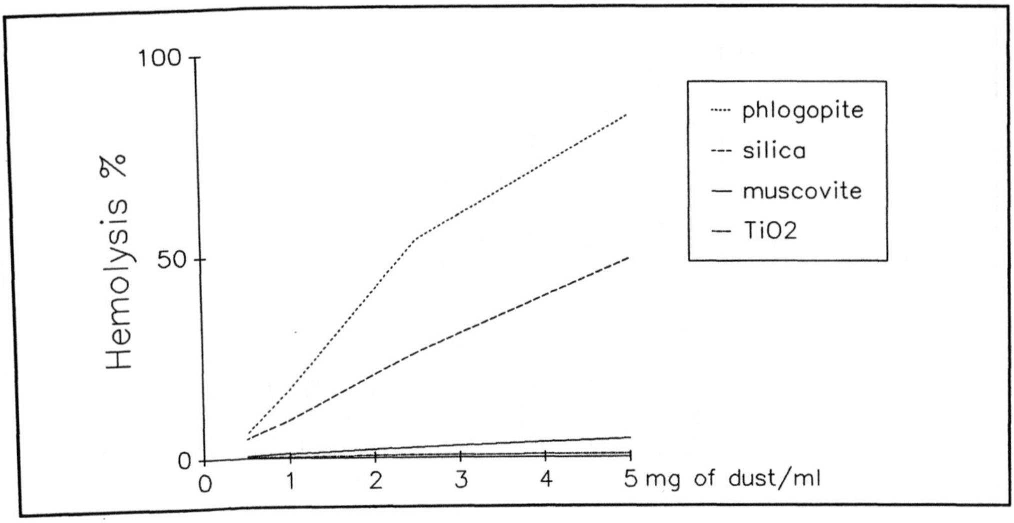

Figure 1. Hemolysis caused by phlogopite, silica, muscovite and TiO_2.

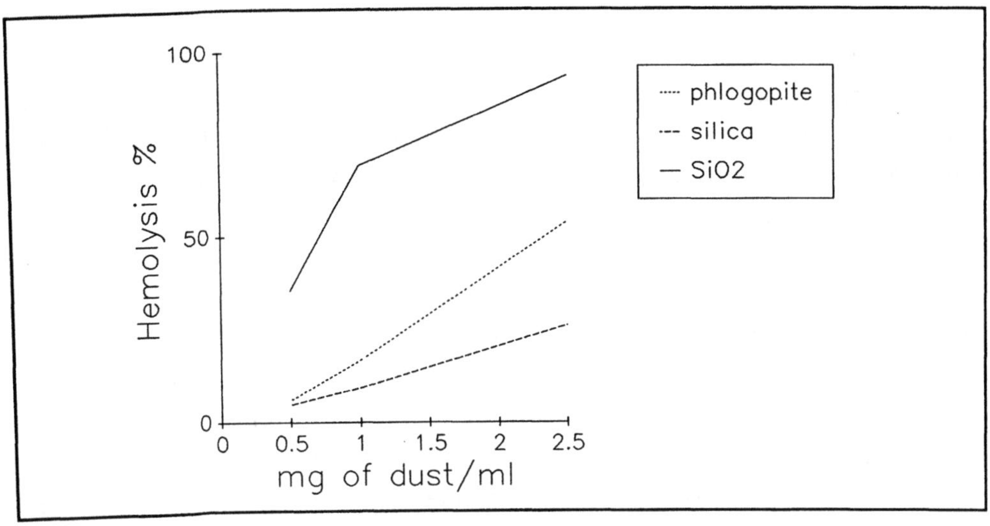

Figure 2. Hemolysis caused by acid treated phlogopite, phlogopite and silica.

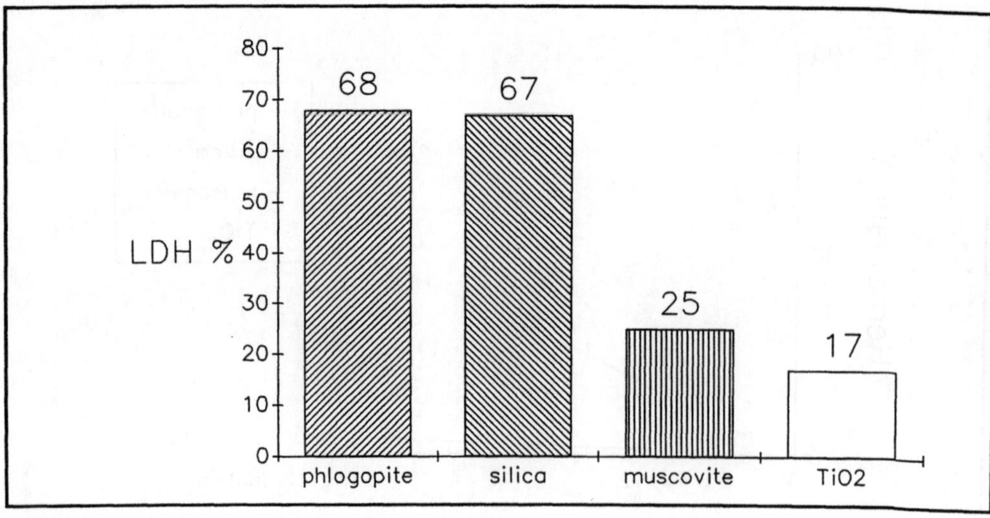

Figure 3. The release of LDH from rat AMs induced by phlogopite, silica, muscovite and TiO_2.

Results

Phlogopite induced 53.9% hemolysis, but muscovite only 2.5% hemolysis when 2.5 mg dose of the dust was used (figure 1). The acid-treated phlogopite was highly hemolytic: a 50% hemolysis was induced by 0.7mg dose of the dust, whereas 2.4 mg of the untreated phlogopite sample was required to induce the 50% hemolysis (Figure 2). In the macrophage tests in Finland, the LDH release was at the level of quartz with phlogopite (68%), but it was lower (25%), when muscovite was used (figure 3). The acid treatment of phlogopite reduced the release of LDH (Figure 4). In the macrophage enzyme assays carried out by NIOSH, the release of LDH was higher with phlogopite (86%), but the release of lysosomal enzymes ß-NAG and ß-GLUC showed comparable levels after exposure to phlogopite and silica (Figure 5).

Phlogopite induced a dose dependent increase in the secretion of hydrogen peroxide from alveolar macrophages. Phlogopite at a concentration of 1 mg/ml induced the secretion of 210 nmol of

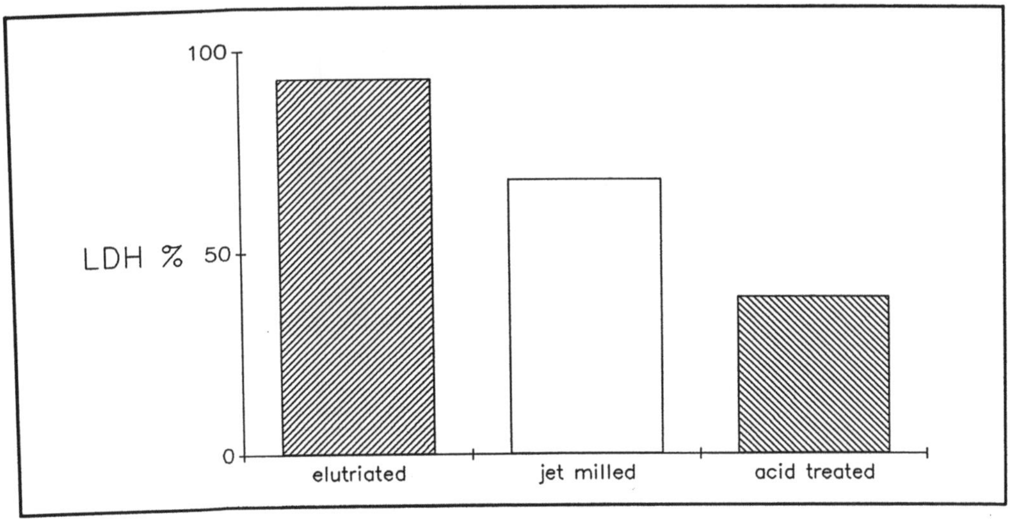

Figure 4. The release of LDH from rat AMs induced by elutriated, jet milled and acid treated phlogopite.

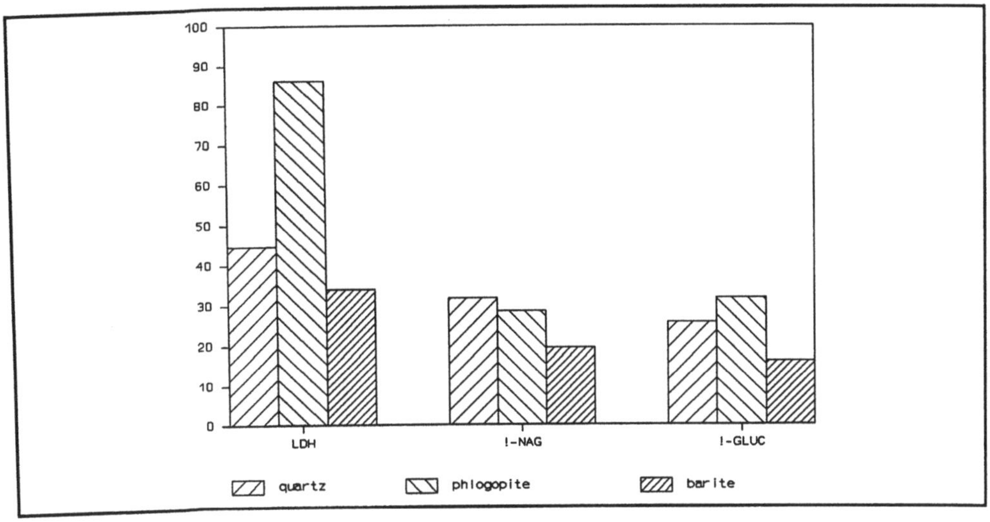

Figure 5. The release of LDH, ß-NAG and ß-GLUC from AMs induced by silica, phlogopite and barite.

hydrogen peroxide per min compared to 89 nmol/min with respirable dust sample of richterite, which is an amphibole mineral from the Siilinjärvi mine (table 3).

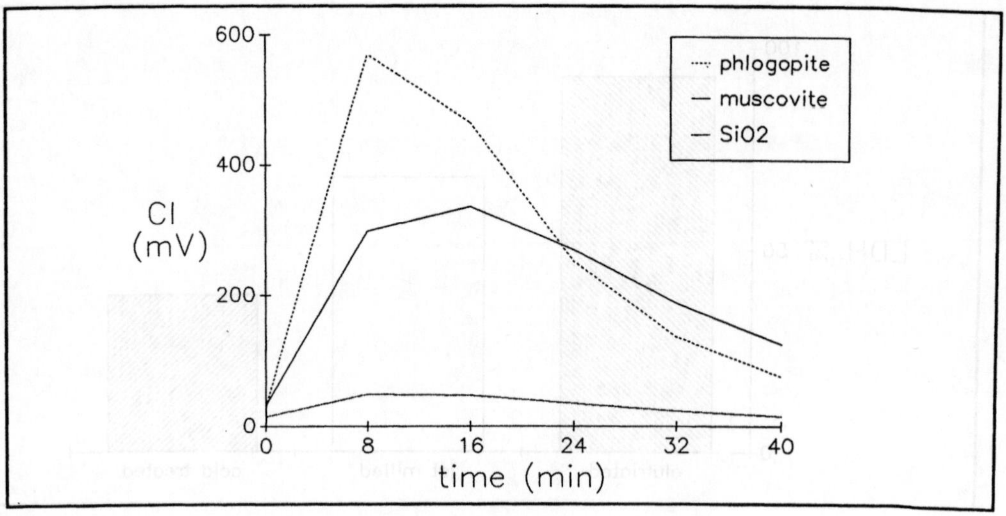

Figure 6. Chemiluminescence response induced by phlogopite, muscovite and acid-treated phlogopite in human polymorphonuclear leukocytes.

The results of CL assay reveal, that phlogopite caused a strong CL response which was completely blunted by acid treatment of the mineral. With muscovite the CL response was between that of those two dusts (figure 6). The dose response curves of the CL tests are shown in figure 7.

Table 3. H_2O_2 production (nmol/min) in rat AMs induced by phlogopite and richterite.

Mineral	H_2O_2 production nmol/min	
mg/ml	Phlogopite	Richterite
0.25	50	41
0.50	85	80
1.0	210	89

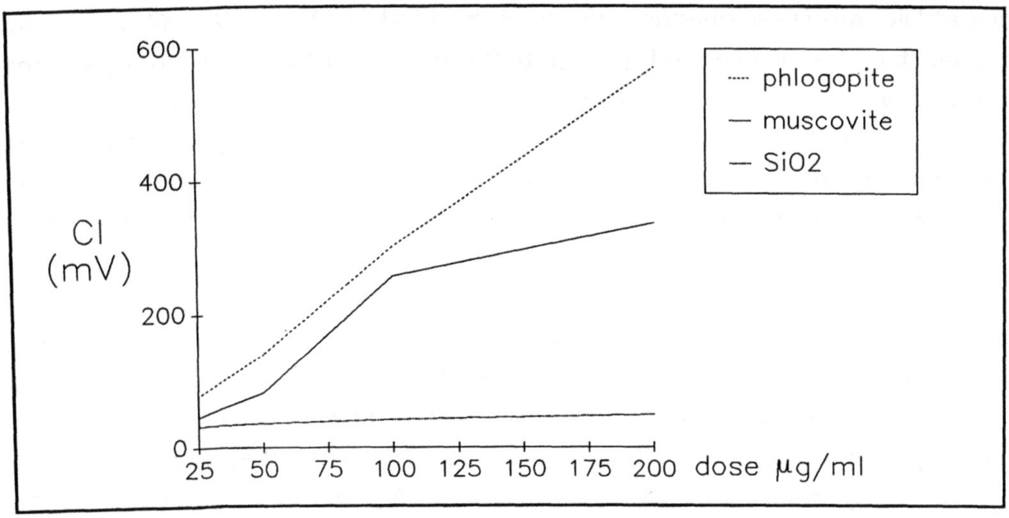

Figure 7. Dose response curves of chemiluminescence induced by phlogopite, muscovite and acid treated phlogopite in human polymorphonuclear leukocytes.

Discussion

The two mica dust samples, phlogopite and muscovite, studied are physically comparable in the particle size distribution and in the shape of the particles. Therefore, the observed differences in the cytotoxicity of these mica samples may be caused not by physical but by the chemical and surface properties of the minerals. The chemical composition of the muscovite and phlogopite samples is different: Phlogopite is rich in MgO (23%) and Fe (6.4%) compared to muscovite (0.4% and 1.2% respectively). Muscovite contains 35% of Al_2O_3, while the content of Al_2O_3 in phlogopite is 11%.

The marked hemolysis caused by the acid-treated phlogopite suggests that the ability of phlogopite to induce hemolysis is not only caused by the metal elements but also by the SiO_2 on the surface of the mineral. The acid treated sample had a strong negative surface charge (zeta potential -20 - -100 mV), as compared to the untreated phlogopite with a positive charge (zeta potential 0 - +45 mV). The negative charge on quartz has shown to be an important determinant in hemolysis induced by that mineral (Nolan et al. 1981), and our results suggest that the same phenomenon seems to be relevant in the case of acid treated phlogopite.

The mechanism in which silicate minerals are toxic to AMs seems to differ from the mechanism in which they are hemolytic. AMs are phagocytic cells, and the phagocytosis itself may influence and modify the effect of minerals on cells. The phagocytic process of AM is a result of cell activation by mineral particles. The activated respiratory burst accompanied by the formation of reactive oxygen metabolites, including hydrogen peroxide, may lead to cell damage, with a subsequent release of enzymes from the phagocytic cell (Brain 1988). Metal elements of a mineral may induce hydroxyl radical production from hydrogen peroxide by the Fenton reaction. The formation of hydroxyl radical, catalyzed by metals of phlogopite, may cause the higher cytotoxicity of the untreated phlogopite to AMs compared to the acid treated sample of that mineral, which was almost pure SiO_2. The phlogopite sample contains 6.4% iron, and the muscovite sample 1.2%. This could explain the high cytotoxicity of phlogopite in macrophage test compared to muscovite.

The CL response of PMN induced by phlogopite was consistent with the results from the other test systems, but the results with acid-treated phlogopite were surprising: this highly hemolytic dust was inert in the CL assay. This result suggests that cationic metal elements of mica are of importance for cell interactions and reactive oxygen metabolite production in PMN.

Retention of the dust in the lower respiratory tract after exposure and the stability of the particles in lung tissue influence to the fibrogenicity of a mineral dust. At the present time there are no definite <u>in vitro</u> tests to estimate the possible fibrogenicity of these mica minerals <u>in vivo</u>. Whereas phlogopite dissolves in weak acids, and to some degree even in neutral solutions, muscovite is a less soluble mineral. In the long term muscovite may therefore be more persistent in lung tissue than phlogopite.

Conclusion

The results reveal that in in vitro tests phlogopite mica was even more cytotoxic than silica. Muscovite, on the other hand, was less cytotoxic in our tests and it was comparable to TiO_2. Extraction of cations from phlogopite by acid treatment converted that mineral more hemolytic but less cytotoxic to AMs, and decreased its potency to induce the formation of reactive oxygen metabolites by PMNs. Hemolysis was correlated with the surface charge of mica, but the release of LDH from AMs was better correlated with metal elements of mica. The pulmonary effects of these mica dusts should be studied in vivo, before any estimate of their fibrogenicity can be given.

Acknowledgements. We thank Ms. Irma Väänänen for her excellent technical assistance.

References

Brain J (1988) Lung macrophages: How many kinds are there? What do they do? Am Rev Resp Dis 137:507-509

Collan Y, Kosma V-M, Anttonen H, Kulju T (1986) Toxicity of richterite in hemolysis test and macrophage cultures. Arch Toxicol Suppl. 9:292-295

Davies D, Cotton R (1983) Mica pneumoconiosis. Br J Ind Med 40:22-27

Dianova AV, Kochetkova TA, Rumjantsev GI (1976) The problem of combined action of the dust of mica and resins used in the production of micanite articles. Gig Sanit 41, 7:33-37. (In Russian)

Dreessen WC, Dallavalle JM, Edwards TI, Sayers RR, Eason HF, Trice MF (1940) Pneumoconiosis among mica and pegmatite workers. Public Health Bulletin no 250. US Public Health Services, Washington DC

Harington JS, Miller K, Macnab G (1971) Hemolysis by asbestos. Environ Res 4:95-117

Hedenborg M, Klockars M (1987) Production of reactive oxygen metabolites induced by asbestos fibers in human polymorphonuclear leukocytes. J Clin Pathol 40:1189-1193

Heimann H, Moskowitz S, Iyer CRH, Gupta MN (1953) Note on mica dust inhalation. Arch Ind Hyg Occup Med 8:531-532

Holopainen M, Collan Y, Kosma V-M, Kalliokoski P, Kulju T, Anttonen H, Tossavainen A, Kauppinen H (1986) Evidence for toxicity of phlogopite in hemolysis and macrophage tests. 2nd International Symposium on Occupational Health and Safety in Mining and Tunneling, September 23-26, Prague

Krasnopejeva LF (1964) Experimental pneumoconiosis caused by the mica dust. Gig Tr Prof Zabol 8, 11:3-8 (In Russian)

Lockard VG, Kennedy RE (1976) Alterations in rabbit alveolar macrophages as result of traumatic shock. Lab Invest. 35:501

Nolan RP, Langer AM, Harington JS, Oster G and Selikoff IJ (1981) Quartz hemolysis as related to its surface functionalities. Environ Res 26:503-520.

Parkes WR (1982) Occupational lung disorders 315-316. Butterworths, London

Reeves WJ, Fimignari GM (1963) An improved procedure for the preparation of chrystalline lactic dehydrogenase from Hog Heart. J Biol Chem 238:3853

Sellinger OZ, Beufay H, Jacques P, Doyan A, De Duve C (1960) Tissue fractionation studies, intracellular distribution and properties of ß-N-acetyl-glucosaminidase and ß-galactosidase in rat liver. Biochem J 74:450

Skulberg KN, Gylseth B, Skaug V, Hanoa R (1985) Mica pneumoconiosis - A literature review. Scand J Work Environ Health 11:65-74

Smith AR (1952) Pleural calcification resulting from exposure to certain dusts. Am J Roentgenol 67:375-382.

KAOLIN GENERATES OH AND CAUSES HEMOLYSIS BY ACTING AS A FENTON REAGENT

A. Ghio, T. Kennedy, M. Baser,[1] W.R. Rawlings,[2] N.V. Rao,[3] J.R. Hoidal[3]
Division of Allergy, Critical Care and Respiratory Medicine
Department of Medicine
Duke University Medical Center
Durham, North Carolina, USA

Kaolin is one of several clays commercially exploited for its content of the nonfibrous, hydrated aluminum silicate kaolinite ($Al_2O_3.2SiO_2.2H_2O$). Industrial uses include the manufacture of adhesives, paper products, refractory materials, ceramics, and fillers for plastics, rubber, and paints. Inhalation of kaolin dust has been well documented to cause pneumoconiosis (Oldham, 1983; Kennedy, 1983; Sepulveda, 1983). Human (Hale,1956) and animal (Policard, 1954) studies provide evidence that kaolin itself is responsible for the lung disease rather than free silica present in the dust.

The biochemical mechanism of kaolin toxicity is unknown. Other silicates including crocidolite, chrysotile, and amosite asbestos have been demonstrated to generate hydroxyl radicals from hydrogen peroxide (Weitzman, 1984). Cytotoxicity from crocidolite and chrysotile has been inhibited by antioxidants (Goodglick, 1987; Shatos, 1987). A candidate reaction for the generation of oxidants by silicates is the Fenton reaction in which iron mediates an electron exchange when reduced from ferric (Fe^{3+}) to the ferrous (Fe^{2+}) state (Cohen, 1985). Because iron is a common structural component or contaminant of silicates causing pneumoconiosis, we tested

[1]Departments of Environmental Health Sciences, Epidemiology, and Medicine, The Johns Hopkins Medical Institutes, Baltimore, Maryland, USA; [2]Surgical Associates of Sandersville, Sandersville, Georgia, USA; [3]Division of Pulmonary, Critical Care and Occupational Medicine, University of Utah, Salt Lake City, Utah, USA

NATO ASI Series, Vol. G 21
Health Related Effects of Phyllosilicates
Edited by J. Bignon

the hypothesis that kaolin is a Fenton reagent that catalyzes generation of reactive oxidant species in the presence of hydrogen peroxide and a reducing substance. We also studied whether *in vitro* hemolysis from kaolin is also explained by kaolin-mediated oxidant injury to the red cell. Results of these studies have been recently reported elsewhere (Kennedy, 1989).

Methods

Kaolin generation of hydroxyl radicals (·OH) from hydrogen peroxide (H_2O_2) was tested using ascorbate-driven Fenton oxidation (Cohen, 1985). Formation of OH was monitored by the reaction of ·OH with dimethyl sulfoxide (DMSO) to yield methane (Repine, 1973). One-milliliter aqueous reaction mixtures (replicates of 5) containing 13 mM DMSO, 1 mM H_2O_2 and kaolin (silica free commercial hydrous kaolin from Geogia with a mean particle diameter of 0.63 ± 0.03 microns and average area per particle of 1.13 microns^3) were incubated in 3-ml sealed glass tubes at 37° C for 30 minutes, with or without 1 mM ascorbate. Generation of methane (CH_4) from DMSO was measured using a Varian gas chromatograph (6 feet by 2 mm stainless steel column packed with 60/80 Carbosieve B; flame ionization detector) and methane standard. In experiments with dimethylthiourea (DMTU), 15 mM of this sulfhydryl ·OH radical scavenger (Dorfman, 1973) was added to the reaction mixture. In experiments using dust pretreated with iron chelators, kaolin was incubated in water with 18 mM deferoxamine for 72 hours or 2 mg/ml transferrin for 6 hours, then washed with double glass distilled iron-free water prior to use.

Human erythrocytes were prepared using fresh whole blood from a single donor. Erythrocytes were washed two times in phosphate-buffered saline (PBS, pH 7.4), centrifuged at 1000 g, and diluted in PBS prior to use. Superoxide anion (O_2^-) generation by erythrocytes was studied by incubation replicates of 10 2-ml suspensions of 2% erythrocytes and 80 micromolar succinylated ferricytochrome c in PBS for 60 minutes at 37 degrees C with or without 200 units of superoxide dismutase (SOD). After centrifugation, SOD-inhibitable absorbance of supernatants at 550 nm, corrected for that of controls containing no cytochrome c or SOD, and an extinction coefficient of 2.1 x 10^4 M^{-1} cm^{-1} were used to calculate O_2^- generation.

Lipid peroxidation of erythrocyte membranes was studied by incubating 2 ml of 2% by volume erythrocytes in PBS with 1 mg/ml kaolin for 12 hours. Cell suspensions were boiled for 20 minutes with thiobarbituric acid, trichloroacetic acid, and 0.01% butylated hydroxytoluene (Buege, 1978). After centrifugation at 2000 g, malondialdehyde content of supernatants was determined by absorbance at 532 nm and an extinction coefficient of 1.56 X 10^5 M^{-1} cm^{-1}. Erythrocyte hemolysis from kaolin was studied by incubating 5 2-ml replicates each of 2% by volume erythrocytes and kaolin (0 to 1.0 mg/ml) in PBS for 1 hour at 37 degrees C with gentle agitation. After centrifugation at 1000 g, percentage lysis was calculated as the ratio of optical density of supernatants at 540 nm compared to that of positive controls (erythrocytes in PBS containing 0.5% Triton X-100). Erythrocytes in 4,4'-diisothiocyano-2,2'-stilbene disulfonate (DIDS) treated experiments were preincubated as a 37.5% suspension in PBS with 1 mM of the anion channel blocker DIDS (Lynch, 1978) for 30 minutes at 37 degrees C, followed by three washes with PBS. Pretreated kaolin was incubated in water with 18 mM deferoxamine for 72 hours or 2 mg/ml transferrin for 6 hours, then washed with double glass distilled iron-free water. Catalase- and N-propyl gallate-treated experiments were performed with 100 units/ml catalase or 6 mM of the ·OH scavenger N-propyl gallate (Dorfman, 1973). Hemolysis curves were compared by multivariant analysis. Values in all studies are presented as means $\pm$ standard error of the mean (SE). Significance was assumed when $p<.05$.

Results

Figure 1 shows that hydrous kaolin is an extremely potent catalyst for generating ·OH. Kaolin generated large quantities of methane from H_2O_2 in the presence of ascorbate (Figure 1). Without ascorbate, kaolin exhibited no catalytic activity. The requirement for a reducing substance such as ascorbate in order for kaolin to generate ·OH is characteristic of Fenton catalysts (Cohen, 1985).

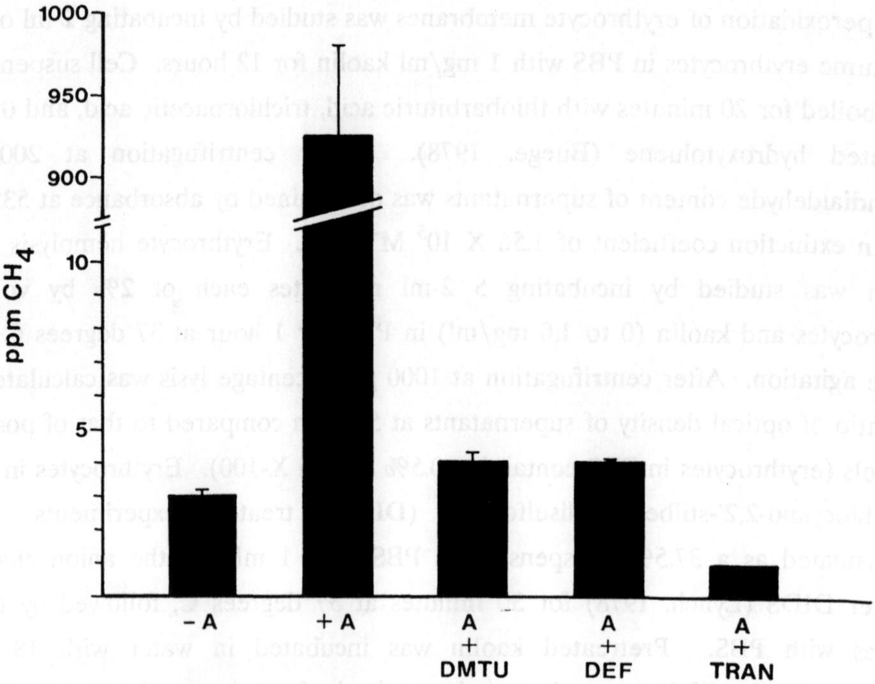

FIGURE 1. Kaolin functions as a Fenton catalyst *in vitro*. In the presence of 1 mM ascorbate (A) as a reducing substance, 1 mg kaolin generates substantial ·OH from 1 mM H_2O_2, measured by evolution of CH_4 from 13 mM DMSO. Production of ·OH, or CH_4, is prevented by the sulfhydryl ·OH scavenger DMTU, or by pretreatment of kaolin with the iron chelators deferoxamine (DEF) or transferrin (TRAN). Reprinted with permission (Kennedy, 1989).

Generation of methane was prevented by ·OH radical scavenger DMTU (4 ± 0.3 ppm CH_4). Fenton activity was also decreased by preincubating kaolin with the iron chelator deferoxamine (4 ± 0.1 ppm CH_4) or transferrin (1 ± 0 ppm CH_4).

To determine whether kaolin might cause hemolysis by acting as a Fenton reagent, we first determined whether erythrocytes could produce a reducing substance, such as O_2^-, to drive the reaction. A 2 ml suspension of 2% erythrocytes incubated with ferricytochrome c generated 1.57 x 10^{-6} mol of O_2^- in 60 minutes. Next we studied whether kaolin-induced hemolysis was accompanied by biochemical evidence of oxidant injury. Kaolin increased malonaldehyde in an erythrocyte suspension to 3.37 ± 0.28 nmol/ml compared to 2.55 ± 0.08 nmol/ml in control incubated with PBS alone ($p < 0.05$).

Finally, we studied kaolin-induced hemolysis with scavengers of reactive oxygen species to determine if selectively interrupting the Fenton reaction at various points attenuated red cell lysis. Figure 2 shows that kaolin caused substantial *in vitro* lysis of human RBCs. Hemolysis was significantly attenuated by pretreatment of erythrocytes with the anion channel blocker DIDS to prevent exit of O_2^- from the red cell into the extracellular medium. Hemolysis was also decreased by the H_2O_2 scavenger catalase, by pretreating silicates with the iron chelators deferoxamine or transferrin, or by adding the lipid-soluble ·OH scavenger N-propyl gallate to the medium. Not shown is the effect of the O_2^- scavenger superoxide dismutase (SOD), which decreased hemolysis at 1000 U/ml (21±1% hemolysis with 1 mg/ml kaolin, $p<.001$ compared to untreated control).

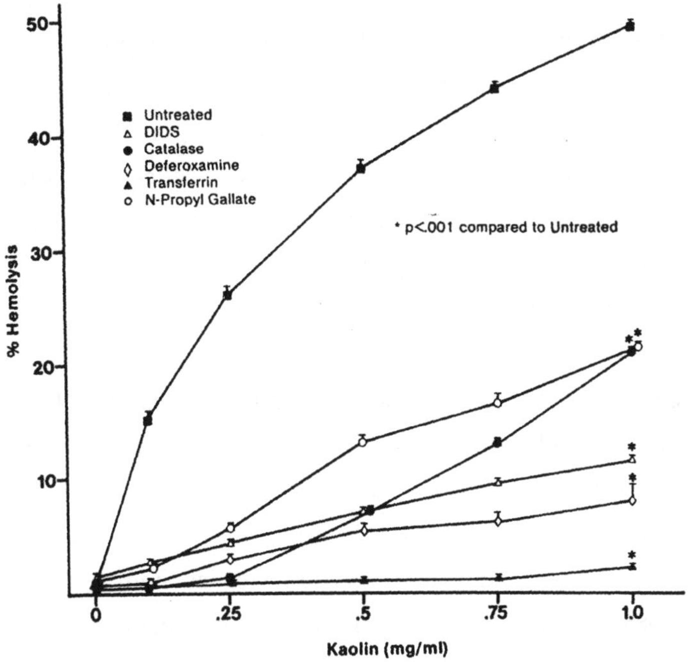

FIGURE 2. *In vitro* hemolysis from kaolin is inhibited by the iron chelators or antioxidant interventions. Kaolin alone caused substantial hemolysis which was significantly attenuated by pretreatment of erythrocytes with the anion channel blocker DIDS to prevent exit of O_2^- derived from hemoglobin autooxidation into the extracellular medium. Hemolysis was also decreased by the H_2O_2 scavenger catalase, by pretreating silicates with the iron chelators deferoxamine or transferrin, or by adding the lipid-soluble ·OH scavenger N-propyl gallate to the medium. Reprinted with permission (Kennedy, 1989).

Discussion

Weitzman and Graceffa demonstrated that crocidolite, chrysotile, and amosite asbestos catalyze generation of toxic $\cdot OH$ from H_2O_2, a normal byproduct of tissue metabolism. The iron chelator deferoxamine inhibited $\cdot OH$ generation indicating a major role for iron or another transitional metal located on the fiber surface as the critical element determining catalytic activity. We have shown the same is true for kaolin, suggesting that its cytotoxicity is explained, in part, by generation of $\cdot OH$. In the presence of a reducing substance such as ascorbate (AH_2), kaolin demonstrated catalytic activity for generating $\cdot OH$ from H_2O_2. This activity was prevented by preincubating kaolin with the iron chelator deferoxamine, demonstrating that, like asbestos, catalytic activity of kaolin for generating $\cdot OH$ depends on a transitional metal adsorbed to the clay surface. Silica and silicates form a coordination complex with metal cations (Dugger, 1964; Schindler, 1976). Adsorption occurs at surface silanol groups:

$$\text{R-Si-OH} + M^{z+} \longrightarrow \text{-(Si-O)}M^{(z-1)+} + H^+$$

Complexed on the surface, Fe^{3+} is capable of catalyzing electron transfer. Ferric iron has frequently been found adsorbed to commercial kaolin (Fordham, 1969). Thus generation of $\cdot OH$ and cell lysis can be explained by the following reactions:

$$\text{Kaolin-}Fe^{3+} + AH_2 \longrightarrow \text{Kaolin-}Fe^{2+} + AH^{\cdot} + H^+$$

$$\text{Kaolin-}Fe^{2+} + H_2O_2 \longrightarrow \text{Kaolin-}Fe^{3+} + {}^{\cdot}OH + OH^-$$

Both O_2^- and H_2O_2 are produced by erythrocytes as hemoglobin autooxidizes (Winterbourn, 1985). H_2O_2 diffuses freely through cell membranes (Ohno, 1985) and O_2^- can permeate erythrocyte membranes rapidly through anion channels (Lynch, 1978). Thus, O_2^- and H_2O_2 could function as the reducing substance and substrate for kaolin-Fe^{3+} mediated $\cdot OH$ generation, causing oxidant hemolysis:

$$\text{Kaolin-}Fe^{3+} + O_2^- \longrightarrow \text{Kaolin-}Fe^{2+} + O_2$$

$$\text{Kaolin-}Fe^{2+} + H_2O_2 \longrightarrow \text{Kaolin-}Fe^{3+} + {}^{\cdot}OH + OH^-$$

In the lung, O_2^- and H_2O_2 are generated from microsomal (Chance, 1979) and mitochondrial (Boveris, 1976) metabolism, or are produced by the respiratory burst of phagocytes such as alveolar macrophages (Hoidal, 1978; Hoidal 1982) during attempts to ingest particles of kaolin. Thus, kaolin could produce pneumoconiosis by generating toxic $\cdot OH$ radicals from lung H_2O_2. Hydroxyl radicals could injure lung cells by lipid peroxidation or oxidation of critical cell proteins (Davies, 1987; Fucci, 1983), and trigger intracellular events by generation of calcium ionophores from oxidation of membrane unsaturated fatty acids (Serhan, 1981; Lebedev, 1982). The tumor antibiotic bleomycin, which chelates iron and oxidatively injures DNA *in vivo* by generating ·OH (Sausville, 1978), produces lung fibrosis clinically in patients and experimentally in animals. Thus, there is biologic precedent for the proposition that kaolin might cause pneumoconiosis by also acting as a Fenton catalyst.

Bibliography

Boveris A, Cadenas E, Stopanni (1976) Role of Ubiquinone in the mitochondrial generation of hydrogen peroxide. Biochem J 156:435-444.

Buege JA, Aust SD (1978) in Methods in Enzymology (Fleischer S and Packer L, Eds.), Vol. 52, pp. 302-310, Academic Press, San Diego.

Chance B, Sies H, Boveris A (1979) Hydroperoxide metabolism in mammalian organs. Physiol Rev 59:527-605.

Cohen G (1985) in Handbook of Methods for Oxygen Radical Research (Greenwald RA, Ed.), pp. 55-64, CRC Press, Boca Raton, FL.

Davies KJA, Goldberg AL (1987) Oxygen radicals stimulate intracellular proteolysis and lipid peroxidation by independent mechanisms in erythrocytes. J Biol Chem 262:8220-8226.

Dorfman LM, Adams GE 1973 Reactivity of the Hydroxyl Radical in Aqueous Solution, National Standard Reference Data System, National Bureau of Standards 46, Washington, DC.

Dugger DL, Stanton JH, Irby BN, McConnell BC, Cummings WW, Maatman RW (1964) The exchange of twenty metal ions with the weakly acidic silanol group of silica gel. J Physic Chem 68:757-760.

Fordham FW (1973) Sorption and precipitation of iron on kaolinite. Factors involved in sorption equilibria. Aust J Soil Res 11:185-195.

Fucci L, Oliver CN, Coon MJ, Stadtman ER (1983) Inactivation of key metabolic enzymes by mixed function oxidation reactions: Possible implication in protein turnover and ageing. Proc Natl Acad Sci USA 80:1521-1525.

Goodglick LA, Kane AB (1987) Role of reactive oxygen metabolites in crocidolite asbestos toxicity to mouse macrophages. Cancer Res 46:5558-5566.

Hale LW, Gough J, King EJ, Nagelschmidt G (1956) Pneumoconiosis in kaolin workers. Br J Ind Med 13:251-259.

Hoidal JR, Repine JE, Beall GD, Rasp FL, White JG (1978) The effect of phorbol myristate acetate on the metabolism and ultrastructure of human alveolar macrophages. Am J Pathol 91:469-482.

Hoidal JR, Niewoehner DE (1982) Lung phagocyte recruitment and metabolic alterations induced by cigarette smoke in humans and in hamsters. Am Rev Respir Dis 126:548-552.

Kennedy TP, Dodson R, Rao NV, Ky H, Hopkins C, Baser M, Tolley E, Hoidal JR (1989) Dusts causing pneumoconiosis generate ·OH and produce hemolysis by acting as Fenton catalysts. Arch Biochem and Biophys 269:359-364.

Kennedy TP, Rawlings W Jr, Baser M, Tockman M (1983) Pneumoconiosis in Georgia kaolin workers. Am Rev Respir Dis 127: 215-220.

Lebedev AV, Levitsky DO, Loginov VA, Sirnov VN (1982) The effect of primary products of lipid peroxidation on the transmembrane transport of calcium ions. J Mol Cell Cardiol 14 Suppl III: 99-103.

Lynch RE, Fridovich I (1978) Permeation of the erythrocyte stroma by superoxide radical. J Biol Chem 253:4697-4699.

Oldham PD (1983) Pneumoconiosis in Cornish China clay workers. Br J Ind Med 40:131-137.

Ohno Y, Gallin JI (1985) Diffusion of extracellular hydrogen peroxide into intracellular compartments of human neutrophils. Studies utilizing the inactivation of myeloperoxidase by hydrogen peroxide and azide. J Biol Chem 260: 8438-8446.

Polichard A, Collet A (1954) Experimental studies of pathological effects. Schewerze Z allg Path 17:320-327.

Repine JE, Eaton JW, Anders MW, Hoidal JR, Fox RB (1979) Generation of hydroxyl radical by enzymes, chemicals, and human phagocytes in vitro. Detection with the antiinflammatory agent, dimethyl sulfoxide. J Clin Invest 64:1642-1651.

Sausville EA, Peisach J, Horwitz SB (1978) Effect of chelating agents and metal ions on the degradation of DNA by bleomycin and iron (II). Biochem 17: 2746-2754.

Schindler PW, Furst B, Dick R, Wolf PU (1976) Ligand properties of surface silanol groups. Surface complex formation with Fe^{3+}, Cu^{2+}, Cd^{2+}, and Pb^{2+}. Journal of Colloid and Interface Science 55:469-475.

Sepulveda MJ, Vallyathan V, Attfield MD, Piacitelli L, Tucker JH (1983) Pneumoconiosis and lung function in a group of kaolin workers. Am Rev Respir Dis 127:231-235.

Serhan C, Anderson P, Goodman PE, Dunham P, Weissman G (1981) Phosphatidate and oxidized fatty acids are calcium ionophores. Studies employing arsenazo III in liposomes. J Biol Chem 256:2736-2741.

Shatos MA, Doherty JM, Marsh JP, Mossman BT (1987) Prevention of asbestos-induced cell death in rat lung fibroblasts and alveolar macrophages by scavengers of active oxygen species. Environ Res 44:103-116.

Winterbourn CC (1985) Free-radical production and oxidative reactions of hemoglobin. Environ Health Perspect 64: 321-330.

Weitzman SA, Graceffa P (1984) Asbestos catalyzes hydroxyl and superoxide radical generation from hydrogen peroxide. Arch Biochem Biophys 228:373-376.

SOME IN VIVO AND IN VITRO STUDIES CARRIED OUT WITH TALC SAMPLES

S. Endo-Capron, J. Fleury-Feith[1], M. Nebut, R. De Neef, M.C. Jaurand

Equipe Pathologie Cellulaire de l'Environnement
INSERM U139, CHU Henri Mondor
94010 Créteil CEDEX, FRANCE

[1]LHPD, UFR de Médecine
Université Paris XII
94010 Créteil CEDEX, FRANCE

Introduction

Talc is a mineral largely used for many industrial and commercial purposes (see Clarke this meeting). Heavy occupational exposure to inhaled talc can result in pulmonary talcosis and several cases of "talcum powder granuloma" induced by the use of talc on surgical gloves have been reported (IARC 1987). No evidence for a carcinogenic potency has been reported. Experimental data obtained up to 1987 have been evaluated by the IARC (IARC 1987). It has been concluded that there was an inadequate evidence of carcinogenicity in experimental animals. In addition, the degree of evidence in short-term tests for genetic activity has been found to be inadequate. In this paper we report data on experiments carried out to evaluate carcinogenic potency and genotoxicity of talc samples. Carcinogenic potency was determined following inoculation of talc into the pleural cavity of rats and genotoxicity by assessing chromosomal effects in cultures of rat pleural mesothelial cells (RPMC).

To investigate chromosomal effects of talc, we used Sister Chromatid Exchanges (SCEs) analysis. SCEs is believed to be sensitive indicator of the carcinogenic and/or mutagenic potential of various substances. Many investigators have used SCEs analysis on human lymphocytes or in mammalian cells for monitoring studies for exposure to environmental mutagenesis (Perry 1980). A positive linear correlation exists between the induction of SCEs and transformation by several carcinogens in Syrian hamster fetal cells (Popescu et al 1985) and similar correlation exists between the SCEs frequencies and gene mutation at the HGPRT locus in Chinese hamster cells (Speit et al 1986). The SCEs assay seems to be a valuable method for predicting mutagenic and carcinogenic potential.

NATO ASI Series, Vol. G 21
Health Related Effects of Phyllosilicates
Edited by J. Bignon

In vivo experiment

Methods

One hundred fifty two Sprague-Dawley rats were studied : 35 animals 10 weeks-old received an intrapleural injection of 20 mg of talc; 40 received an intrapleural injection of saline, 38 were without any injection and 39 received Canadian chrysotile (90% of fibres <8µm of length determined by transmission electron microscopy, TEM). Animals were sacrified when moribund by an intraperitoneal injection of pentobarbital. After opening the thoracic wall, the pleural cavity was observed and the lungs were fixed in situ by an intrabronchial perfusion of formalin, according to a procedure described elsewhere (Jaurand et al. 1987).

Large sections of each lung and/or tumor when observed, were embedded on paraffin. 5 µm sections stained by hematein eosin were observed under light microscopy.

Talc was obtained from Luzenac (mean size 2.6 ± 2.3 µm determined by TEM).

Results

No pleural tumour was observed in the talc-treated group. In contrast, 25.6% of rats injected with chrysotile fibres developed mesothelioma. No tumour was observed in the control groups.

In vitro experiments

Materials and methods

* *Cell cultures.*

Rat pleural mesothelial cells (RPMC) were obtained as described elsewhere (Jaurand et al, 1981). Cells were routinely grown in medium Ham F10 (Flow) supplemented with 10 % fetal bovine serum (Boehringer), 100 U/ml penicillin (Flow), 50 µg/ml streptomycin (Flow), 2 mM glutamine (Gibco), 1 mM vitamine C (Merck) and 10 mM Hepes (Intermed).

* *Particles and chemicals.*

A sample of European talc (mean size = 4.2 ± 3.8 µm determined by TEM) was tested. Two reference particles, anatase (size < 2 µm; gift from P.Sebastien, CERCHAR) and attapulgite (mean length = 0.7 ± 0.4 µm) were also studied. Furthermore, mitomycin C (Choay) and K_2CrO_4 (Aldrich) were used as positive references.

Particles were suspended in complete medium by sonication for 5 minutes (20 KHz, 3 watts), at a concentration of 560 µg/ml. K_2CrO_4 and 2-bromodeoxyuridine (BrdU) were dissolved directly in complete medium. Mitomycin C was dissolved in distilled water.

* *Cell viability.*

Viability of talc-treated rat pleural mesothelial cells was determined by Trypan blue exclusion. Non cytotoxic concentration of talc were used to detect sister chromatid exchanges.

* *Sister chromatid exchanges analysis.*

RPMC between 8th and 20th passage were used for this assay. Cells were plated at densities of $2x10^6$ in $75cm^2$ flasks. Twenty four hours later, cells were treated with particles, chemicals or references and 3 µg/ml BrdU for 48 hours. Two hours before harvesting cells, 0.2 µg/ml colchicine (Sigma) was added. Chromosome preparations were made according to a technique described by Dutrillaux (Dutrillaux 1975). The last fixation step lasted for one night. The cell suspension was dropped on ice-cold slides.The slides were stained by technique fluorescence plus Giemsa (Perry and Wolff 1974).

Thirty metaphases having 37-42 chromosomes were counted per assay.

* *Statistical analysis.*

The comparison between SCEs rates in treated and untreated cultures was made by Mann Whitney test.

Results

Number of SCEs and the standard deviation for differents treatments are shown in figures 1 to 5. Results obtained with control particles, attapulgite and anatase are reported in figures 1 and 2 respectively. No significant enhancement of SCEs was found. RPMC treated with 2 ng/ml mitomycin (Fig 3) and 0.5 µg/ml K_2CrO_4 (Fig 4) showed a frequency of SCEs highly increased ($p < 0.005$). Figure 5 reports data obtained with talc. No statistically significant enhancement of SCEs was observed in cultures treated with talc, even at the highest concentration (15µg/cm^2).

Discussion

These studies indicate that talc tested in vivo had no carcinogenic potency. In contrast, chrysotile fibres were carcinogenic and the rate of mesothelioma production was comparable to that reported elsewhere (Jaurand et al. 1987). In vitro studies showed that talc sample used in this experiment was not clastogenic. In contrast, SCEs analysis showed that RPMC was sensitive to mitomycin C, a known clastogen (Galloway et al.1979 ; Murli et al.1988) and K_2CrO_4, another clastogen (Latt et al.1982; Majone and Levis 1979).

SCEs frequency in cells treated with 2 ng/ml of mitomycin C increased 3 fold compared to untreated cells. This data indicates that RMPC respond to very low concentration of this substance compared to other cellular systems such as human lymphocytes (Latt 1974) and chinese hamster cell line (Kato and Shimada 1975; Perry 1975).K_2CrO_4 is weaker inducer of sister-chromatid exchanges than mitomycin C. It was reported that Cr^{6+} compounds are weak inducers of sister-chromatid exchanges, in spite of greatly increasing the frequency of chromosomal aberrations. This is also true for other compounds such as caffeine and bleomycin (Majone and Levis 1979).

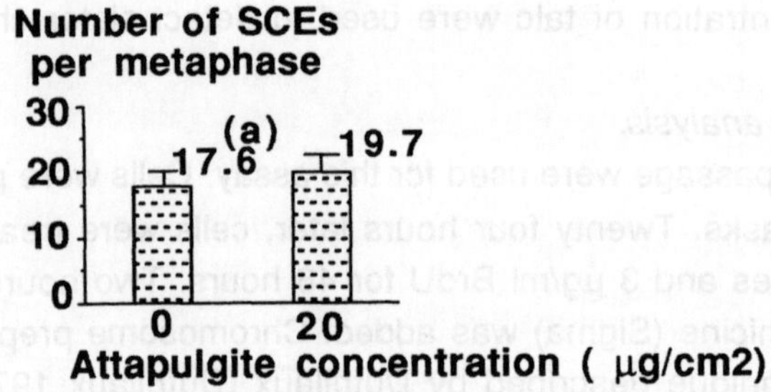

(a) : Numbers represent mean number of SCEs from 3 experiments

fig. 1. Number of SCEs per metaphase after treatment with attapulgite

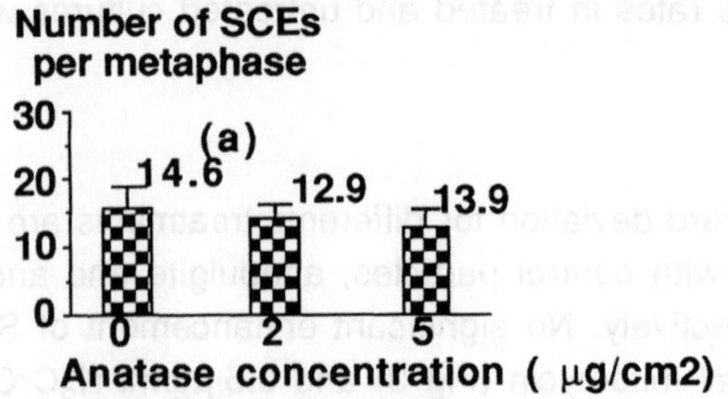

(a) : Numbers represent mean number of SCEs from 9 experiments

fig.2. Number of SCEs per metaphase after treatment with different concentrations of anatase

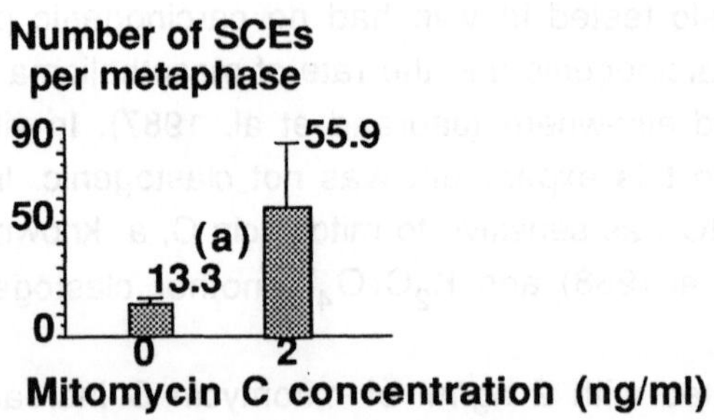

(a) : Numbers represent mean number of SCEs from 8 experiments

fig.3. Number of SCEs per metaphase after treatment with mitomycin C. Comparison of SCEs frequencies between control cells and mitomycin C-treated cells showed a significant difference for each experiment ($p<0.005$)

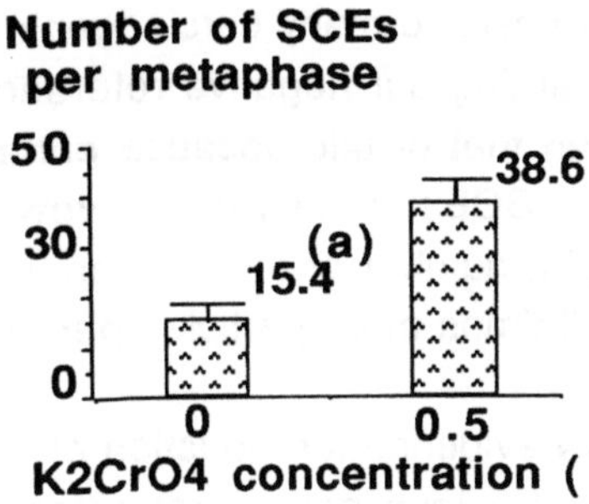

(a) : Numbers represent mean number of SCEs from 8 experiments

fig.4. Number of SCEs per metaphase after treatment with K2CrO4 treated cells showed a significant difference for each experiment ($p<0.005$)

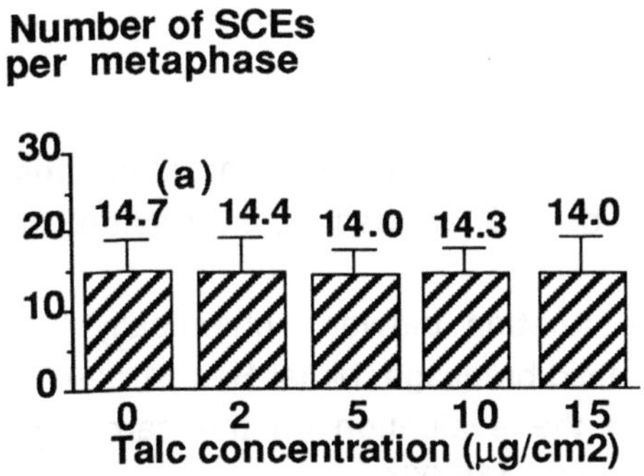

(a) : Numbers represent mean number of SCEs from 8 experiments

fig. 5. Number of SCEs per metaphase after treatment with different concentration of talc

Our results show that RPMC are very sensitive cells to many carcinogenic substances and very good cells for SCEs analysis. Concerning our negative reference, anatase, we used lower concentrations of anatase than that of talc because enhancement of anatase concentration would have impede SCEs observation. However, these particules are very small and at a given weight, number of anatase particles is much higher than the one of talc (there is about 100 times more particles per unit weight of anatase than of talc).

All of 8 experiments performed did not show evidence for induction of SCEs by this talc sample. This data is in agreement with previous papers reporting absence of chromosomal effects observed in mammalian cells in vitro and in vivo (IARC 1987).

Conclusion

One talc sample used in the present experiments did not show indication of tumourigenic potency after intrapleural injection in rats and the other sample did not enchance SCEs on cultures of rat pleural mesothelial cells.

References

Dutrillaux B (1975) Sur la nature et l'origine des chromosomes humains. Monographies des annales de génétiques. L'expansion scientifique, Paris p 90

Galloway SM, Wolf S (1979) The relation between chemically induced sister-chromatid exchanges and chromatid breakage. Mutat Res 61 : 297-307

IARC (1987) IARC monographs on the evaluation on the carcinogenic risk of chemicals to humans. IARC Lyon 42 : 185-224

Jaurand MC, Bernaudin JF, Renier A, Kaplan H, Bignon J (1981) Rat pleural mesothelial cells in culture. In vitro 17 : 98-105

Jaurand MC, Fleury J, Monchaux G, Nebut, M, Bignon J (1987) Pleural carcinogenic potency (asbestos, attapulgite) and their cytotoxicity on cultured cells. J Natl Cancer Inst 49 : 797-803

Kato H, Shimada H (1975) Sister chromatid exchanges induced by mitomycin C :a new method of detecting DNA damage at chromosomal level. Mutat Res 28 : 459-464

Latt SA (1974) Sister chromatid exchanges, indices of human chromosome damage and repair : detection by fluorescence and induction by mitomycin C. Proc Nat Acad Sci, USA 71 : 3162-3166

Latt SA, Schreck RR, Sahar E, Paika IJ, Kittrel C (1982) Basis and biological significance of sister chromatid exchange. In : Sugimura T, Konda S, Takabe H (eds). Proceeding of the 3rd International Conference on

Environmental Mutagens. University of Tokyo Press, Tokyo and Alan R. Liss, Inc. New York, .431-441

Majone F, Levis AG (1979) Chromosomal aberrations and sister chromatid exchanges in chinese hamster cells treated in vitro with hexavalent chromium compounds. Mutat Res 67 : 231-238

Murli H, Galloway SM, Ivett JL, Parry DM, Mulvihill JJ (1988) Repair of sister chromatid exchange-inducing lesion in mutagen-treated cultures of human whole blood and purified fresh or frozen lymphocytes. Mutat Res 202 :125-132

Perry PE and Wolff S (1974) New giemsa method for the differential staining of sister chromatid. Nature 251 : 156-158

Perry PE, Evans HJ (1975) Cytological detection of mutagen-carcinogen exposure by sister chromatid exchange. Nature 258 : 121-125

Perry PE (1980) Chemicals mutagens and sister-chromatid exchange. In: De Serres FJ, Hollaender A (eds) Chemical mutagens, Principles and Methods for their detection, Vol 6, Plenum Press, New York and London, 1-39

Popescu NC, Amsbaugh SC, DiPaolo JA (1985) Persistence of sister-chromatid exchanges and in vitro morphological transformation of Syrian hamster fetal cells by chemical physical carcinogens. Carcinogenesis 6 : 1627-1630

Speit G (1986)The relationship between the induction of SCEs and mutation in Chinese hamster cells. Mutat Res. 174 : 21-26

Slate Dust Toxicity

Qamar Rahman, M.F. Khan and S.V. Singh
Industrial Toxicology Research Centre
Post Box No. 80, Mahatma Gandhi Marg
Lucknow-226001, India

Introduction

Chronic exposure to slate dust causes pneumoconiosis (Hunter 1969). Surveys conducted at Mandsour (India), where slate pencil Industries are situated, revealed serious health problems among workers and general public exposed to slate dust (Jain et al, 1980; Chakarvarti, 1981). As there was no available data on the experimental aspects of slate dust toxicity, we have, therefore, conducted a series of experiments (Singh et al, 1982; Khan et al 1983; Singh et al 1983; Singh & Rahman 1987) to elucidate the toxicity of slate dust to which large number of people are exposed. In the present paper we have studied hemolytic and cytotoxic behaviour of slate dust along with biochemical parameters related to fibrogenic responses under stress conditions, induced by slate dust exposure.

Materials & Methods

Dust samples

Slate dust was collected from Mandsour, M.P., India, and reduced to particle size below 5 μm according to Zaidi (1969).

Chemicals

All the chemicals utilised during the study were of Analar or GR grade either from BDH or E. Merck.

Hemolytic studies

The hemolytic studies were conducted in human RBCs according to Rahman et al (1974).

Animal experimentation

Female albino rats bred in ITRC animal colony weighing 150-200 g, were used. The dry dust and 0.15 M NaCl were separately autoclaved

NATO ASI Series, Vol. G 21
Health Related Effects of Phyllosilicates
Edited by J. Bignon

at a pressure of 15 lbs for 15 min. The dust was suspended in 0.15M NaCl just before innoculation. Rats were intratracheally instilled with 50 mg of dust suspended in 1.0 ml of 0.15 M NaCl according to the procedure of Zaidi (1969). The corresponding control animals received 1.0 ml of normal saline. All the animals were maintained on a laboratory stock diet, supplied by Hindustan Lever Ltd. Bombay, India. Six animals from each group were sacrificed at 1,2,4,8, 16, 30,60,90,120 and 150 days after instillation.

Preparation of lung lavage, homogenate and mitochondria

The washing of lungs for lavage collection and preparation of cellular and acellular fractions were carried out as described by Khan et al (1983). For the preparation of homogenate, lungs were minced and then homogenized in sucrose-mannitol medium (0.15 M sucrose, 0.15M Mannitol, 1 mM Tris-HCl, 1 mM EDTA, pH 7.4). The homogenate was filtered through muslin cloth (mesh size 150x100 μ) and made upto 10% concentration (w/v) with the same medium. Isolation of mitochondria was performed according to Mustafa (1974).

Acid phosphatase assay

Acid phosphatase (EC 3.1.3.2) was assayed according to Oser (1965).

Mitochondrial enzyme assays

Monoamine oxidase (EC 1.4.3.4) was assayed according to Green and Haughton (1961), cytochrome c oxidase (EC.1.9.3.1) according to Cooperstein and Lazarow (1951), and glutamate dehydrogenase (EC.1.4.3.2) according to Olson and Anfinsen (1952).

Estimation of lung tissue constituents

A weighed portion of lung was dried at 110°C to a constant weight and powdered. Fifty mg of dry tissue was first hydrolysed with 6N HCl and then neutralized with 6N NaOH. Using suitable aliquots, collagen, hexosamine and sialic acid contents were assayed according to methods of Stegmann (1958), Ashwell (1957), Whitehouse & Zilliken (1963) respectively.

Protein

Protein was estimated according to Lowry et al. (1951), using crystalline bovine serum albumin as standard.

Phospholipids

The total lipids were extracted according to Folch et al (1957) and total inorganic phosphorus was estimated according to Wagner et al (1962). Lipid pi multiplied by 25 (based on dipalmitoyl lecithin, $C_{40}H_{78}O_9$ NP, mol. wt. 748.1; 4.1% phosphorus) gave the amount of phospholipids.

Statistical analysis

All values are the result of at least six determinations per point. Standard deviations were less than 10% of the mean value.

Results & Discussion

Hemolytic index of slate dust

In Figure 1, the lytic activity of slate dust is recorded. Hemolytic activity increased up to 4.5 mg, beyond this there was a slow decline.

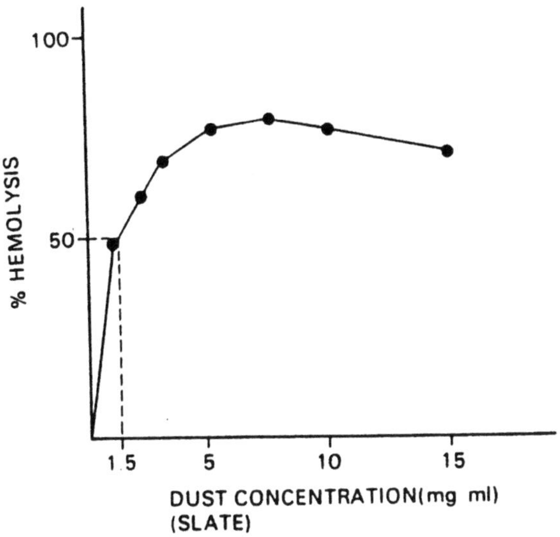

Fig. 1. Effect of varying slate dust concentration on human erythrocyte lysis. The incubation time was fixed for two hours.

From these data, it follows that 1.5 mg dust caused 50% lysis. Calculated from this hemolytic index, the lytic activity of slate dust actually was 666 units/g dry dust, which is comparable with the most powerful hemolytic mineral fiber chrysotile, where lytic index is 833 units/g dry dust (Singh et al 1983).

Cytotoxic effects of slate dust

Numerical alterations in the free cell population and changes in the acid phosphatase activity in lung lavage is reported in Table 1. Slate dust stimulated free cell population even after 24 hours of exposures as a defensive mechanism reaches maximum at four days when almost five fold increase recorded. Subsequently as a result of cytotoxicity the free cell number, decreases with the progress of toxic conditions. The increase in acid phosphatase activity in the acellular fraction and the gradual decrease by detergent activation in the cellular fraction clearly indicated the cytotoxic action of slate dust *in vivo*. There is another interesting observation that changes in the free activity was correlated with the gradual loss of latency of the cellular fraction. The latency is not totally broken even at the most advanced state since a doubling of activity occurs on detergent treatment. Apparently, slate dust caused partial macrophage membrane damage *in vivo* which was observed with asbestos also.

Biochemical & fibrogenic effects of slate dust

Changes in the lung tissue constituents is recorded in Table 2. Protein and phospholipid contents of lung showed a maximum increase of 23 and 34%, respectively, at 150 days. One of the salient features of the development of slate dust toxicity is the increase in collagen content which starts after 90 days of exposure and about 45% higher ($P<0.001$) at 150 days. This clearly indicates that fibrosis commences as a result of slate dust exposure. The concurrent increase in hexosamine and sialic acid in the lungs of experimental animals may lead to the accumulation of mucopolysaccharides. Mucopolysaccharide accumulation has been reported in various types of pneumoconiosis, such as in the early proliferative phase of silicosis (Singh et al, 1975), and in asbestosis (Viswanathan et al, 1978; Misra et al, 1976). As in the case of asbestosis (Singh et al 1976), phospholipid content of lungs also increased as a result of slate dust exposure.

Effect of slate dust on the mitochondrial enzymes

The changes in cytochrome c oxidase, monoamine oxidase and glutamate dehydrogenase activities in the lung mitochondria after intratracheal instillation of slate dust has been recorded in Table 3. The cytochrome c oxidase activity in the treated animals was generally higher at all stages, being 85% ($P<0.01$) higher at 150 days. Monoamine oxidase, the marker for outer membrane of mitochondria, was unaffected at most of the stages and only a 20% increase was observed at 150 days. The free glutamate dehydrogenase as well as total activity in experimental animals was high as compared to respective controls.

The biochemical changes produced by slate dust in mitochondria is of great physiological significance. Our findings, like increase in cytochrome c oxidase activity on slate dust exposure indicate the involvement of mitochondria in the pathogenesis of slate dust-induced toxicity, which is in agreement with silica (Singh et al 1977), and asbestos (Beg et al 1973, Das et al 1983) toxicity. The enhanced terminal oxidation will help in furnishing additional energy needed for biochemical changes like collagen deposition through the process of oxidative phosphorylation. Another line of evidence for the involvement of mitochondria is the increasing trend of glutamate dehydrogenase activity. It may be mentioned here that glutamate dehydrogenase activity in mitochondria has long been used as a marker for the intactness of mitochondria (Endahl & Kochakia, 1957), and that the integrity of the mitochondrial membrane is the most vital pre-requisite for their proper functioning in such processes as oxidative phosphorylation(Mitchell 1977).

Conclusion

In conclusion, our studies indicate that slate dust is not only potential hemolytic agent, but also cytotoxic to macrophages, and induces marked biochemical and fibrotic changes in the lung of experimental animals. Further, biomembranes are the target of slate dust toxicity. On the basis of these observations it is, therefore, suggested that the workers exposed to slate dust needs all the precautions as required for silica and asbestos exposures.

Table 1

Number of free cells and acid phosphatase activity in the lung lavage of control and slate dust exposed rats[a]

	1 day	2 days	4 days	8 days	16 days	30 days	90 days
No. of free cells x 10^6 per lung							
Control	9.72	9.01	7.76	7.50	9.58	6.02	6.18
Treated	24.18	28.19	41.53	25.27	17.82	12.55	11.14
Free acid phosphatase activity in lung lavage:							
Cellular fractions*							
Control	31.74±5.71	38.19±4.89	40.35±6.0	35.57±4.21	43.93±7.35	50.99±7.07	32.51±5.93
Treated	42.30±1.38	51.26±5.81	80.13±4.09	106.09±2.51	75.51±3.62	90.05±4.20	78.15±3.56
Acellular fraction**							
Control	11.43±1.64	14.18±1.63	17.35±2.45	16.67±2.01	18.06±0.84	15.72±2.67	14.44±1.51
Treated	14.54±3.26	16.45±0.98	25.72±1.65	35.08±0.96	49.63±1.64	70.19±1.47	85.14±0.51
Activation with Triton X 100 in cellular fraction***							
Control	3.02±0.14	3.50±0.07	3.74±0.045	3.91±0.12	3.22±0.095	3.07±0.12	3.15±0.15
Treated	3.22±0.155	3.38±0.12	3.43±0.097	3.15±0.095	2.95±0.16	2.48±0.085	1.95±0.105

*Values expressed as µg Pi liberated/h at $37^{o}C/10^{6}$ cells.

**Values expressed as µg Pi liberated/h at $37^{o}C$/total lavage fluid.

***Expressed as activity with detergent/activity without detergent.

[a]Values are arithmetic mean±S.D. of six determinations.

Table 2

Changes in the lung tissue constituents of slate-dust exposed rats[a]

No. of days after administration	Collagen (mg per 100 mg of dry tissue)	Hexosamine (µg per g fresh tissue)	Sialic acid (µg per g fresh tissue)	Phospholipids (mg per g fresh tissue)	Protein (mg per g fresh tissue)
15	3.46±0.17 (4.31±0.30)	718.98±49.61 (791.92±72.19)	673.69±28.69 (660.00±14.08)	20.11±0.47 (19.84±0.46)	96.60±3.56 (95.53±2.87)
30	3.16±0.09 (3.91±0.46)	823.18±64.79 (792.00±34.03)	728.24±18.63 (644.09±19.91)	20.40±0.58 (19.98±0.30)	98.25±1.71 (93.45±2.94)
60	4.39±0.13 (4.47±0.22)	896.12±38.04 (844.02±26.90)	725.51±25.97 (651.87±18.63)	21.06±0.86 (20.19±0.46)	108.00±3.26 (97.49±2.44)
90	4.85±0.19 (4.55±0.23)	1156.62±43.38 (922.17±35.59)	902.80±47.45 (695.51±27.27)	22.36±0.63 (20.31±0.26)	115.72±3.82 (98.75±4.22)
120	5.47±0.13 (4.62±0.14)	1443.11±62.23 (937.80±53.80)	1052.81±37.26 (709.15±8.90)	23.76±1.05 (20.47±0.36)	121.79±3.00 (98.75±2.05)
150	6.71±0.24 (4.63±0.14)	1318.19±123.14 (963.86±70.93)	1418.30±92.14 (714.60±25.97)	27.60±0.45 (20.57±0.15)	121.60±2.94 (98.75±1.22)

[a]Values are arithmetic mean ± SD of six animals. Control values are given in parentheses.

Table 3

Mitochondrial protein, phospholipids and enzyme activities in normal and slate dust exposed rats[a]

No.of days after administration	Protein (mg per g fresh tissue weight)	Phospholipid (mg per g fresh tissue weight)	Monoamine oxidase activity*	Cytochrome c oxidase activity**	Glutamate dehydrogenase activity***	
					Free	Total
15	10.35±1.23 (10.80±0.61)	3.15±0.13 (3.11±0.10)	1420±116 (1459±45)	864±30 (824±32)	410±54 (400±27)	770±25 (785±27)
30	8.48±1.49 (8.66±1.18)	4.04±0.26 (3.18±0.26)	1400±64 (1415±79)	1004±28 (985±40)	385±16 (312±22)	730±25 (705±32)
60	12.67±0.46 (11.60±0.46)	4.15±0.44 (3.15±0.15)	1598±94 (1460±85)	1011±60 (903±55)	458±52 (398±14)	814±12 (754±23)
90	13.57±0.77 (11.96±0.56)	4.87±0.26 (3.18±0.15)	1657±102 (1523±27)	1099±143 (824±101)	567±69 (400±31)	850±12 (711±14)
120	14.10±1.70 (12.14±0.50)	4.82±0.35 (3.28±0.23)	1765±67 (1578±32)	1374±32 (942±55)	603±59 (410±34)	886±41 (748±20)
150	15.98±0.89 (11.87±0.89)	4.51±0.39 (3.35±0.30)	1953±114 (1617±111)	1885±157 (1020±115)	675±60 (446±50)	832±32 (743±65)

*n moles of substrate utilized per min per g fresh tissue weight

**n moles of cytochrome c oxidized per min per g fresh tissue weight.

***n moles of NADH oxidized per min per g fresh tissue weight.

[a]Values are arithmetic mean ± SD of six animals. Respective control values are given in parentheses.

References

Ashwell G (1957) Colorimetric analysis of sugars. In: Colonick S.P., Kaplan N.O. (eds) Methods in Enzymology Vol 3, Academic Press, New York, p 73.

Beg MU, Rahman Q, Viswanathan PN, Zaidi SH (1973) The effects of asbestos dust on mitochondrial enzymes of rat lung. Environ Physiol Biochem 3: 185-191.

Chakarvarti P (1981) Silicosis-A fatal disease among industrial workers. Sci Reporter 18: 299-306.

Cooperstein SJ, Lazarow A (1951) A microspectrometric method for the determination of cytochrome c oxidase. J Biol Chem 189: 665-670.

Das B, Misra V, Viswanathan PN, Rahman Q (1983) Lung mitochondria in experimental asbestosis. Environ Res 31: 390-399.

End:hl GL, Kochakia CD (1957). Ionic strength of homogenizing medium or glutamic dehydrogenase activity of tissues. Proc Soc Exptl Biol Med 94: 192-193.

Folch J, Lees M, Stanleg GHS (1957) A simple method for the isolation and purification of total lipids from animal tissues. J Biol Chem 226: 497-509.

Green AL, Havghton TM (1961) A colorimetric method for the estimation of monoamineoxidase. Biochem J 78: 172-175.

Hunter D (1969) Diseases of occupations, 4th edn English Universities Press.

Jain SM, Sepaha GC, Khare KC, Dubey VS (1980) Ventilatory functions and sputum cytology in slate pencil workers silicosis. Ind J Chest Dis Allied Sci 22: 103-109.

Khan MF, Jaffery FN, Ali S, Rahman Q (1985). Biochemical studies on the toxicity of slate mine dust. Environ Hlth Perspect 51: 305-310.

Lowry OH, Rosebrough NJ, Randall RJ (1951). Protein measurement with Folin phenol reagent. J Biol Chem 195: 265-275.

Misra V, Rahman Q, Viswanathan PN (1978). Biochemical changes in guinea pig lungs due to amosite asbestos. Environ Res 16: 55-61.

Mitchell P (1977). Vectorial chemosmotic processes. Ann Rev Biochem 46: 996-1005.

Mustafa MG (1974). Augmentation of mitochondrial oxidative metabolism in lung tissue during recovery of animals from acute ozone exposure. Arch Biochem Biophys 165: 531-538.

Olson JA, Anfinsen CB (1952). The crystallization and characterization of L-glutamic acid dehydrogenase. J Biol Chem 197:67-79.

Oser BL (1965). Determination of serum phosphatase activity In: Oser B (ed) Hawk's Physiological Chemistry 14th ed. McGraw Hill, New York, p 1118.

Rahman Q, Kaw JL, Narang S, Zaidi SH (1974). Asbestos induced hemolysis in relation to its silica solubility. Environ Physiol Biochem 4:284-288.

Singh J, Kaw JL, Viswanathan PN, Zaidi SH (1975). Changes in free and bound amino acids in experimental silicosis. Environ Physiol Biochem 5: 165-171.

Singh J, Beg MU, Kaw JL, Viswanathan PN, Zaidi SH (1976). Biochemical changes and pulmonary response of guinea pigs to asbestos dust. Acta Pharmacol Toxicol 39: 77-86.

Singh J, Viswanthan PN, Pandey SD, Zaidi SH (1977). Changes in mitochondrial enzymes activity of rat lung during the development of silicosis. Life Sci 20: 367-374.

Singh SV, Das B, Rahman Q, Viswanathan PN, Murti CRK (1982). Indust Hlth 20: 139-145.

Singh SV, Viswanathan PN, Rahman Q (1983). Interaction between erythrocyte plasma membrane & silicate dusts. Environ Hlth Perspect 51: 55-60.

Singh SV, Rahman Q (1987) Interrelationship between hemolysis and lipid peroxidation of human erythrocytes induced by silicic acid and silicate dusts. J Appl Toxicol 7:91-96.

Stegmann J (1958) Microdetermination of hydroxyproline with chloramine T and p-dimethylaminobenzaldehyde. Hoppeseylers Z Physiol Chem 311: 41-45.

Viswanathan PN, Dogra RKS, Shanker R, Zaidi SH (1973). Pulmonary fibrogenic response of guinea pigs to amosite dust. Inter Arch Arbeitsmed 31: 51-59

Wagner H, Lissau A, Holzl J, Horhammer L (1962) The incorporation of ^{32}P into the inositol phosphatides of rat brain. J Lipid Res 3:177-180.

Whitehouse MW, Zilliken F (1963). Isolation and determination of sialic acid. Meth Biochem Anal 8: 215-216.

Zaidi SH (1969). Experimental Penumoconiosis. John Hopkins Press, Bultimore p. 33.

BIOLOGICAL EFFECTS OF MICA DUST IN EXPERIMENTAL ANIMALS

ANAND P. SAHU
Scientific Commission for Continuing Studies on Effects of Bhopal Gas Leakage on Life Systems
Second Floor, Sardar Patel Bhavan
New Delhi-110 001
INDIA

ABSTRACT

Earlier survey of Mica miners of Bihar province in India revealed a high incidence of silicatosis in workers handling mica. Studies dealing with the biological effects of muscovite variety of Indian mica on experimental animals were done. Pulmonary fibrotic lesions and many cholesterol cleft like structures were observed at 210 days after intratracheal injection of 5 mg of mica dust in mice. The lymph nodes showed marked density of mica in cortico-medullary zone and medullary cords where swollen macrophages were fully packed with mica. Intraperitoneal injection of 5 mg of mica dust in mice at 90 days produced thick dense reticulination together with excessive lymphoid hyperplasia in omental tissue. In lymph node slight reticulinosis was seen in the medullary region corresponding to deposition of particles. In rats, 25 mg of mica dust by intratracheal injection at 330 days, produced small compactly arranged fibrotic lesions in lung parenchyma and lymph nodes were markedly increased in size with fibrosis in the medullary region which involved reticulin and collagen fibres. In guinea pigs 75 mg of mica dust after intratracheal injection provoked fibrotic lesions with thick reticulin fibres in the lymph nodes at 365 days. The poor fibrogenic response of mica dust may not produce severe adverse effects in early stage. However, prolonged inhalation may complicate with the exposure to other environmental chemicals and/or infections. When combined with mica choline produced in rats severe fibrotic reaction, collagen formation and adenocracinoma of bronchiolar epithelium and marked lymphadenopathy with abnormal cells in the lymph nodes.

INTRODUCTION

Micas are complex aluminium silicates combined with either potassium, magnesium or iron. The more common varieties are muscovite or white mica [$K_2O.\ Al_2O_3.\ 6\ SiO_2.\ 2H_2O$] biotite [($KH_2)_2\ MgFe)_2\ (Al\ Fe)_2\ SiO_4)_3$] and phlogopite or

NATO ASI Series, Vol. G 21
Health Related Effects of Phyllosilicates
Edited by J. Bignon

amber mica $[(KH)_3\ Mg_3\ Al\ (SiO_4)_3]$. The muscovite or white mica is transparent and colourless in thin sheet and is distinguished by its highly perfect clevage. It is one of the important and indispensable mineral used as insulating material in electrical and electronic industries. By viture of its thermal and dieletric properties it is used in Telestar, Transmission Satellites, Space Research Programme, Aeroplane industries and many other industries. During operational processes of crude mica, workers associated with either mining operation or their use in industry, inhales mica dust and varying degrees of pulmonary silicatosis have been observed (Selikoff, 1972). The present communication describes the biological effects of muscovite variety of Indian mica in experimental animals.

METHODS

Samples of mica dust used in the investigations was obtained from Bihar province of India. The particle size below 5 um was made according to the procedure described by Zaidi (1969). The chemical composition of mica dust is given in Table 1. The experimental protocol is given in Table 2.

Table 1. Chemical Composition of Mica dust

Constituents	Per cent
Silica	46.29
Alumina	34.70
Potassium Oxide	11.48
Loss on Ignition	2.97

The animals were inoculated and maintained under normal laboratory conditions and killed at various periods as mentioned in Table 2. The left and right tracheobronchial lymph nodes were carefully excised and fixed in Bouin's solution. The lungs were fixed *in situ* by 10 per cent formol saline. Paraffin imbedded blocks were prepared and histological procedures were used for studies.

Table 2. Experimental Protocol

Animal	Number	Route of Adminis-tration	Quantity of dust in 0.15 M NaCl (mg/ml) per animals (sterilized)	Period of Study
Mice	80	IT[a]	5.0/1.0	7, 15, 30, 60, 90, 120, 150 and 210 days
	40	IT (control)	0.0/0.1	
	20	IP[b]	5.0/0.1	24 hrs, 3, 7, 15, 30, 60 and 90 days
	20	IP (control)	0.0/0.1	
Rats	10	IT	25/1.0	90, 180 and 330 days
	10	IP, Choline Chloride (CC)[c]	-	
	10	IT + IP (CC)	25/1.0	
	10	Control		
Guinea Pigs	36	IT	75/1.5	24 hrs, 7, 15, 30, 60, 90, 120, 180, 300 and 365 days
	4	Control	-	

[a] Intratracheal administration

[b] Intraperitoneal administration

[c] Intraperitoneal administration of 40 injections of 25 mg Choline Chloride (CC) in 1 ml strile distilled water given for 5 days/ week/animals

RESULTS

In the lungs of mice, after intratracheal inoculation of 5 mg of mica dust produced, acute inflammatory reaction at early periods, and later many discrete foci of dust laden macrophags together with marked perivascular and peribranchiolar lymphoid hyperplasia. In the lymph node at an early stage extracellular deposition of dust particles were observed, and later marked density of birefringent particles in cortico-medullary zone and medulla together with little reaction. At 210 days postinoculation, there was shrinkage of pulmonary lesions and many cholesterol cleft like structures were found with increased fibroblastic activity and infiltration of lymphocytes. In the lymph nodes, marked density of mica in the cortico-medullary zone and medullary cords where swollen dust laden macrophages were fully packed with mica dust and their nuclei stained faintly were observed. Intraperitoneal injection of 5 mg of mica dust in mice produced lymphoid hyperplasia around the dust mass together with active proliferation of fibroblasts in the omental tissue. At 90 days there was increased fibroblastic activity together with lymphoid hyperplasia which showed thick dense reticulation. In the lymph node at 90 days, the macrophages were fully packed with dust particle, and retained their normal histological characteristics.

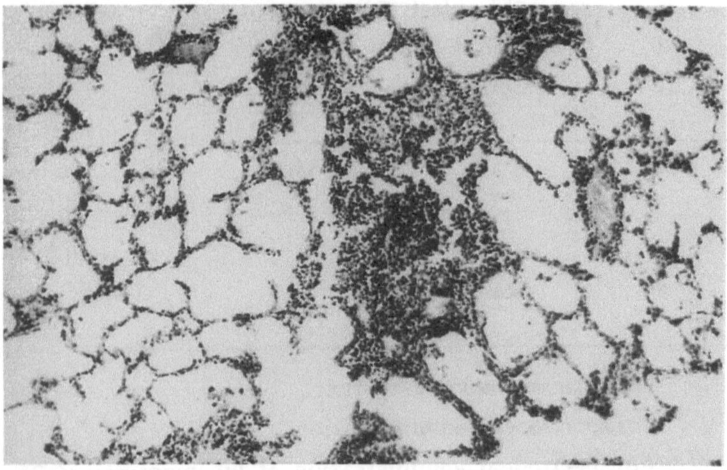

Figure 1. Section of rat lung at 330 days after intratracheal inoculation of Mica dust, fibrotic lesions. Hematoxylin and eosin x 200

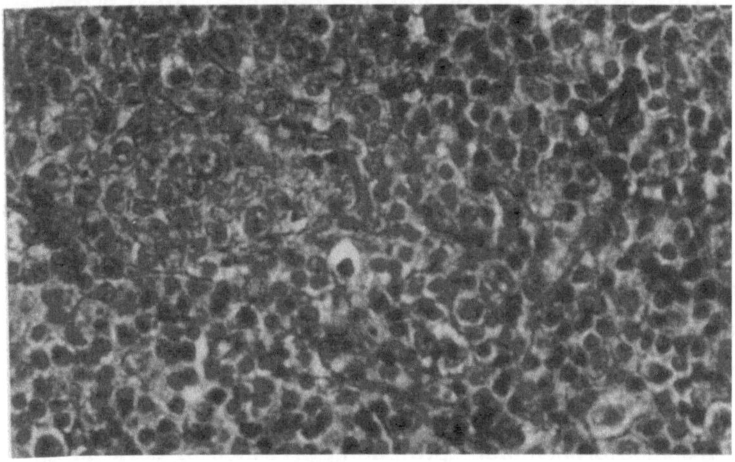

Figure 2. Section of tracheobronchial lymph node of rat at 180 days after intratracheal inoculation of Mica dust. Dust cell reaction in the paracortical region. H & E x 800.

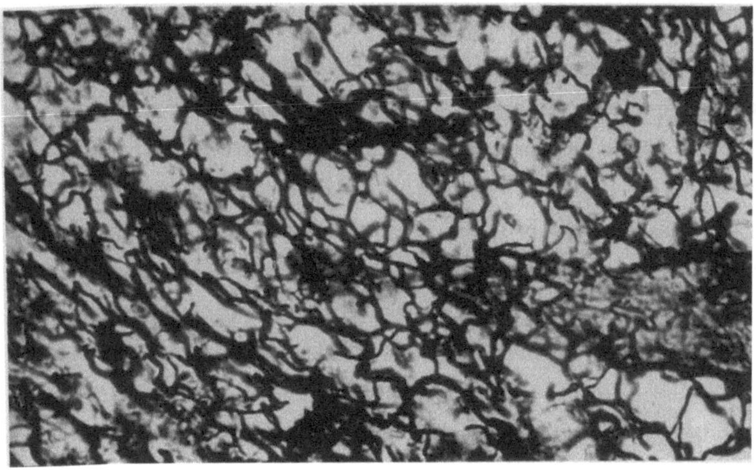

Figure 3. Tracheobronchial lymph node section of rat lung at 330 days after intratracheal inoculation of Mica dust. Reticulin and collagen fibres. Silver inpregnation x 800.

In rats, 25 mg of mica dust by intratracheal inoculation produced characteristic dust cell reaction with thin reticulin fibres in the lungs at 90 days. At 180 days small compactly arranged cellular lesions were found in the lung parenchyma (Figure 1). However at 330 days, small compactly arranged cellular lesions which contained homogenous, acellular eosinophilic areas. In lymph nodes at 180 days small to large dust foci were observed in the paracortical and medullary cord regions containing degenerating dust laden macrophages along with marked cellular reaction (Figure 2). Lymph nodes at 330 days showed marked cellular reaction, fibrosis in the medullary region which involved reticulin and collagen fibres (Figure 3).

In the combined group with mica dust and choline, at 90 days the lung showed hypertrophied bronchiolar epithelium, presence of dark basophilic lymphoid cells around bronchioles and blood vessels. At 180 days, the adenomatoid changes were found around bronchiolar lumen. At 330 days the bronchiolar epithelium continued to exhibit adenomatoid changes, and adenocarcinoma were also observed. The lymph nodes at 180 days showed an occasional blue-grey sheet of abnormal cells in the subcapsular region in Giemsa stained sections. Small dust foci and plasma all reactions were very prominent in cortex and medullary cord. Silver impregnation showed several epitheloid cell granulomata to contain remnants of large dark abnormal cells entrapped in reticulin fibres. At 330 days the lymphs node showed predominently histiocytoid cell reaction, blue grey sheets of large abnormal aggregated cells and thickned capsule. Reticulin fibres were increased especially in the medullary cord and paracortical region.

Intratracheal inoculation of 75 mg mica dust in guinea pigs, lymph nodes at 30 days showed small foci of mica dust and pigment laden macrophages along with some extracellular deposits of birefringent particles. At 90 days very little evidence of cytotoxicity was observed along with loose network of reticulin fibres. At 300 and 365 days there was overall increase in the density of birefingent particles in the cortico-medullory zone with marked proliferation of lymphocytes and fibroblasts. The swollen dust laden macrophages retained their normal structure and were associated with the formation of thick reticulin fibres though the density of birefringent particles remained high.

DISCUSSION

From the present studies, it is evident that mica dust produces poor fibrogenic response but definite fibrotic reaction at various stages after

Intratracheal inoculation in mice, rats and guinea pigs. Mica dust administered in mice by intratracheal and intraperitoneal route and in rats by intratracheal route, there was considerable transport of dust particles from lungs to regional lymph nodes in comparison to that in guinea pigs, the slow transport of dust particles was observed. May particulate pollutants are known to alter the structure of cell membrances and causing damage (Kilroe - Smith, 1974). Choline is an essential component of all membrances and interaction between it an particulate pollutant could be pathogenic. In rats when combine with mica dust by intratracheal route and choline by parenteral administration induced severe fibrotic reaction, collagen formation and adenocarcinoma of bronchiolar epithelium and marked lymphadenopathy with abnormal cells in the lymph nodes (Sahu et al, 1984, 1985).

Morrow (1972) stated that mere fibrotic replacement of nodal tissue due to chemical substances should not only be considered as harmful effects, there could still be more dangerous effects and sequalae. Maigetter and his associate (1976) have pointed out that the exposure to particulate dust in mice alters the resistance to bacterial and viral pneumonias. Earlier, animal experiment show that tuberculous infection of the lungs combined the inhalation or intratracheal injection of various dusts and coal can induce extensive fibrosis (Zaidi et al. 1955). The tuberculous infection or other infections combined with the exposure of mica dust can produce massive fibrosis in mica miners of India.

REFERENCES

Kilroe-Smith TA (1974) Peroxidative action of quartz in relation to membrane lysis. Environ Res 7:110-116.

Maigetter R Z, Erlich R, Fenters J D, Gardner D E (1976) Potentiating effects of manganese dioxide on experimental respiratory infections. Environ Res 11:386-391.

Morrow P E (1972) Lymphatic drainage of the lung in dust clearance. Ann N Y Acad Sci 200:46-65.

Sahu A P, Shukla L J, Krishna Murti C R (1984) Effect of mica dust and choline on the lymph nodes of rats. Brit J exp Path 65:533-541.

Sahu A P, Singh K P, Shukla L J, Shanker R (1985) Choline and Mica dust induced pulmonary lesions in rats. Indust Health 23:135-144.

Selikoff I J (1972) Occupational Lung diseases. In : D H K Lee (ed) Environmental Factors in Respiratory Disease, Academic Press, New York p.199-217.

Zaidi S H (1969) Experimental Pneumoconiosis, The Johns Hopkins Press, Baltimore, p.35-41.

Zaidi S H, Harrison C V, King E J, Mitchison DA (1955) Experimental infective pneumoconiosis. Massive pulmonary fibrosis produced by coal-mine dust and isoniazid-resistant tubercle bacilli of low virulence. Brit J exp Path 36:553-559.

ASSESSMENT OF THE FIBROGENICITY OF ATTAPULGITE

R Bégin, S Massé and P Sébastien
Unité de recherche pulmonaire
Université de Sherbrooke
Sherbrooke (QC) J1H 5N4
and
McGill School of Occupational Health
Montréal (QC)
Canada

INTRODUCTION

Attapulgite, also named palygorskite, is a fibrous clay mineral extracted from the soil in several countries, mainly USA, France, Spain, India and USSR for a world consumption in excess of one million tons per year. The material is an hydrated magnesium aluminum silicate with unique chain structure that gives it unusual colloidal and sorptive properties. Its colloidal properties find applications in paints and drilling mud whereas its sorptive properties find uses as decolorating and clarifying agents, in water purification, animal litter, pesticide carrier, pharmaceutical antacid, etc...(Haden and Schwint, 1967; Hénin and Caillère, 1975). In recent years, it has been increasingly used as asbestos replacement, particularly in friction materials (Washabaugh, 1983) and in several other applications which resulted in a forcasted world consumption that could triple in the coming years (Anonymous, 1985).

Because the mineral is a short thin fibrous mineral which could be inhaled and reach the alveolar space, its potential toxicity has been considered. Attapulgite has been tested in vitro and found to be cytotoxic and a carcinogen binder as asbestos (Bignon et al, 1980; Jaurand et al, 1979; Harvey et al, 1984). In vivo, the potential of the material to induce pleural mesothelioma has been documented in animals (Pott et al, 1976). Furthermore, the fibrogenic lung activity of the material has been questioned in an attapulgite mine engineer with lung fibrosis (Sors et al, 1979).

NATO ASI Series, Vol. G 21
Health Related Effects of Phyllosilicates
Edited by J. Bignon

In this study, we compared the lung toxicity of american attapulgite, UICC canadian chrysotile and short chrysotile fibers.

MATERIALS AND METHODS

Animals. Sixty-four sheep weighing 25-40 kg were enrolled in this study. They were prepared and accustomed to the pulmonary techniques as previously reported (Bégin et al, 1983; 1985A).

Experimental design. The flock was divided into a group of 16 sheep exposed once to phosphate buffered saline (PBS) only (saline exposed sheep), a group of 16 sheep exposed once to 100 mg of UICC canadian chrysotile in 100 ml PBS (UICC asbestos exposed sheep), a group of 16 sheep exposed once to 100 mg U. of Sherbrooke short chrysotile in 100 ml PBS (short asbestos exposed sheep) and a group of 16 sheep exposed once to 100 mg of american attapulgite in 100 ml PBS (attapulgite exposed sheep). Exposures were carried out through bronchoscopic canulation and a slow infusion in the tracheal lobe. The animals were studied prior to exposure day 0 and at day 12, 24, 40 and 60 after exposure by analyses of cellularity and biochemistry of bronchoalveolar lavage (BAL) and histopathology of the tracheal lobe at day 60.

Fibrous minerals. The UICC canadian chrysotile samples were kindly provided by Dr. V. Timbrell of the Medical Research Council, Pneumoconiosis Research Unit, Penarth. These fibers have been well characterized (Timbrell et al, 1978) nearly 100% being longer than 2 µm, 42% > 5 µm, 21% > 10 µm and 8% > 20 µm. The short chrysotile fibers were kindly obtained from Dr. C. Jolicoeur of the Department of Chemistry, U. of Sherbrooke after separation with selective aqueous sedimentation technique (Jolicoeur et al, 1981), with 50% of fibers shorter than 0.5 µm, 98% < 3 µm and 100% < 8 µm (mean length = 0.92 µm, diameter 0.08 µm). The attapulgite samples were obtained from Dr. P. Sébastien of the Dust Disease Unit, McGill University. The latter have also been well characterized with a mean length of 0.8 µm (range 0.25 - 4 µm), diameter of 0.02 µm (fig. 1) and originated from the northern part of Florida, USA.

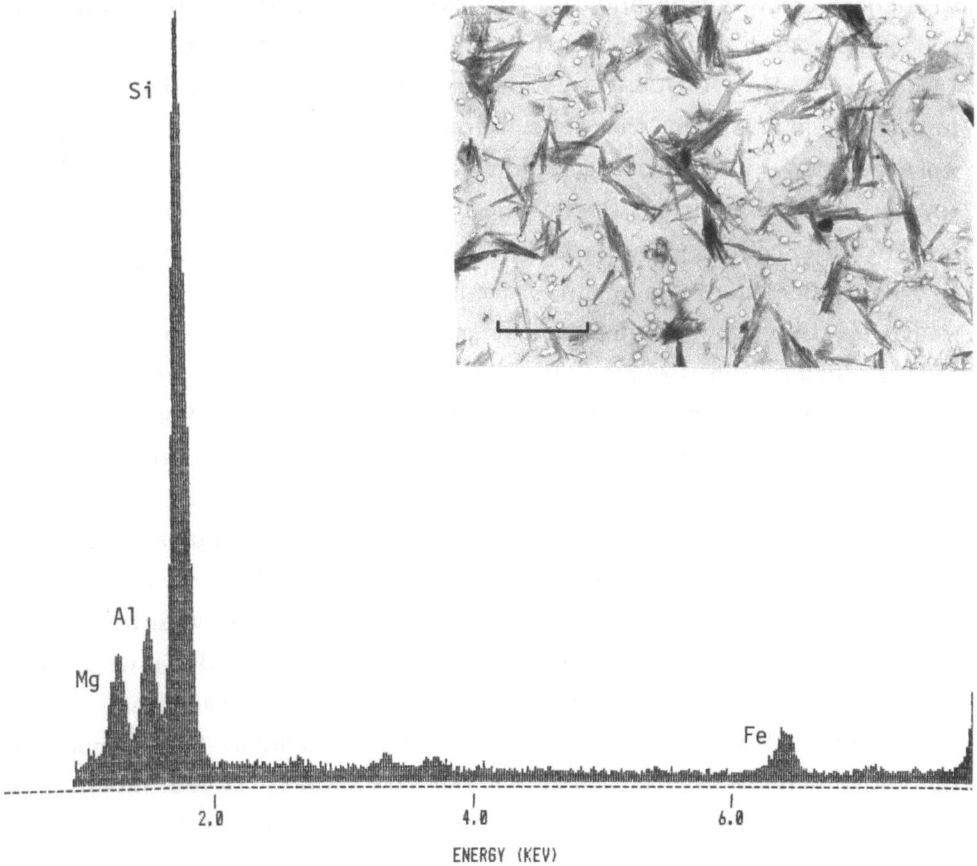

Figure 1. Analysis of the elemental chemistry of individual attapugite fibres by energy-dispersive spectrometry of X-rays.

Bronchoalveolar lavage and fluid analysis. Most of the techniques in BAL procedures and analyses have been previously described (Bégin et al, 1983). The BAL effluent was passed through four layers of cheesecloth to remove mucus and the cells were pelletized by centrifugation. Cells were counted in a hemocytometer and cell viability was determined by the trypan blue exclusion technique. Cytocentrifuge smears served to identify the cellular populations recovered with the Wright-Giemsa stain. In the supernatant, albumin, immunoglobulin IgG and fibronectin were measured by the immunochemical methods of Killingsworth and Savory (1972), using the Behring laser nephelometer instrument (Behring LN modular system, Hoechst Behring, Frankfurt, WG). For sheep albumin and IgG, specific antiserum raised in rabbits was obtained commercially (Cappell Lab. Inc, Downingtown, PA). Sheep BAL fibronectin was purified by affinity column followed by chromatography and antisheep fibronectin antibodies prepared in rabbits as described (Engvall and Ruoslahti, 1977; Rennard et al, 1980). In the supernatant, the activity of lactate dehydrogenase (LDH) and β-glucuronidase (β-G) were measured by standard methods (Wacker et al, 1956; Fishman et al, 1967). Procollagen III was measured as type III procollagen N-terminal peptides by radioimmunoassay as reported by Low (1983) based on the method originally described by Rohde (1979). All results were expressed per milliliter of BAL fluid.

Histopathology. At day 60 of the study, all sheep were sacrificed and the lungs removed from the chest cavity. The tracheal lobe was identified and 9 samples of the lobe were obtained each time for microscopic examination. The lung samples were processed as routinely done for human lung tissue.

Analysis of the data. In the presentation of the results, values of the data for each group of sheep are followed by the standard error as an index of dispersion. The data were evaluated by analysis of variance for experiments having repeated measurements on the same subjects. When a significant effect was detected, a Kruscal Wallis test was used to determine which group means were significantly different (Snedecor and Cochran, 1967). Differences with $P < 0.05$ were considered significant.

RESULTS

Bronchoalveolar lavage cellularity. In the saline exposed sheep, total BAL cells and subpopulations did not vary significantly over time, total cells averaging 22 ± 3 x 10^4/ml, macrophages 16 ± 3 x 10^4/ml (73%), lymphocytes 5 ± 2 x 10^4/ml (23%) and neutrophils 1 ± 1 x 10^4/ml (4%) ($P > 0.05$). In the UICC asbestos exposed sheep, there were significant sustained increases in total BAL cells averaging 50 ± 10 x 10^4/ml, macrophages 25 ± 5 x 10^4/ml (50%), lymphocytes 6 ± 2 x 10^4/ml (12%) and neutrophils 19 ± 8 x 10^4/ml (38%). In the short asbestos exposed sheep, total BAL cells and subpopulations did not differ from saline exposed sheep ($P > 0.05$). In the attapulgite exposed sheep, total BAL cells/ml and subpopulations increased significantly above control at day 12, 24, 40 but by day 60, they were back to control levels.

Bronchoalveolar lavage biochemistry. In the saline exposed sheep, BAL albumin averaged 95 ± 10 µg/ml, procollagen III 1.1 ± 0.3 ng/ml, fibronectin 20 ± 3 µg/ml, LDH 7 ± 2 mUI/ml and β-G 0.9 ± 0.3 ng/min/ml and did not vary significantly over time ($P > 0.05$). In the UICC asbestos exposed sheep, albumin increased significantly after exposure, averaging 124 ± 12 µg/ml (1.5 x control), procollagen III increased to a peak of 4.6 ± 2 ng/ml (4 x control) at day 40, fibronectin increased after exposure to a sustained level of 56 ± 5 µg/ml (3 x control), LDH increased similarly to a level of 19 ± 3 mUI/ml (3 x control) and β-G similarly to an averaged 2.2 ± 0.4 ng/min/ml. In the short asbestos exposed sheep, all biochemical parameters of BAL were comparable to saline exposed sheep. In the attapulgite exposed sheep, albumin and procollagen III did not differ from control level, whereas fibronectin, LDH and β-G levels were significantly above control after exposure and closer to the UICC asbestos exposed group than saline exposed sheep.

Histopathology. In the saline exposed sheep, the histopathology of the tracheal lobe remained normal. In the UICC asbestos exposed sheep, at day 60, the tracheal lobe infiltrate was composed of macrophages and neutrophils predominantly located in the peribronchiolar areas with early peribronchiolar fibrosis and distortion of the small airways. In the short asbestos exposed sheep, the infiltrates were predominantly alveolar, largely composed of macrophages with some peribronchiolar accumulation. Overall, the profusion of these infiltrates were at 15-25% those of sheep exposed to UICC asbestos. Nonetheless, areas of early fibrosis were seen. In the attapulgite exposed sheep, the infiltrates were also predominantly alveolar and quite comparable to those of short asbestos except that peribronchiolar lesions were seen more often and in 3 of the sheep of that group, these early peribronchiolar lesions were quite similar to those of the UICC asbestos exposed sheep although less intense (figure 2).

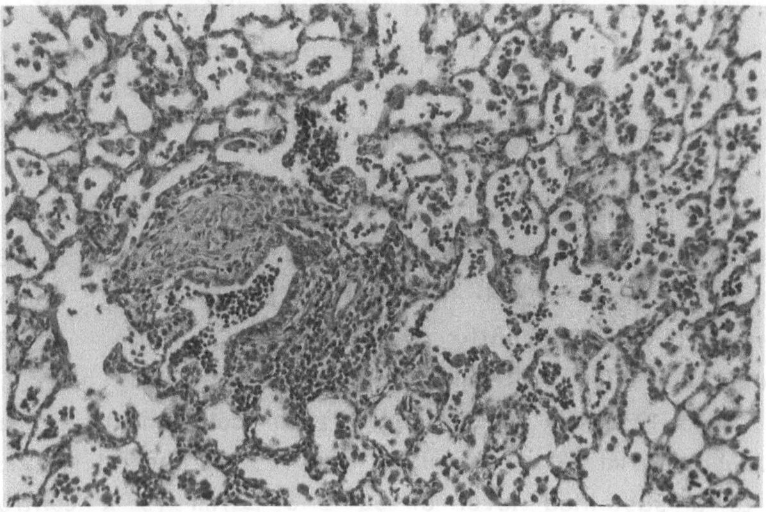

Figure 2. Histopathology of one of the attapulgite exposed sheep with peribronchiolar and alveolar alveolitis (hematoxylin-eosin x 160).

DISCUSSION

The present study of the lung biological activity clearly documented that american attapulgite was biologically active and induced lung lesions comparable to those of short chrysotile fibers in all instances and in some cases, reproduced an early peribronchiolar fibrosing alveolitis. By and large, the attapulgite induced lung lesions were 75-85% less intense than those of UICC asbestos but were accompanied by similar increases in BAL fibronectin, LDH and β-G. Because the cellular and biochemical changes in the BAL of attapulgite exposed sheep were quite similar to those reported in early fibrosing alveolitis in several animal models (Bégin et al, 1985A; Schoenberger et al, 1981; Fahey et al, 1982; Smith et al, 1982) and in humans (Bégin et al, 1985B; 1985C; 1986), we believe that the attapulgite induced lung lesions should have a fibrotic evolution as reported for other short fibers (Lee et al, 1981). This should be further assessed by longer follow-up of attapulgite exposed animals.

These experiments were reported in more details recently (Bégin et al, 1987).

REFERENCES

Anonymous (1985) Attapulgite output to treble. Industrial Minerals 209: 14

Bégin R, Rola-Pleszczynski M, Massé S, Lemaire I, Sirois P, Boctor M, Nadeau D, Drapeau G, Bureau MA (1983) Asbestos induced lung injury in the sheep model: the initial alveolitis. Environ Res 30: 195-210

Bégin R, Rola-Pleszczynski M, Massé S, Berthiaume Y, Drapeau G (1985A) Selective exposure and analysis of the sheep tracheal lobe to study rapidly in vivo the toxicity of respirable particles. Environ Res 36: 389-404

Bégin R, Massé S, Rola-Pleszczynski M, Drapeau G (1985B) Bronchoalveolar and lung tissue analyses in asbestos exposed humans and sheep. IN: In vitro effects of mineral dusts (Beck and Bignon, Eds) Springer-Verlag, Heidelberg, p. 359-367

Bégin R, Cantin A, Berthiaume Y, Boileau R, Bisson G, Lamoureux G, Rola-Pleszczynski M, Drapeau G, Massé S, Boctor M, Breault J, Péloquin S, Dalle D (1985C) Clinical features to stage the alveolitis in asbestos workers. Am J Indust Med 8: 521-536

Bégin R, Martel M, Desmarais Y, Drapeau G, Boileau R, Rola-Pleszczynski M, Massé S (1986) Fibronectin and procollagen 3 levels in bronchoalveolar lavage of asbestos-exposed human subject and sheep. Chest 89: 237-243

Bégin R, Massé S, Rola-Pleszczynski M, Geoffroy M, Martel M, Desmarais Y, Sébastien P (1987) The lung biological activity of american attapulgite. Environ Res 42: 328-339

Bignon J, Sébastien P, Gaudichet A, Jaurand MC (1980) Biological effects of attapulgite. IN: Biological Effects of Mineral Fibers. Vol. 1. Ed. JC Wagner. IARC Scientific Publication # 30, IARC, Lyon, 1980

Engvall E, Ruoslahti E (1977) Binding of soluble form of fibroblast surface protein, fibronectin, to collagen. Int J Cancer 20: 1-5

Fahey PJ, Utell MJ, Mayewski RJ, Wandtke JD, Hyde RW (1982) Early diagnosis of bleomycin pulmonary toxicity using bronchoalveolar lavage in dogs. Am Rev Respir Dis 126: 126-130

Fishman WH, Kato K, Anstiss CL, Green S (1967) Human serum β-glucuronidase: its measurement and some of its properties. Clin Chim Acta 15: 435-447

Haden WL Jr, Schwint IA (1967) Attapulgite: its properties and applications. Indust Eng Chem 59: 58-69

Harvey G, Pagé M, Dumas L (1984) Binding of environmental carcinogens to asbestos and mineral fibres. Brit J Indust Med 41: 396-400

Hénin S, Caillère S (1975) Fibrous Minerals. IN: Soil Components, ed. Gieseking JE, Springer-Verlag, N.Y. vol. 2, pp 335-349

Jaurand MC, Bignon J, Magne L, Renier A, Lafuma J (1979) Interaction des fibres avec les globules rouges et les macrophages alvéolaires in vitro. Rev Fr Mal Resp 7: 717-722

Jolicoeur C, Roberge P, Fortier JL (1981) Separation of short fibers from bulk chrysotile asbestos fiber materials: analysis and physico-chemical characterization. Can J Chem 59: 1140-1148

Killingsworth LM, Savory J (1972) Manual nephelometric methods for immunochemical determination of immunoglobulins IgG, IgA and IgM in human serum. Clin Chem 18: 335-339

Lee KP, Barras CE, Griffith FD, Waritz RS, Lapin CA (1981) Comparative pulmonary responses to inhaled inorganic fibers with asbestos and fiberglass. Environ Res 24: 167-191

Low RB, Cutroneo KR, Davis GS, Giancola MS (1983) Lavage type III procollagen N-terminal peptides in human pulmonary fibrosis and sarcoidosis. Lab Invest 48(6): 755-759

Pott F, Dolgner R, Friedrichs KH, Huth F (1976) L'effet oncogène des poussières fibreuses. Ann Anat Pathol 21: 237-246

Rennard SI, Berg R, Martin GR et al (1980) Enzyme-linked immunoassay for connective tissue components. Anal Biochem 104: 205-214

Rohde H, Vargas L, Hahn E, Kalbfleisch H, Bruguera M, Timpl R (1979) Radioimmunoassay for type III procollagen peptide and its application to human liver disease. Eur J Clin Invest 9: 451-459

Schoenberger C, Hunninghake G, Gadek J, Crystal R (1981) Inflammation and asbestosis: Characterization and maintenance of alveolitis following acute asbestos exposure. Chest 80: 70S-71S

Smith TJ, Beck BD, Brain JD, Hinds WC, Baron SG, Weil L (1982) Prediction of pneumoconiosis risk by bioassays of particulate from occupational exposures. Ann Occup Hyg 26: 435-448

Snedecor GW, Cochran WC (1967) Statistical methods. Iowa: Iowa State University Press

Sors H, Gaudichet A, Sébastien P, Bignon J, Even P (1979) Lung fibrosis after inhalation of fibrous attapulgite. Thorax 34: 695-696

Timbrell V, Gibson JC, Webster I (1978) UICC reference samples of asbestos. Int J Cancer 3: 406-408

Wacker W, Ulmer DD, Vallee BL (1956) Metalloenzymes and myocardial infarction. II. Malic and lactic dehydrogenase activites and zinc concentrations in serum. N Engl J Med 255: 449-456

Washabaugh FJ (1983) (to Engelhard Corporation, Edison, N.J.), U.S. Patent 4,373,037 (Feb. 8)

PULMONARY TOXICITY OF KAOLIN IN RATS EXPOSED BY INHALATION

A. Wastiaux & H. Daniel
CERCHAR
B.P. N° 2
60550 Verneuil en Halatte
France

1 INTRODUCTION

By contrast with some other phyllosilicates which can be associated with free silica, kaolin is an industrial mineral generally used in its pure form. It has numerous industrial uses, mostly in the manufacture of paper, rubber, plastics, ceramics and pharmaceutical products.

Kaolin dust has been described to be fibrogenic for the human lung.

Peribronchiolar macules, nodules and also massive fibrosis and interstitial fibrosis can be observed in lungs of kaolin workers (Lynch and Mac Iver, 1954 ; Hale et al, 1956 ; Lapenas et al, 1984 ; Wagner et al, 1986 ; Morgan et al, 1988 ; Green and Churg, 1988). It is now accepted that the long term inhalation of high quantities of kaolin dust can lead to the development of pulmonary fibrosis. The prevalence of this disease among exposed workers seems to be low however (Hale et al, 1956 ; Sepulveda et al, 1983 ; Kennedy et al, 1983 ; Green et Chung, 1988 ; Morgan et al, 1988).

In the animal studies, kaolin dust was moderately fibrogenic by tracheal or peritoneal injection (King et al, 1948 ; Policard et Collet, 1954 ; Attygalle et al, 1955 ; Zaidi, 1969 ; Martin et al, 1977 ; Sahu et al, 1978). However, it can exhibit acute pulmonary toxicity after a single intratracheal injection (Daniel et al, 1988).

In that context, we found it useful to test the pulmonary fibrogenic potential of kaolin dust in a long term inhalation study in rats.

NATO ASI Series, Vol. G 21
Health Related Effects of Phyllosilicates
Edited by J. Bignon

2 METHODS

The experiment was carried out on 45 female Wistar rats exposed to 300 mg/m3 respirable Cornish kaolin dust, 6 hours a day, 5 days a week, for 3 months. X-ray diffraction and infra red spectroscopy did not detect any free silica in the sample. The control group was housed in the same conditions but without dust exposure. Aerosol generators and inhalation facilities are described in details elsewhere (Le Bouffant L., 1961).

The pulmonary response was assessed on 8 animals at month 3, 6, 12, 24 and 28 after the beginning of the experiment. At these times, animals were killed and the lungs removed.

The macroscopical findings and the weight of fresh lungs were recorded for each animal. Left lobes were used for histopathological examination. They were perfused under 25 cm H_2O pressure and fixed in 10 % neutral buffered formalin. Sections were stained with hematoxylin eosin and silver impregnation for reticulin (Gordon Sweets). The stained sections were also examined under polarized light.

In each group, remaining fragments of lung tissue were pooled, dried and analyzed for collagen, lipids and lung dust burden. Collagen was measured by the method of Stegeman (1958). Lung dust was extracted by low temperature ashing, ash suspension and filtration through a polycarbonate membrane filter. The quantity of kaolin on the membrane was deduced from aluminum concentration measured by X-ray fluorescence.

3 RESULTS

Five animals died during the course of the experiment, respectively at month 16, 16, 22, 25 and 26.

Before month 24, no significant gross changes were noticed in the lungs. Macroscopic tumorous lesion were present in one animal at month 24 and in three animals at month 28. Tracheobronchial lymph nodes were slightly enlarged after 6 months.

At month 3, the mean weight of fresh lungs was slightly increased, and this difference persisted until month 28 (Table I below).

Table I. Mean weight (g) of fresh lungs (standard deviation) after a 3 month inhalation period in female Wistar rats

Groups	Month 3	Month 6	Month 12	Month 24	Month 28
Kaolin	1.80 (0.13)	2.05 (0.09)	1.71 (0.07)	1.90 (0.16)	2.22 (0.14)
Controls	1.04 (0.02)	1.16 (0.02)	1.09 (0.03)	1.54 (0.08)	1.28 (0.04)

At month 3, the main histopathological lesions were lipoproteinosis and alveolitis with numerous dust-laden macrophages in the alveolar spaces (Figure 1). Few macules were located in peribronchiolar perivascular areas. Under polarized light birefringent particles were visible in the macrophages.

At month 6, proteinosis and alveolitis were still present, and in the brownish macules a very thin reticuline positive network was observed, but without collagen formation. In addition to these lesions, particulate-filled macrophages distended the interstitium (Figure 2).

At month 12, epithelial hyperplasia and bronchiolar metaplasia were visible associated with interstitial fibrosis. Multinucleated giant cells granulomas were also encountered (Figure 3).

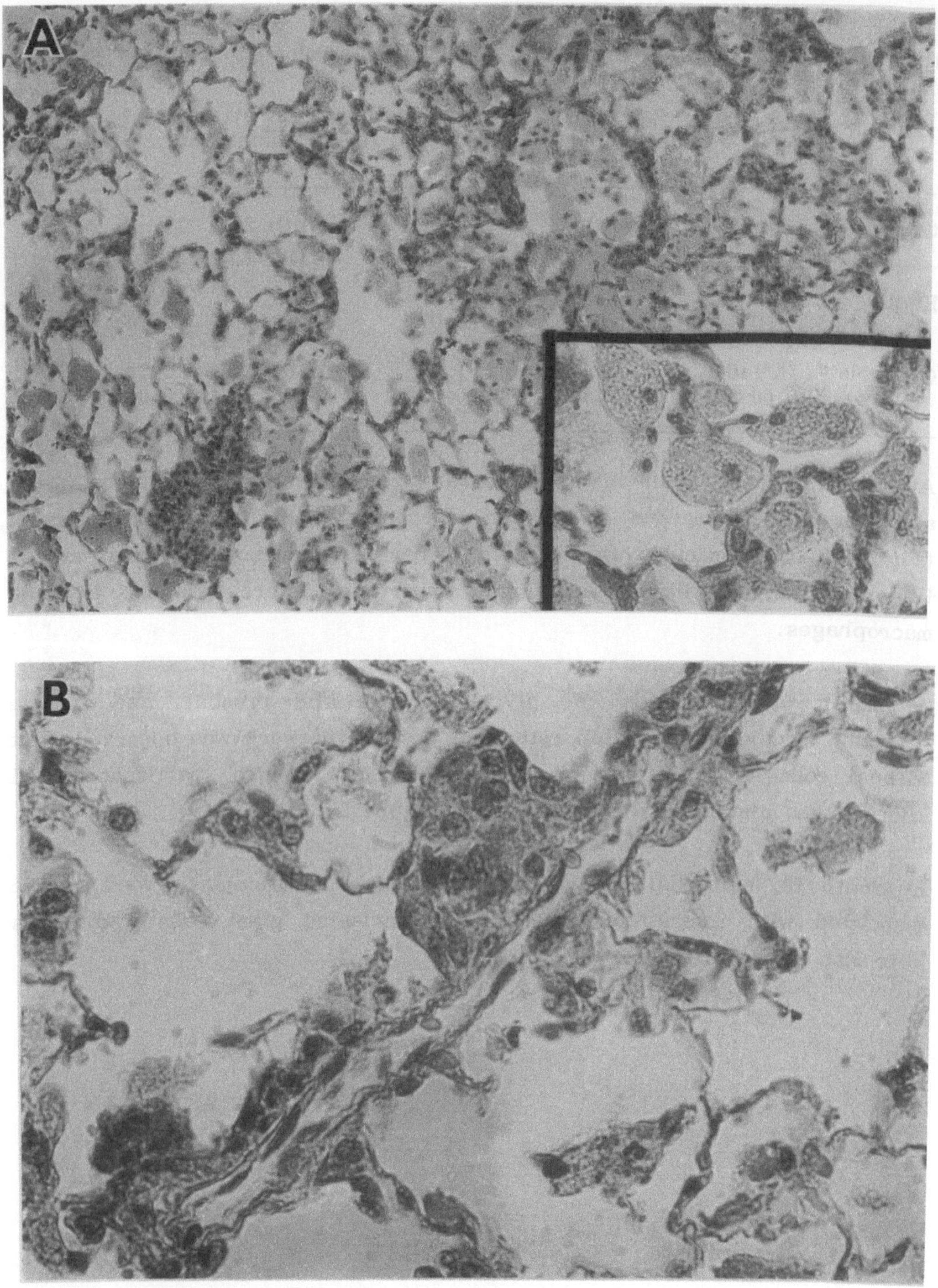

Figure 1. Histopathological features at month 3. Alveolitis with dust-laden macrophages in the alveolar space (A X 125). Perivascular macules (B X 400). Hematoxylin and eosin stain.

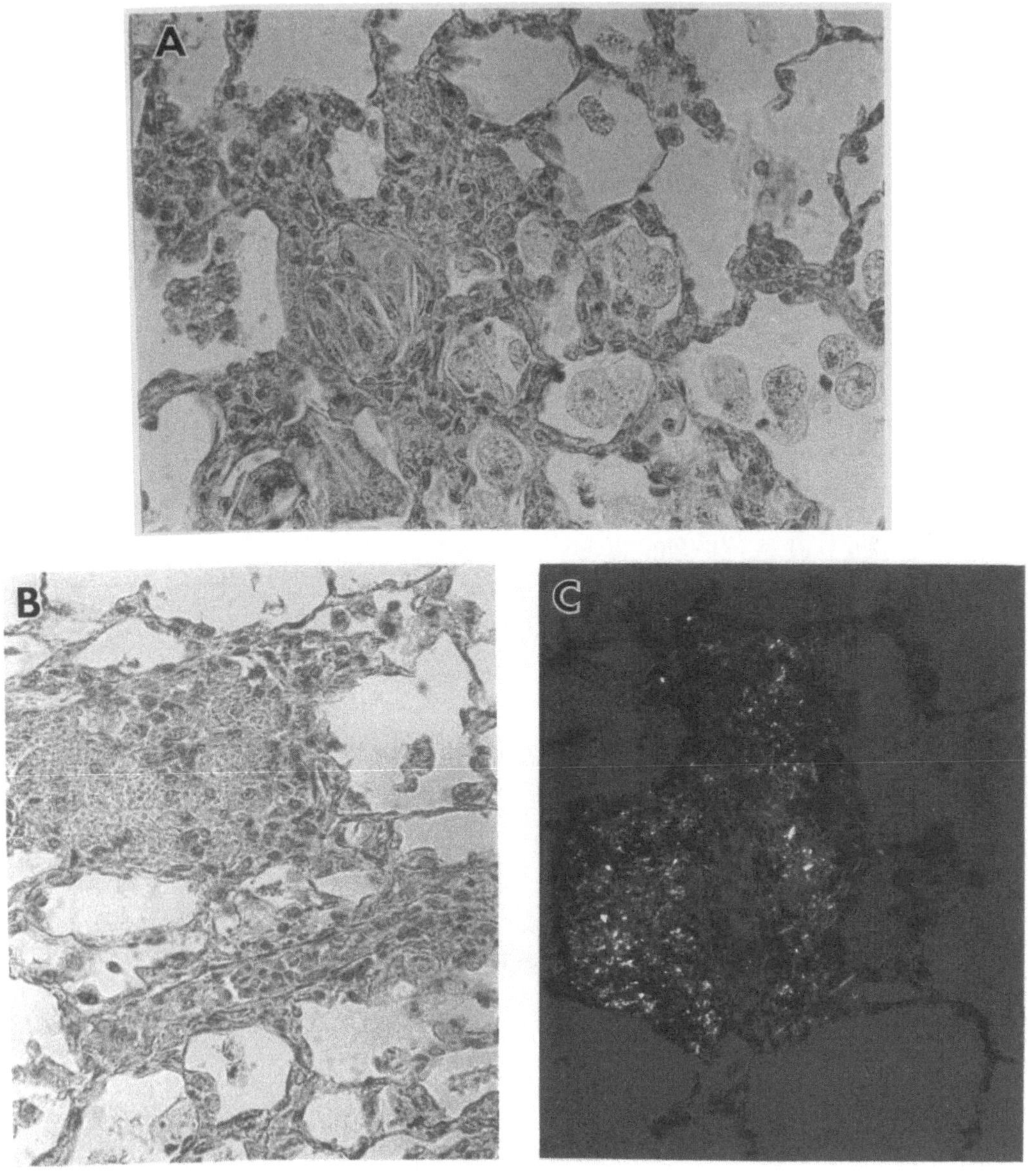

Figure 2. Histological features at month 6.
Dust laden macrophages in the alveolar spaces and in the interstitium (A, x 310). Macules are larger than at month 3 (B, X 310). Birefringent material is visible by sopolarized light in the macules (C X 310).

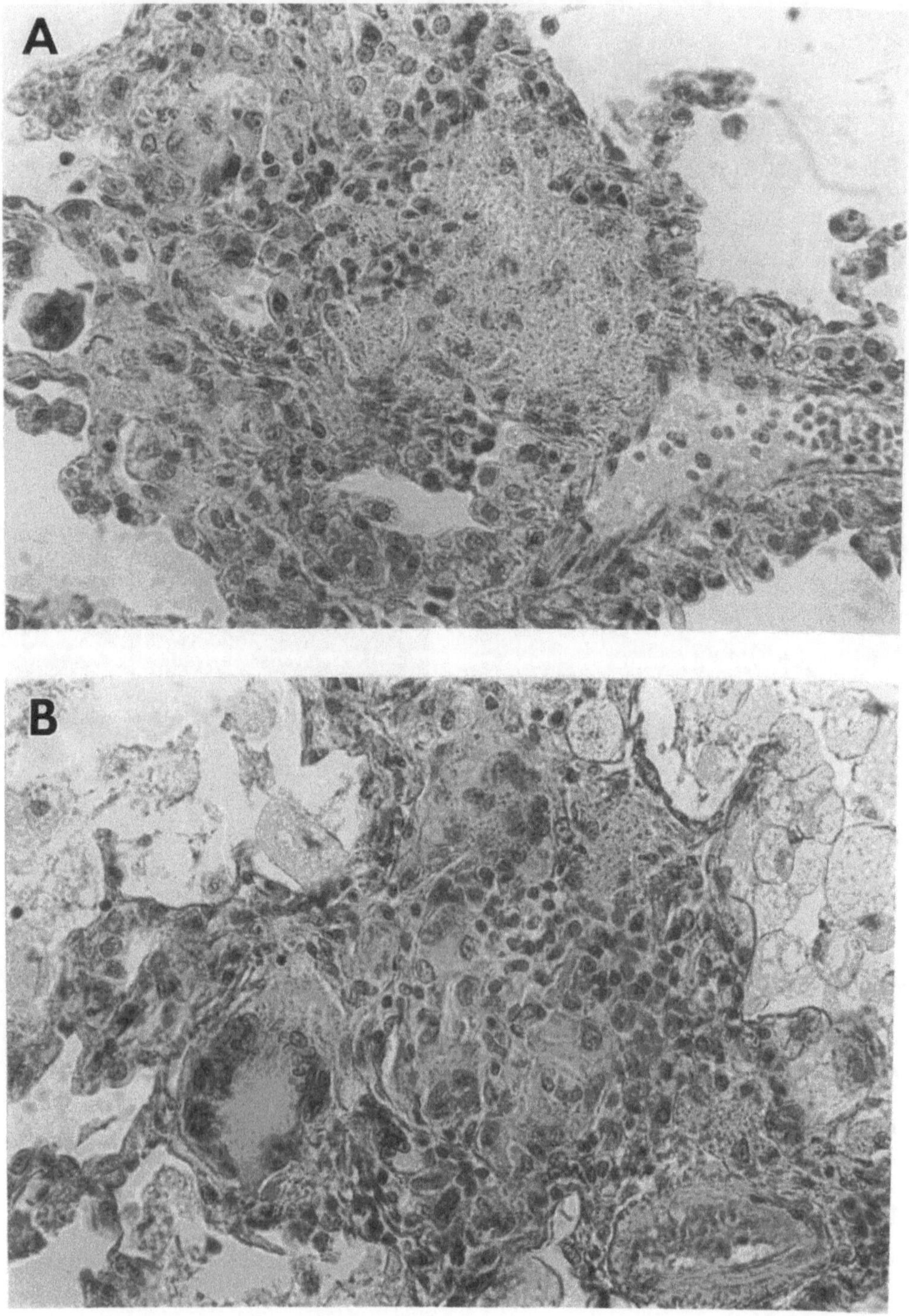

Figure 3. Histological features at month 12.
Macules are larger than at month 6 (A, X 310).
Multinucleated giant cell granuloma (B, X 310).
Hematoxylin and eosin stain.

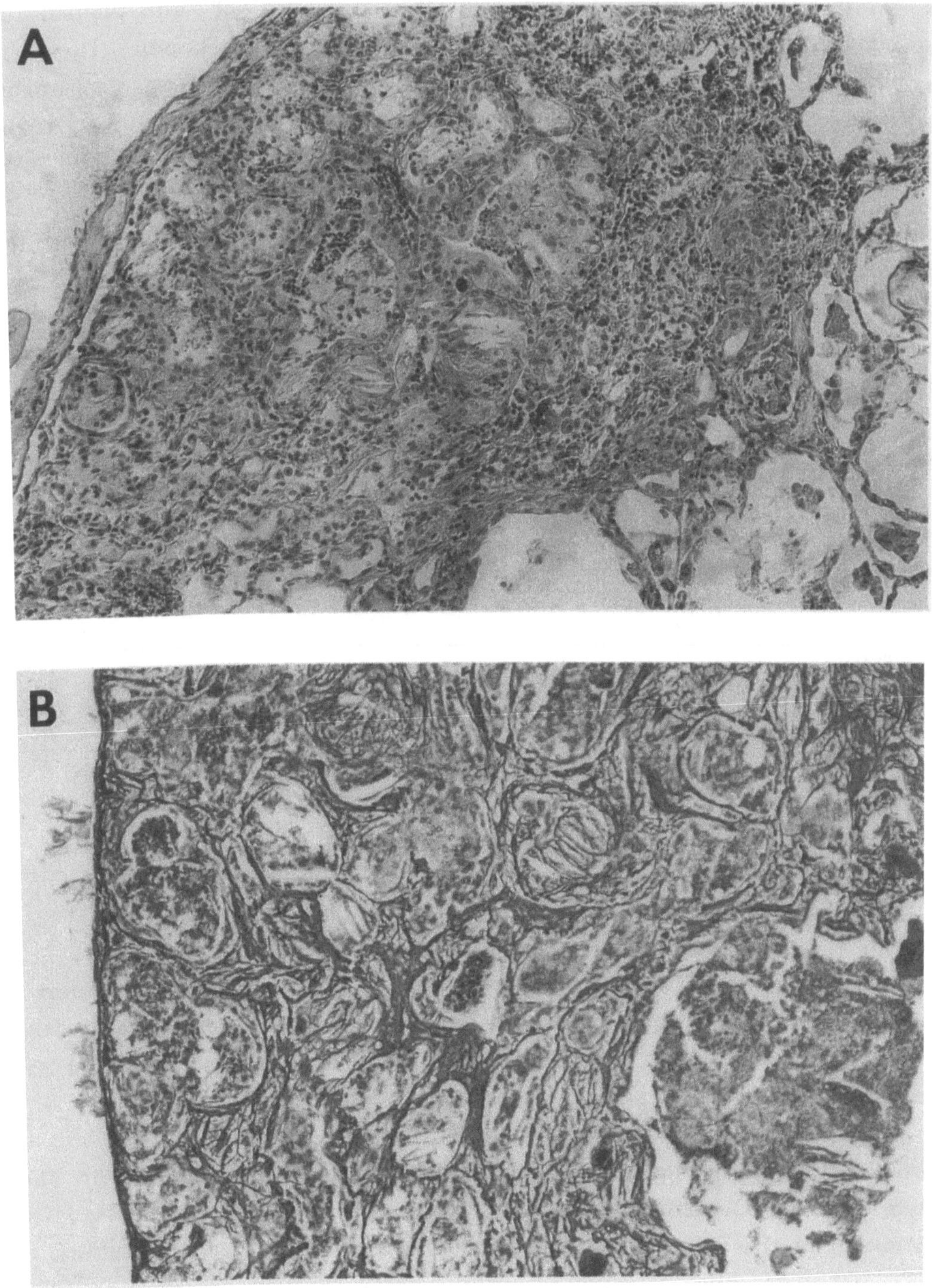

Figure 4. Histological features at month 24.
Granulomas with collagen bands. (A : hematoxylin and eosin X 125 ; B : Gordon Sweets, X 125).

At month 24, the main lesion was the presence of sub-pleural or parenchymal granulomas traversed by spaced collagen bands (fig. 4). Facing those peripheral lesions, local pleural thickening was sometimes observed. Similar lesion were present at month 28, but in addition, three squamous carcinomas were found among the nine surviving animals.

The measurements of pulmonary collagen and dust retained in the lung are reported in the Table II below. During the course of the experiment, an overall increase of the pulmonary collagen was noticed, the increase rate being higher for the first six months. At the end of the exposure period, the amount of kaolin retained in the lung averaged 20.7 mg per rat, with very slow clearance afterwards. At month 28, an average of 17.9 mg was measured.

Table II. Excess of pulmonary collagen and lung dust burden

	Month 3	Month 6	Month 12	Month 24	Month 28
Formed collagen (mg)	8.9	17.8	14.5	19.9	21.7
Dust burden of the lung (mg)	20.7	19.0	15.3	15.9	17.9

4 DISCUSSION

In this long term inhalation experiment, kaolin exhibited moderate fibrogenicity. The first microscopical lesion observed was an an alveolitis, probably responsible for the difference in lung weight between treated and control animals.

Alveolar proteinosis was present until month 12 of this experiment. This lesion has been described, in both human and experimental pathology, as a reaction of the alveolar epithelium to exceptionally high concentrations of fine particulate silica (Craighead et al, 1988). Similar observations after kaolin exposure are not so frequent (Sahu et al, 1978). Martin et al (1977) did not observe alveolar proteinosis after an intratracheal injection of 50 mg of kaolin in rats.

Brownish macules were visible since month 3 with progressive reticulin formation. The presence of particle-laden macrophages in the alveolar spaces is always described in the early stages of dust-related pulmonary disease (Martin et al, 1977 ; Sahu et al, 1978 ; Lapenas et al, 1984 ; Daniel et al, 1988). Using polarized light microscopy, birefringent particles are visible in the macrophages (Parkes, 1982 : Craighead et al, 1988 ; Green and Churg, 1988). Macules are often described in human and experimental pathology (Martin et al, 1977 ; Sahu et al, 1978 ; Morgan, 1983 ; Lapenas et al, 1984 ; Craighead et al, 1988 ; Green and Churg, 1988 ; Morgan et al, 1988). They are collections of dust-laden macrophages located predominantly in peribronchiolar and perivascular areas. Macules differ from nodules in their collagen content.

Giant cell granulomas are lesions larger than macules in which collagen bundles are intermingled with reticulin and inflammatory cells. Common reaction after talc exposure, kaolin granulomas are seldom described. In this experiment, difficulties arose regarding the differential diagnosis between interstitial fibrosis and collagenised granulomas (see for example Figure 4).

Nodules as mentioned in the literature are dense and collagenised but quite different from silicotic nodules. A precise histological definition is lacking for the term nodule which, in fact can cover several morphologically different forms. For F. Green and A. Churg (1988), the nodule may represent a fibrotic stage of the granuloma. These authors claimed also that the morphologic separations between macule, interstitial fibrosis and nodules may in some instance be arbitrary and can depend on the plane of section.

If a silicates pneumoconiosis is generally admitted, no specific pathological feature is described for kaolin and the pulmonary lesions are common with other silicates. According to Zaïdi (1969), no animal model have proved to be satisfactory for kaolin because the pulmonary response is limited pathologically to reticulin formation and macrophage reactivity. The observations made in this long term inhalation experiment in rats were helpful to confirm the pulmonary pathogenicity of pure kaolin and to understand the progression of the lesions from macules to interstitial fibrosis and collagenised granulomas. The late tumoral development will not be discussed here because the experiment was not designed to explore the carcinogenicity of kaolin.

REFERENCES

Attygalle D., Harrison C.V., King E.J., Mohanty G.P. (1955). Infective pneumoconiosis. The influence of dead tubercule bacilli (BCG) on the dust lesions produced by anthracite coal mine dust, and kaolin in the lung of rats and guinea pigs. Br. J. Ind. Med. 11 : 245-254

Craighead J.E. and members of the Silicosis and Silicate Disease Committee (1988). Diseases associated with exposure silica and non fibrous silicate minerals. Arch. Pathol. Lab. Med., 112 : 673-720

Daniel H., le Bouffant L., Wastiaux A., Sébastien P. (1988). Pulmonary toxicity of illite and kaolin dusts. VIIth International Pneumoconiosis Conference. Pittsburg USA.

Hale L.W., Gough J., King E.J., Nagelschmidt G. (1956). Pneumoconiosis of kaolin workers. Br J. Indust. Med., 13 : 251-259

Kennedy T., Rawlings W.. Baser J.R.M., Tockman M. (1983). Pneumoconiosis in Georgia kaolin workers. Am. Rev. Respir. Dis., 127 : 215-220

King E.J.M., Harrison V., Nagelschmidt G. (1948). The effects of kaolin on the lung of rats. J. Path. Bact., 60 : 435-440.

Lapenas D., Gale P., Kennedy T., Rawlings W., Dietrich P. (1984). Kaolin pneumoconiosis. Am. Rev. Respir. Dis., 13 : 282-288.

Le Bouffant L. (1961). Etude quantitative de l'épuration pulmonaire chez le rat. Comparaison entre poussières inertes et poussières nocives. In : Davis C.N., ed. Inhaled particles and vapors. Oxford : Pergamon Press : 369-383.

Le Bouffant L., Daniel H., Martin J.C., Bruyère S. (1982). Effect of, impurities and associated minerals on quartz toxicity. Ann. Occup. Hyg., 26 : 625-634.

Lynch K.M., Mc Iver F.A. (1954). Pneumoconiosis from exposure to kaolin dust : kaolinosis. Am. J. Pathol. 30 : 1117-22.

Martin J.C., Daniel H., Le Bouffant L. (1977). Short-and long-term experimental study of the toxicity of coal-mine dust and of some of its constituents. International Symposium organized by British Occupational Hygiene Society, Edinburgh, 22-26 September 1975. Inhaled Particles IV, pp 361-370 .

Morgan W.K.C. (1983). Kaolin in the lung. Am. Rev. Resp. Dis., 127 : 141-142.

Morgan W.K.C., Donner A., Higgins I.T.T., Pearson M.G., Rawlings W. (1988). The effect of kaolin in the lung. Am. Rev. Respir. Dis., 138 : 813-820.

Parkes W.R. (1982).Occupational Lung Disorders, 2nd Ed. Butterworths, London, Boston, Sydney, Wellington, Durban, Toronto, 310-313.

Policard A., Collet A. (1954) : Etude expérimentale des effets pathologiques de kaolin. Pathologie und Bakteriologie, 17 : 320-325.

Sahu A.P., Shanker R., Zaidi S.H. (1978) : Pulmonary response to kaolin, mica and talc in mice. Exp. Pathol., 16 : 276-282.

Sepulveda M.J., Vallyathan V., Attfield M.D., Piacitelli L., Tucker S.H. (1983) : Pneumoconiosis and lung function in a group of kaolin workers. Am. Rev. Respir. Dis., 127 : 231-235.

Stegeman H. : Mikrobestimmung von hydroxyprolin mit chloramin-t und p-dimethylaminobenzaldehyd. Hoppe-Sayler's Zeitschrift für Physiologische Chemie, 311 : 41-45 (1958).

Wagner J.C., Pooley F.D., Gibbs A., Lyons J., Sheers G., Moncrieff C.B. (1986) : Inhalation of China stone and China clay dusts : relationship between the mineralogy of dust retained in the lungs and pathological changes. Thorax, 41 : 190-196 .

Zaïdi S.H. : Experimental pneumoconiosis (1969). Baltimore : Johns Hopkins Press. 144-224.

PART V

IMPLICATIONS IN INDUSTRIAL HYGIENE AND IN PUBLIC HEALTH

THE SCIENTIFIC BASIS FOR CONTROLLING EXPOSURES TO PHYLLOSILICATE DUST AT THE WORKPLACE

P Sébastien
Groupe "Santé au Travail,
Hygiene et Toxicologie Industrielles"
CERCHAR
BP # 2
60550 Verneuil en Halatte
France

1 INTRODUCTION

This paper will only consider the risk of non malignant pulmonary disease ("Silicatosis") associated with the inhalation of phyllosilicate dust at the workplace. Other sources of exposure were dealt with by Drs BOUTIN and VIALLAT at this conference and evaluation for carcinogenicity is the role of the International Agency for Research on Cancer (IARC). Among the non fibrous phyllosilicates, only talc has been evaluated for carcinogenicity (World Health Organization 1978). According to IARC, there is inadequate evidence for the carcinogenicity to humans of talc not containing asbestiform fibres, while there is sufficient evidence for the carcinogenicity to humans of talc containing asbestiform fibres.

The difficulties involved to proper identify asbestiform fibres in mineral products must be pointed out. The concept of asbestiform habit in geology refers to criteria of physical properties such as visual appearance or tensile strength, which cannot be readily assessed under a microscope (Zoltai 1978). Moreover, such fibres often occur in the mineral product just as traces difficult to detect. For these two reasons, analysis of fibres in the final products is perhaps not the best way to attack the problem. An alternative approach could be for the geologists to state on the degree of contamination of individual deposits. The approach would probably be more reliable and would avoid exposures associated with the mining and processing of contaminated minerals.

Many conferences have addressed the questions related to the health effects of exposure to asbestos, silica or coal dust. This conference is the first one dealing with phyllosilicates. It could be a very rough indication

NATO ASI Series, Vol. G 21
Health Related Effects of Phyllosilicates
Edited by J. Bignon

that overall the risks associated with phyllosilicates are rather low. But pathologists at this conference convinced us on the reality of a specific lung reaction to mineral dusts called "silicatosis" (Craighead 1988). It is certainly legitimate to want to control this risk at the workplace. But the scientific basis is rather weak. Data are especially lacking on the quantitative aspects of exposure response relationships, preventing us to make proper scientifically based recommendations. In this context, the objective of this paper will only be to draw attention on several points which may be eventually considered for the risk evaluation.

2 SOME ASPECTS OF THE SILICATOSIS RISK EVALUATION

There is some indications that phyllosilicate exposures in the past were substantial, but similar to those occurring at that time in other mining industries. Studies that provided information on occupational exposures to talc were reviewed by IARC experts in monograph 42 (World Health Organization 1987). Only two equivalent studies were identified for kaolin (Sepulveda et al 1983, Alterkruse et al 1984). Some of these data are reported in Table I below.

Table I. Mean dust concentrations in some mining industries

	Midget Impinger, mpcf			Resp. dust mg/m³ Recent
	-1945	-1965	-1969	
St Lawrence County Talc milling. (Kleinfeld et al, 1974)	154	37	28	
Montana vermiculite milling. (McDonald et al, 1986)		29	8	
Québec chrysotile milling. (Gibbs & Lachance, 1972)		30		
Talc mines and mills. (NIOHS, 1979)				1.2
Kaolin mining and processing. (Alterkruse et al, 1984)				1.5
Coal mining in France. (Bruyet et al, 1988)				2.3

Before 1970, an air sampling instrument called the midget impinger was commonly used for dust measurement in mining industries. Airborne particles were collected by impingment in a solution and then counted at low magnification under an optical microscope. Only particles larger than about 0.75 micron were analyzed by this technique (Ayer et al 1965). There was no identification criteria. Concentrations were expressed in million of particles per cubic foot of air sampled (mpcf). Table I shows some mpcf values obtained before 1970 in the talc and in the vermiculite mining industries. They are similar to those measured by Gibbs & Lachance (1972) in the Québec asbestos mining industry. After 1970, hygienists started to use portable pumps and cyclones to collect "respirable" dust particles on membrane filters. Gravimetric concentrations were measured. For this recent period, data are available for some talc and kaolin mining industries. Concentrations are lower than those measured for example in the French coal mines. It should be mentioned at this stage that coal miners are substantially exposed to phyllosilicates. Some coal mine dusts may contain up to 60% of phyllosilicate minerals such as kaolin, muscovite or illite (Figure 1).

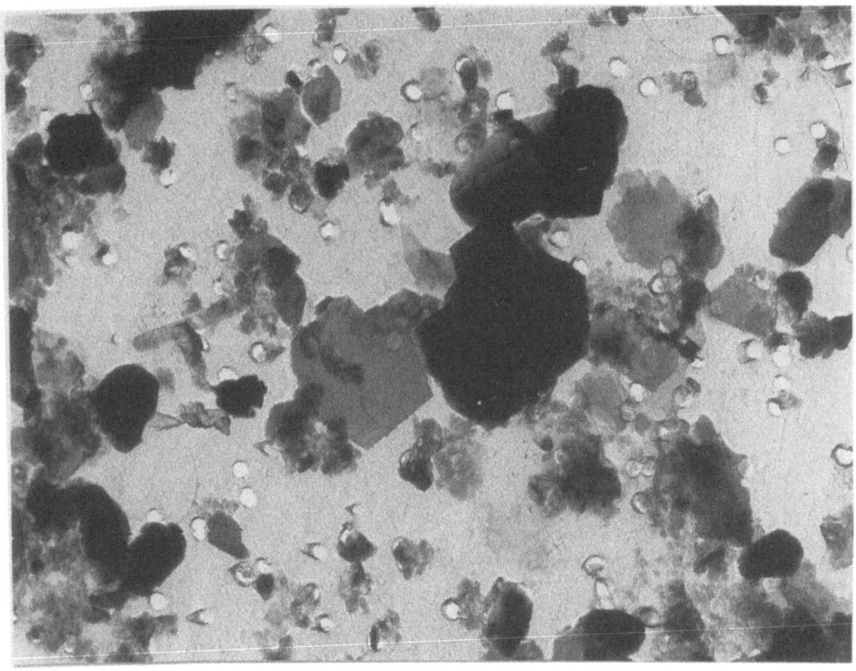

Figure 1. Phyllosilicate particles in coalmine dust.
Transmission electron microscopy x 12,000.

Measurements of dust in human lung at autopsy provide indirect information on past exposure. Several studies documented the presence of talc, kaolin or mica dust particles in the human lung (Berner et al 1981, Brambilla et al 1979, Davies & Cotton 1983, De Vuyst et al 1987, Friedrichs 1987, Hale et al 1956, Kleinfeld et al 1963, Lapenas et al 1984, Miller et al 1971, Val Vallyathan & Craighead 1981, Wagner et al 1986). But only three studies provided quantitative data on mineral content (Hale et al 1956, Davies & Cotton 1983, Wagner et al 1986), and only one (Hale et al 1956) reported measurement results in terms of total amount of dust retained in the lung. This seems to be the proper way of reporting retention data. Reporting concentrations is misleading since the pathological reaction may modify the tissue density and so affect the denominator component of the concentration values. Data on the kaolin retention by Hale et al (1956) are compared in Figure 2 with other lung measurement results, essentially obtained in coalminers. Kaolin retention is located in the middle range of coal retentions with low quartz contents. Wagner et al (1986) measured kaolinite and quartz in the lung of 22 deceased china clay workers. Unfortunately, they reported concentrations in mg of dust per g of dry lung. Converting their data by assuming a mean lung weight of 200 g yields to a location in Figure 2 very close to that of Hale et al (1956). To our knowledge, there is no similar data for talc retention in human lung.

The data reported above give a very rough indication of exposures experienced by workers in several mining industries dealing with phyllosilicates. The growing interest for the subject will possibly lead to an increasing effort by occupational hygienists to more regularly monitor and control exposures. The gravimetric concentration of respirable dust will be the measurement criteria retained for now. Is there an urgent need to be more specific by measuring also the proportions of each type of mineral present in the dust ? Such a mineralogical identification will be costly, and will require that proper analytical methods be developed. Bearing in mind our poor understanding of the biological significance of the mineral composition of mixed dust at present, it would not be reasonable to impose such analyses in routine occupational surveys. But they must definitely be part of future epidemiological studies on exposure response relationships. Another problem which coul be tackled in these studies is the biological significance of several concentration parameters : mass, number or surface area. This has been long regarded as an academic question. The more recent research on the mechanisms of action of asbestos fibres, clearly indicates that this is not the case (NIEHS 1989).

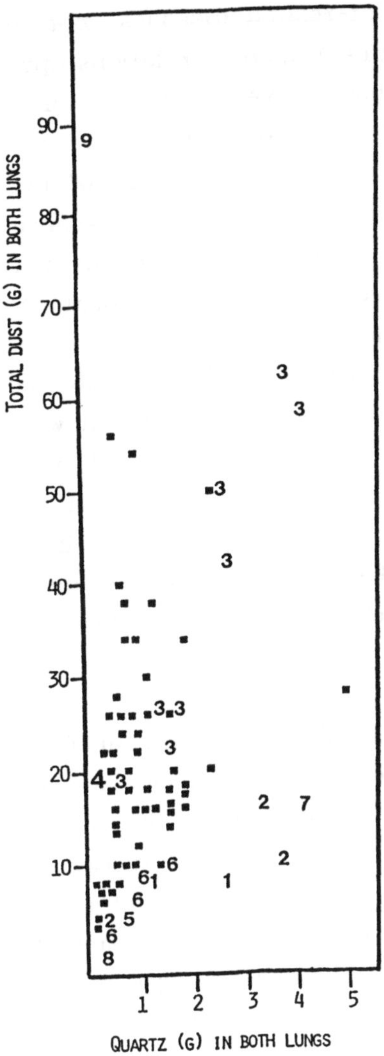

Authors	Number of cases
(■) Coal workers	
Badham and Taylor (1938, 1941)	24
Bergman and Casswell (1972)	314
Davis et al. (1977)	74
Faulds et al. (1959)	33
King et al. (1956)	65
Nagelschmidt et al. (1963)	18
Rivers et al. (1960)	45
Rossiter (1972)	145
Rossiter et al. (1967)	88
Ruckley et al. (1981)	387
	1,193
Others	
(1) Badham and Taylor (1938, 1941) Australian metal miners and tunnel workers	26
(2) Dobreva et al. (1977) Bulgarian workers exposed to dust with high quartz content	24
(3) Faulds and Nagelschmidt (1962) Hematite miners from Cumberland	72
(4) Hale et al. (1956) Kaolin worker	1
(5) King et al. (1956) Tin miners and granite workers	15
(6) McLaughlin and Harding (1956) Foundry workers	65
(7) Nagelschmidt (1960) Foundry workers	2
(8) Verma et al. (1982) Hard rock miners	6
(9) Watson et al. (1959) Carbon electrode makers	2
	213

Figure 2. Published data on the amounts of quartz and total dust measured in both lungs at autopsy. Results reported in the graph are average values for different subgroups of the populations studied.

It has to be realized that in the field of dust-related diseases, not so many studies addressed the question of exposure-response relationships. Most of them dealt with asbestos or coal dust; there is very little on silica (McDonald & Oakes 1984). The troublesome conclusion from a review of the results of these studies is the specificity of the risk to the nature of the mineral product or to the type of industry. Not a single exposure-response curve was able to accommodate the data from the several occupational groups studied. This has been observed with asbestos dust (McDonald & McDonald 1986), as well as with coal dust (Hurley et al 1982). In French coal mines for example, the risk of developing simple pneumoconiosis or progressive massive fibrosis is high in collieries in north of France but was never detected in collieries in south of France (Amoudru 1987). Similar observations have been done in other coal producing countries. It is generally accepted that mining coal of high rank does carry a higher risk, but the significant physical and/or chemical characteristics of the dust particles have not been yet identified, in spite of many years of research on the subject (Hamilton 1985).

A major difficulty with coal mine dusts is that they are in fact complex mixtures of coal, quartz and clay particles. The proportions and characteristics of each component may vary from one deposit to the other. There is some experimental evidence for some beneficial or detrimental interaction between the several components of the dust (Le Bouffant et al 1988, Daniel et al 1988). For example, in rats exposed by intratracheal injection, a mixture of quartz and kaolin produced far more collagen in the lung than did a mixture of quartz and illite (Figure 3). This concept of complex mixture certainly applies to phyllosilicates. Their characteristics vary with the geological origin and they are almost always found in association with other minerals.

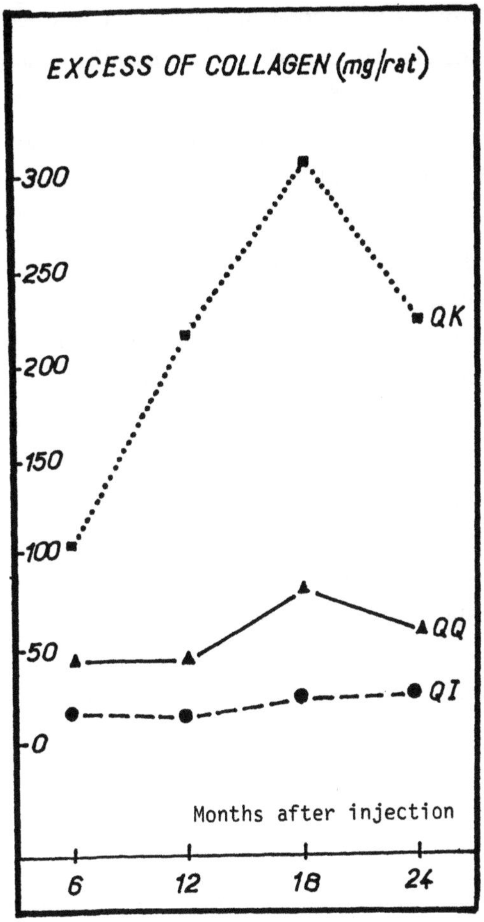

Figure 3. Pulmonary collagen in rats a different times
after an intratracheal injection of :
12.5 mg of quartz (QQ),
12.5 mg of quartz + 37.5 mg of illite (QI),
12.5 mg of quartz + 237.5 mg of kaolin (QK).

Phyllosilicates is a generic term for a large variety of minerals. Because the biological activity of inhaled particles is so related to their chemical and physical characteristics, it would certainly be an error not to distinguish between the several types of minerals or not to assess their toxicity individually. Such a mistake was done in the early days of research on asbestos. Kaolin, talc and mica are presently the three mineral names the more often cited in the literature on the subject. There is not enough data to establish a gradient of toxicity between these minerals.

But the phyllosilicates share some common physical characteristics, such as the sheet structure. It could be asked if this peculiarity may be of biological significance. The resulting effect could be called the "phyllo effect", by analogy to the "fibre effect" which has been - and still is - offered as a explanation for the biological activity of asbestos and other mineral fibres. At least two elements could support the hypothesis of a phyllo effect. First, the shape of the particles will influence their aerodynamic behavior, by reducing their aerodynamic diameter. Although this has not been specifically studied, it is probable that this phenomenon could bring to the alveolar region of the lung some particles of relatively large dimensions (Cheng et al 1988). The pulmonary deposition of airborne particles is essentially related to their aerodynamic diameter, whereas there is much evidence that the biological reaction "in situ" is related to the geometrical diameter. The peculiar pulmonary reactions described by pathologists in silicatosis are perhaps related to this special relationship between the two types of diameter for the phyllosilicates. Another component of the hypothetical phyllo effect could come from the chemical functionalities present at the edges of the particles, as indicated by Dr Pézerat at this conference.

3 CONCLUSIONS

Pathologists have identified a non malignant pulmonary reaction called silicatosis. Overall their observations cannot be explained by peculiar circumstances of extraordinary exposures. Although limited, the available information on the subject indicates that exposures to phyllosilicate dust in the past may have been substantial, but similar to those occurring at that time in other mining industries. It has to be concluded that some phyllosilicates may be reactive when deposited in the lung. The concern for the biological activity of phyllosilicates was apparently enough to initiate this conference.

Industrial dusts containing phyllosilicates are complex mixtures. Coal mine dust is an example. Because of this complexity, it will probably take a long time to the toxicologists to identify the determinants of the biological activity. These are still unknown for coal mine dusts.

More epidemiological studies on exposure-response relationships are needed to feed a scientific data base usable to recommend some control limit values. In the absence of such data, no sound recommendation can be made. May be that experience with coal mine dust could serve as a reference in that field. Coal mine dusts do contain a lot of phyllosilicates and the two types of dust exhibited some similar reactions in several toxicological models. In several countries the control limit value for coal dust is at 2 mg/m^3 (respirable fraction, time weighed average). Some people may think that the implementation of such a value did not completely eliminate the pneumoconiosis risk, especially in high rank coal mining. It will certainly be interesting to compare results of the future epidemiological studies on exposure-response relationships in the field of phyllosilicates with those already available in the field of coal. It will also be interesting to see if the silicatosis risk is dependent on the nature of the mineral and/or the industrial process.

Before getting these data, it is advisable to be practical, to maintain dust exposures as low as possible and to control them by regular industrial hygiene practices. In fact the problem arose because of the dust produced by some engineering techniques. We can be confident that engineers will be smart enough to modify such techniques in order to reduce dust exposures.

REFERENCES

Alterkruse EB, Chaudhary BA, Pearson MG & Morgan WKC (1984). Kaolin dust concentrations and pneumoconiosis at a kaolin mine. Thorax, 39, 436-441.

Amoudru C (1987). Les pneumoconioses - Dimension actuelle des problèmes. Colloque INSERM, 155, 3-40.

Ayer HE, Lynch JR & Fanney JH (1965). A comparison of impinger and membrane filter techniques for evaluating air samples in asbestos plants. Ann NY Acad Sci, 132, 274-287.

Badham C & Taylor HB (1938). Annual Report, Department of Public Health, NSW, 1936, p100. As quoted by Nagelschmidt in Br J Ind Med (1960)17, 247-259.

Badham C & Taylor HB (1941). Annual Report, Department of Public Health, NSW, 1939, p 1. As quoted by Nagelschmidt in Br J Ind Med (1960)17, 247-259.

Bergman I & Casswell C (1972). Lung dust and lung iron contents of coal workers in different coalfields in Great-Britain. Br J Ind Med, 29, 160-168.

Berner A, Gylseth B & Levy F (1981). Talc dust pneumoconiosis. Acta path microbiol scand, 89, 17-21.

Boutin C & Viallat JR (1989). Health effects related to therapeutic, cosmetic and addict uses of talc. Presented at the First International Conference on Health Related Effects of Phyllosilicates, NATO Advanced Workshop, Colloque INSERM, Paris, March 16-18, 1989.

Brambilla C, Abraham J, Brambilla E, Benirschke K & Bloor C (1979). Comparative pathology of silicate pneumoconiosis. Am J Pathol, 96, 149-170.

Bruyet B, Courbon P & Sébastien P (1988). Personal exposures to respirable dust in french coal mines. Presented at the International Symposium on Pneumoconioses, Shenyang,China, 30 May - 2 June, 1988, proceedings in press.

Cheng YS, Yeh HC & Allen MD (1988). Physical characterization and aerodynamic behavior of plate-like particles. Ann occup Hyg, 32, 369-378.

Craighead JE (1988). Diseases associated with silica and nonfibrous silicate minerals. Arch Pathol Lab Med, 112, 673-720.

Daniel H, Le Meur M & Wastiaux A (1988). Characteristics and toxicity of coal dust. Presented at the International Symposium on Pneumoconioses, Shenyang, China, 30 May - 2 June, 1988, proceedings in press.

Davis MG (1977). The effect of quartz and other non-coal dust in coalworkers' pneumoconiosis. In "Inhaled Particles IV", Walton WH ed, Pergamon Press, Oxford-New York-Toronto-Sydney-Paris-Frankfurt, pp 691-702.

Davies D & Cotton R (1983). Mica pneumoconiosis. Br J Ind Med, 40, 22-27.

De Vuyst P, Dumortier P, Léophonte P, Vande Weyer R & Yernault JC (1987). Mineralogical analysis of bronchoalveolar lavage in talc pneumoconiosis. Eur J Respir Dis, 70, 150-156.

Dobreva M, Burilkov T, Koley K & Lalova P (1977). Characteristics of lung dusts and their relation to dust exposure and pathological findings in the lungs. In "Inhaled Particles IV", Walton WH ed, Pergamon Press, Oxford-New York-Toronto-Sydney-Paris-Frankfurt, pp 717-725.

Faulds JS, King EJ & Nagelschmidt G (1959). The dust content of the lungs of coalworkers from Cumberland. Br J Ind Med, 16, 43-50.

Faulds JS & Nagelschmidt GS (1962). The dust in the lungs of haematite miners from Cumberland. Ann occup Hyg, 4, 255-263.

Friedrichs KH (1987). Electron microscopic analyses of dust from the lungs and the lymph nodes of talc-mine employees. Am Ind Hyg Assoc J, 48, 626-633.

Gibbs GW & Lachance M (1972). Dust exposure in the chrysotile asbestos mines and mills of Québec. Arch Environ Health, 24, 189-197.

Hale LW, Gough J, King EJ & Nagelschmidt G (1956). Pneumoconiosis of kaolin workers. Br J Ind Med, 13, 251-259.

Hamilton RJ (1985). Hygiène industrielle dans les mines. Commission des Communautés européennes, rapport EUR 9253 FR, Office des publications officielles des Communautés européennes, Luxembourg.

Hurley JF, Burns J, Copland L, Dodgson J & Jacobsen M (1982). Coalworkers' simple pneumoconiosis and exposure to dust in 10 British coalmines. Br J Ind Med, 39, 120-127.

King EJ, Maguire BA & Nagelschmidt G (1956). Further studies of the dust in lungs of coalminers. Br J Ind Med, 13, 9-23.

Kleinfeld M, Giel CP, Majeranowski JF & Messite J (1963). Talc pneumoconiosis. A report of six patients with postmortem findings. Arch Environ Health, 7, 101-115.

Kleinfeld M, Messite J & Zaki MH (1974). Mortality experiences among talc workers : a follow-up study. J occup Med, 16, 345-349.

Lapenas D, Gale P, Kennedy T, Rawlings Jr W & Dietrich P (1984). Kaolin pneumoconiosis. Radiologic, pathologic, and mineralogic findings. Am Rev Respir Dis, 130, 282-288.

Le Bouffant L, Addison J, Bolton RE, Bruch J, Bruyet B, Daniel H, Davis JMG, Degueldre G, Demarez J, Dodgson J, Gormley IP, Hadden GG, Kovacs MP, Martin JC, Reisner MTR, Robertson A & Rosmanith J (1988). Compared in-vitro and in-vivo toxicity of coalmine dusts. Relationship with mineral composition. Ann occup Hyg, 32, 611-620.

McLaughlin A & Harding HE (1956). Pneumoconiosis and other causes of death in iron and steel foundry workers. AMA Arch Ind Health, 14, 350-378.

McDonald JC & McDonald AD (1986). Epidemiology of asbestos-related lung cancer. In "Asbestos-related Malignancy", Antmann K & Hisner J eds, Grove & Stratton Inc, pp 57-79.

McDonald JC, McDonald AD, Armstrong B & Sébastien P (1986). Cohort study of mortality of vermiculite miners exposed to tremolite. Br J Ind Med, 43, 436-444.

McDonald JC & Oakes D (1984). Exposure-response in miners exposed to silica. In "VIth International Pneumoconiosis Conference 1983", International Labour Office, pp 114-123.

Miller A, Teirstein AS, Bader ME, Bader RA & Selikoff IJ (1971). Talc pneumoconiosis. Significance of sublight microscopic mineral particles. Am J Med, 50, 395-402.

Nagelschmidt G (1960). The relation between lung dust and lung pathology in pneumoconiosis. Br J Ind Med, 17, 247-259.

Nagelschmidt G, Rivers D, King EJ & Travella W (1963). Dust and collagen content of lungs of coalworkers with progressive massive fribrosis. Br J Ind Med, 20, 181-191.

National Institute for Occupational Safety and Health (1979). Mining surveillance : potentially toxic occupational exposures. Morgantown, West Virginia, USA.

National Institute of Environmental Health Sciences (1989). Workshop on fiber toxicology research needs. Research Triangle Park, North Carolina, USA.

Pézerat H (1989). Surface properties of phylosilicates. Presented at the First International Conference on Health Related Effects of Phyllosilicates, NATO Advanced Workshop, Colloque INSERM, Paris, March 16-18, 1989.

Rivers D, Wise ME, King EJ & Nagelschmidt G (1960). Dust content, radiology, and pathology in simple pneumoconiosis of coalworkers. Br J Ind Med, 17, 87-108.

Rossiter CE (1972). Relation between content and composition of coalworkers' lungs and radiological appearances. Br J Ind Med, 29, 31-44.

Rossiter CE, Rivers D, Bergman I, Caswell C & Nagelschmidt G (1967). Dust content, radiology and pathology in simple pneumoconiosis of coalworkers. In "Inhaled Particles and Vapors II", Davies CN ed, Pergamon Press, Oxford-London-Edinburgh-New York-Toronto-Sydney-Paris-Braunschweig, pp 419-437.

Ruckley VA, Chapman JS, Collings PL, Douglas AN, Fernie JM, Lamb D & Davis JMG (1981). Autopsy studies of coalminers' lungs. Phase II. Final report on CEC contract 7246-15/8/001, Institute of Occupational Medicine, Edinburgh, UK.

Sepulveda MJ, Vallyathan V, Attfield MD, Piacitelli L & Tucker JH (1983). Pneumoconiosis and lung function in a group of kaolin workers. Am Rev Respir Dis, 127, 231-235.

Val Vallyathan N & Craighead JE (1981). Pulmonary pathology in workers exposed to nonasbestiform talc. Hum Pathol, 12, 28-35.

Verma DK, Muir DCF, Stewart ML, Julian JA & Ritchie AC (1982). The dust content of the lungs of hard-rock miners and its relationship to occupational exposure, pathological and radiological findings. Ann occup Hyg, 26, 401-409.

Wagner JC, Pooley FD, Gibbs A, Lyons J, Sheers G & Moncrieff CB (1986). Inhalation of china stone and china clay dusts : relationship between the mineralogy of dust retained in the lungs and pathological changes. Thorax, 41, 190-196.

Waston AJ, Black J, Doig AT & Nagelschmidt G (1959). Pneumoconiosis in carbon electrode makers. Br J Ind Med, 16, 274-285.

World Health Organization (1987). International Agency for Research on Cancer, IARC monographs on the carcinogenic risk of chemicals to humans, Silica and Some Silicates. Volume 42, IARC, Lyon, France.

Zoltai T (1978). History of asbestos-related mineralogical terminology. In "Proceedings of the workshop on asbestos : definitions and measurement methods", Gravatt CC Lafleur PD & Heinrich KFJ eds, National Bureau of Standards Special Publication 506, Us Government Printing Office, Washington, pp 1-18.

CONTROL LIMIT VALUES IN DIFFERENT INDUSTRIALIZED COUNTRIES: USA

Richard W. Niemeier, Ph.D.
Division of Standards Development and Technology Transfer
National Institute for Occupational Safety and Health
4676 Columbia Parkway
Cincinnati, Ohio 45226 (USA)

In December 1970, Congress enacted the Occupational Safety and Health Act (Public Law 91-596) "to assure safe and healthful working conditions for working men and women; by authorizing enforcement of the standards developed under the Act; by assisting and encouraging the States in their efforts to assure safe and healthful working conditions; by providing for research, information, education, and training in the field of occupational health". The Act created two separate federal agencies and provided a legislative mandate to each with specific responsibilities:

The Occupational Safety and Health Administration (OSHA), within the Department of Labor (DOL), promulgates and enforces safety and health standards in the workplace and educates workers.

The National Institute for Occupational Safety and Health (NIOSH), within the Department of Health and Human Services, plans, directs, and coordinates the national program effort to develop and establish recommended occupational safety and health standards and conducts research, training, and related activities to assure safe and healthful working conditions for every working man and woman. These responsibilitites include:

(1) conducting experimental research or demonstrations directly or supported through grants and contracts;

(2) developing innovative methods and approaches for dealing with occupational safety and health problems;

(3) producing criteria dealing with toxic substances and physical agents which describe exposure levels that are safe for various periods of employment, including but not limited to the exposure levels at which no employee will suffer impaired health or

NATO ASI Series, Vol. G 21
Health Related Effects of Phyllosilicates
Edited by J. Bignon

functional capacities or diminished life expectancy as a result of work experience. Emphasis is placed on ways to discover latent disease, establishing causal relationship between diseases and work conditions;

(4) serving as a principal focus for training programs to increase the number and competence of personnel engaged in the practice of occupational safety and health;

(5) publishing an annual listing of all known toxic substances and the concentrations at which such toxicity is known to occur;

(6) developing and coordinating the appropriate reporting procedures which assist in accurately describing the nature of the national occupational safety and health problems;

(7) conducting and publishing industry-wide studies of the effect of chronic or low-level exposure to industrial materials, processes, and stresses on the potential for illness, disease, or loss of functional capacity in workers;

(8) consulting with DOL, other federal, state and local government agencies, industry and employee organizations, and other appropriate individuals, institutes, and organizations with regard to promoting occupational safety and health.

Under the Federal Mine Safety and Health Act of 1977 (Public Law 91-164), which reflects changes to the Federal Coal Mine Health and Safety Act of 1969 (Public Law 91-173), NIOSH is also responsible for:

(1) surveying toxic substances and hazardous physical agents in mines and recommending worker exposure limits to the Mine Safety and Health Administration (MSHA), DOL;

(2) administering the X-ray surveillance program for coal miners;

(3) testing and certifying respirators and hazard measuring devices.

In order for the Department of Labor to carry out its Congressional mandate through MSHA and OSHA, it encourages the occupational safety and health community, industry, labor, academia, and the public to participate in the regulatory process. Both regulatory agencies within DOL have developed similar administrative and legal processes for promulgating occupational safety and health standards. The process usually includes:

(1) notification of interested parties by publication in the Federal Register the intent to promulgate a regulation that includes a detailed description of the proposed rule

(2) submission of comments, advice and data by interested parties who choose to participate in the process

(3) if requested, an informal public hearing that includes submission of testimony by interested parties and questioning of the participants

(4) submission by the participants of summary statements, additional data, and clarifications in the form of post-hearing comments and briefs

(5) publication of the final rule in the Federal Register

Topics for occupational safety and health regulations stem from a number of sources including:

NIOSH criteria for recommended standards;
DOL initiatives, advisory committees, and task forces;
legal challenges from industry, labor or public interest groups;
referrals from other Federal agencies, such as the Environmental Protection Agency; and,
other miscellaneous sources including public requests.

NIOSH, when preparing recommendations for the prevention of selected occupational safety and health hazards, compiles, analyzes and critiques existing detailed scientific and technical information. The most complex form of these policy statements take the form of Criteria Documents and contain the following elements: Recommended Exposure Limit (REL), monitoring and recording exposures, medical surveillance, employee notification of hazards, work practices, engineering controls, personal protective equipment including respirator selection and use, worker training, sanitation, and recordkeeping. The recommendations contained in criteria documents are based on an objective scientific analysis of all available data, qualitative and quantitative risk assessments, considerations of technical feasibility, and the availability of suitable analytical methods for measuring the hazard. During peer review, NIOSH solicits comments on the draft criteria documents from experts as well as tripartite review from interested parties, including industry, labor and academia. After careful review of comments, revisions to the document are made and a response prepared for each comment that is transmitted to DOL with the final version of the criteria document.

Another form of these recommendations is the NIOSH testimony to DOL during hearings on proposed standards. The testimony is formulated to support scientific and technical considerations. This testimony is developed and coordinated by a team of experts within NIOSH who have particular scientific expertise on the topic of the regulation. One of the most recent activities involved more than 10 person years of effort expended in less than four months to support the OSHA Air Quality Standards rulemaking. Current scientific and technical information was reviewed and summarized on over 600 chemicals with Interim Recommended Exposure Limits (I-REL) transmitted to OSHA.

Additional mechanisms of transmitting occupational safety and health information to the public are the Current Intelligence Bulletin (CIB) and the Alert. The CIB disseminates new scientific information about occupational hazards and may focus on a previously unrecognized hazard or may report new data on a known hazard. It capsulizes background information about the hazard and outlines recommended action for controlling exposures. The Alert which is a short narrative of new findings published with the intent to reduce injuries, disease or fatalities through specific prevention strategies. Alerts are targeted to those in a position to directly intervene in the workplace to quickly eliminate the problem or reduce the risk.

In order to present the most current U.S. regulations (OSHA) and those recommendations of NIOSH on Phyllosilicates for this First International Conference of Health Related Effects of Phyllosilicates, a mineralogical classification was used as a guide to define members of this class and groups of minerals as shown in Table 1. The derivation of the name of this group implies (Greek: "phyllon" - meaning leaf) most of its many members have a platy or flaky habit and one prominent cleavage. The phyllosilicates are generally soft, of relatively low specific gravity and show flexibility or even elasticity of the cleavage plate. These physical properties result from the indefinitely extended sheet of silicate tetrahedra. Members of the phyllosilicates include the following groups of minerals: serpentine, clay mineral, mica and chlorite. It is obvious from this table that additional research and subsequent recommendations are needed to fill the gaps in knowledge and regulation.

Table 1

U.S. Regulations or Recommendations on
Phyllosilicates based on Mineralogical Classifications[a]

	OSHA (PEL)	NIOSH (REL)	NIOSH (I-REL)
Serpentine Group			
Antigorite	--	--	--
Lizardite	--	--	--
Chrysotile	200,000 f/m^3 (TWA)[b] 100,000 f/m^3 (Action Level) 1,000,000 f/m^3 (30 min ceiling)	100,000 f/m^3[b]	--
Clay Mineral Group			
Kaolinite (Kaolin)	10 mg/m^3 (total) 5 mg/m^3 (respirable)	--	same as PEL
Talc (containing no asbestos)	2 mg/m^3 (respirable)	--	same as PEL
Pyrophyllite	--	--	--
Vermiculite	--	--	--
Mica Group			
Muscovite Phlogopite Biotite Lepidolite	3 mg/m^3 (> 1% quartz)	--	same as PEL
Margarite	--	--	--
Chlorite Group			
Chlorite	--	--	--
Other closely related minerals			
Apophyllite	--	--	--
Prehnite	--	--	--
Chrysocolla	--	--	--

[a] mineralogical classification adopted from: Klein, C, Hurlbut, C (1985) Phyllosilicates. In: Manual of Mineralogy, Twentieth Edition, Chapter 11, Systematic Mineralogy Part IV: Silicates. John Wiley & Sons, New York, pp 418-436

[b] fibers > 5 ≤m in length with aspect ratio of ≥ 3:1

Limit Values: the European Dimension

Dr. M. DE SMEDT
COMMISSION OF THE EUROPEAN COMMUNITIES
Directorate General 'Employment, Social Affairs, Education
Health and Safety Directorate
Industrial Medicine and Hygiene Unit

Jean Monnet Building
Plateau du Kirchberg

L - 2920 LUXEMBOURG

The framework Directive 80/1107/EEC on agents at work, adopted by the Council in November 1980 (1), sets out two objectives:

- the elimination or limitation of exposure to chemical, physical and biological agents
- the protection of workers who are likely to be exposed to these agents

This Directive requires that exposure of workers to chemical substances is as low as reasonably achievable, together with the measures required to achieve this.

Based on Article 118A of the EEC Treaty, concerned with the minimal requirements for the protection of the health and safety of workers, the Council adopted on 16th December 1988 a new directive (2) amending the above mentioned directive.

In this amending directive the Council argued that, with regard to the agents referred to in Annex I to Directive 80/1107/EEC the compulsory limit values would continue to be laid down by Council directives. For agents other than mentioned in Annex I (like phyllosilicates), limit values will be established and laid down by Commission directives.

(1) OJ N° L327, 3.12.80, p. 8
(2) OJ N° L356, 24.12.88, p. 74

NATO ASI Series, Vol. G 21
Health Related Effects of Phyllosilicates
Edited by J. Bignon

As mentioned in its action programme (1), the Commission will carry out studies to collect and evaluate toxicological and health data for individual agents and their absorption pathways. Ideally, the level of exposure of workers to the chemical agent of interest should be used as a criterion for setting priorities, but such data are often limited, necessitating surveys in industry of the extent of exposure.

When toxicological data are available, they can be used to determine the risk to workers by examining the magnitude, level of significance and consistency of these data.

This should result in higher priority being given to those agents producing higher risks to health.

From 10-12 May 1989 a workshop "On Harmonization of Criteria Documents used for the Establishment of Recommended Occupational Exposure Limits" will be organized in The Hague (the Netherlands) by the Commission of the European Communities and the Dutch Directorate General of Labour (Ministry of Social Affairs and Employment). The objectives of this workshop are to suggest a format and framework of harmonised criteria documents as a basis for setting occupational exposure limits and to suggest a mechanism for international cooperation in the preparation of harmonised criteria documents. In order to prepare the proposals for limit values, the Commission intends to set up a committee responsible for the scientific evaluation of the data available for the various agents.

For some phyllosilicates, as kaolin, mica and talc, various Member States adopted limit values either based on their own toxicological data or based on limit values set by ACGIH.

These data will be taken into account by the committee above mentioned.

---

(1) OJ N° C28, 3.2.88, p. 3

KAOLIN	CAS No 01332-58-7	mg/m^3
BELGIE	ACGIH	10
DENMARK	Graensevaerdier for stoffer og materialer	/
BRD	Senatskommission zur Prüfung der Arbeitsstoffe	/
ELLAS		
ESPAÑA	ACGIH	10
FRANCE	Circulaire	10
IRELAND	UK limits or ACGIH	10
ITALIA	ACGIH	10
LUXEMBOURG	BRD or FRANCE or ACGIH	10
NEDERLAND	MAC Lijst '89	10 (5: respir. dust)
PORTUGAL	ACGIH	10
UNITED KINGDOM	HSE	

TALC	not asbestos fiber containing CAS no 14807-96-6	mg/m^3
BELGIE	ACGIH	2
DANMARK	Graensvaerdier for stoffer og materialer	2
BRD	Senatskommission zur Prüfung der Arbeitsstoffe	2
ELLAS		
ESPAÑA	ACGIH	2
FRANCE	Circulaire	
IRELAND	UK limits or ACGIH	10 (tot. inhal.) 1 (resp. dust)
ITALIA	ACGIH	2
LUXEMBOURG	BRD or FRANCE or ACGIH	2
NEDERLAND	MAC lijst '89	6 (tot. inhal.) 3 (resp. dust)
PORTUGAL	ACGIH	2
UNITED KINGDOM	HSE	10 (tot. inhal.) 1 (resp. dust)

MICA	CAS no 12001-26-2	mg/m³
BELGIE	ACGIH	3
DANMARK	Graensevaerdier for stoffer og materialer	
BRD	Senatskommission zur Prüfing der Arbeitsstoffe	6 (total fein ''dust quartz < 1%)
ELLAS		
ESPAÑA	ACGIH	3
FRANCE	Circulaire	
IRELAND	UK limits or ACGIH	3
ITALIA	ACGIH	3
LUXEMBOURG	BRD or FRANCE or ACGIH	3
NEDERLAND	MAC lijst '89	5 (2,5 voor respirable dust)
PORTUGAL	ACGIH	3
UNITED KINGDOM		10 (total inhal.) 1 (respir. dust)

OCCUPATIONAL EXPOSURE LIMITS FOR PHYLLOSILICATES IN THE UNITED KINGDOM

N G West
Health and Safety Executive
403 Edgware Road
LONDON NW2 6LN
United Kingdom

Introduction

Apart from the special case of the asbestiform mineral chrysotile, there are only three categories of phyllosilicate which have received much attention in the United Kingdom regarding their potential health effects. These are talc, mica and kaolin which are used widely, and sometimes interchangeably, in a range of industries notably paper, paint and cosmetics manufacture. Talc usuage is some 100,000 tonnes per annum, mainly imported, the sole United Kingdom source being in the Shetland Islands. Importation of talc is via a small number of primary suppliers and cosmetic grade talcs in particular are subject to strict quality checks. There is only one major processor of mica in the United Kingdom though many end users. The crude imported muscovite is prepared in various grades by wet and dry grinding. The production and processing of china clay or kaolin is a major industry in the south-west of England. Production is approximately 3 million tonnes per annum two thirds of which goes for export.

Attempts to establish the potential of the phyllosilicates to cause lung diseases have often been confounded by the presence of mineral impurities (Skulberg et al., 1985). The situation is particularly confusing as regards 'talc' since the term is commonly used to describe a range of products from cosmetic talc containing close to one hundred percent of the mineral talc to industrial talcs with major concentrations of other

NATO ASI Series, Vol. G 21
Health Related Effects of Phyllosilicates
Edited by J. Bignon

minerals such as magnesite and chlorite. (Hamer et al., 1976) The present epidemiological and toxicological evidence points to the fact that the cytotoxic and fibrogenic potential of mineralogically pure talc, mica and kaolin is low and that there is no evidence of carcinogenic effects. The development of pneumoconiosis is in many cases linked to the presence of quartz and fibrous amphiboles in talc products and the association of quartz with mica and kaolin (Parkes, 1982).

Occupational Exposure Limits

In the United Kingdom exposure limits are set following consideration by an advisory committee having representatives from the Confederation of British Industry, the Trades Union Congress and the Health and Safety Executive, together with independant experts. Limits are established primarily on consideration of potential health effects although the practicalities of industry controlling to the limits are also taken into account.

Ideally the limits should be set on the basis of well established exposure - response curves; once a decision has been taken on the risk to health 'acceptable' to society for a given substance then the appropriate limit can be calculated. Unfortunately very few such exposure - response curves are available and decisions usually have to be based on inadequate information inevitably involving some subjective judgement. For minerals, which are natural products and are rarely pure in a chemical sense, the situation is made more difficult by misuse of mineralogical terminology. This is particularly true of the medical literature where often there is inadequate characterisation of the minerals studied in epidemiological surveys.

Exposure limits are in place for only two phyllosilicates, talc and mica, as shown in the table.

Occupational Exposure Standards (HSE, 1989)

	Total inhalable dust (mgm^{-3})	Respirable dust (mgm^{-3})
Talc	10	1
Mica	10	1

Respirable dust measurements are the more important in terms of assessing the potential hazard to health from talc and mica. The china clay industry also operates to a voluntary limit for kaolin of 4.3 mgm^{-3} respirable dust and the advisory committee intends to review occupational exposure to kaolin when all the results of an on-going epidemiological survey become available. Measurement of respirable dust concentration is a relatively simple procedure and industry in the United Kingdom has generally found the limits to be workable. The results emerging from the kaolin survey (Ogle et al., 1989) suggest that long-term exposure to dust concentrations below ~ 2 mgm^{-3} does not produce significant respiratory effects. It is interesting to note the recent suggestion that there could be a common (generic) phyllosilicate health effect (Sebastien, 1989). If such a thesis was established in the case of kaolin, talc and mica this could indicate that the three minerals should have similar exposure limits with the limits for talc and mica coming more into line with the American Threshold Limit Values (ACGIH, 1987).

In dealing with the problem of mineral impurities associated with talc, mica and kaolin the United Kingdom approach is to make recourse to the appropriate exposure limits as and when necessary, rather than attempting to classify bulk mineral samples as containing more or less than a certain level of a contaminant. Thus if a consignment of mica was thought to contain quartz, the quartz exposure limit would be brought into play or if a sample of talc was suspected of containing asbestiform tremolite then the airborne asbestos fibre limit would be invoked. The alternative of assessing the potential health hazard on the basis of the bulk composition is flawed in that the airborne dust composition may differ considerably from the bulk composition. In addition the analytical procedure for determining traces of contaminants in bulk materials are often complex and expensive. However, there are circumstances in which the assessment of bulk composition may be necessary, for example, for the purposes of prohibition and labelling.

For the future, consideration is being given to a review of the health effects of other phyllosilicates particularly the clay minerals. With the recent introduction of a major new legislation for the Control of Substances Hazardous to Health (COSHH Regulations) attention is being focused as never before on the health risks associated with chemicals usuage. The Health and Safety Executive is commited to reducing the incidence of occupational disease in United Kingdom industry through a combination of information, advice and enforcement.

References

ACGIH (1987) Threshold Limit Values and Biological Exposure Indices for 1987-1988. American Conference of Governmental Industrial Hygienists, Cincinnati

Hamer D H, Rolle F R, Schelz J P (1976) Characterisation of talc and associated minerals. Am Ind Hyg Assoc J 37:296-304

HSE (1989) Occupational Exposure Limits 1989, Guidance Note EH40/89. Health and Safety Executive, London

Ogle C J, Rundle E M, Sugar E T (1989) China clay workers in the south west of England: analysis of chest radiograph readings, ventilatory capacity and respiratory symptoms in relation to type and duration of occupation. B J I M 46: 277-285

Parkes (1982) Occupational Lung disorders, 2nd edn. Butterworths, London

Sebastien P (1989) Proceedings of the first international conference on health related effects of phyllosilicates.

Skulberg K R, Gylseth B, Skaug V, Hanoa R (1985) Mica pneumoconiosis - A literature review. Scand J Work Environ Health 11: 65-74

OCCUPATIONAL EXPOSURE LIMITS IN FRANCE

B. HERVE - BAZIN
Institut National de Recherche et de sécurité
Centre de Recherche, BP n°27
54501 VANDOEUVRE CEDEX

1. ELABORATION AND DEFINITIONS.

A scientific committee gathers experts, personnally appointed by the French Ministry of Labour, and administration representatives. This group critically reviews published data (industrial experience, experimental toxicological data, physicochemical properties, sampling and analysis capabilities, exposure/response relationship), discusses possibilities (taking technical possibilities and socio-economical demands into account), and selects a value for proposition to the "Conseil Supérieur de la Prévention des Risques Professionnels" (High Committee for Occupational Health Hazards Prevention).
An OEL is the concentration level of a pollutant in the air that can be inhaled by a worker during a given period without incurring any health hazard, even if reversible physiological effects are sometimes accepted. No excessive annoyance or uncomfort is allowed for, however; no irreversible organ or function alteration is reasonably to be expected.
In practice, experience shows that new pathologies continue to be discovered; this is the reason why industrial practices should reduce exposure levels to lowest reasonably possible levels : **OELVs are to be considered as minimal objectives.**

Two kinds of values have been defined :
- a short-term OEL (VLE, Valeur Limite d'Exposition), whose respect protects from immediate or short-term effects. It is in fact a ceiling value (i.e. never to be exceeded), for which the control period is not to exceed 15 minutes, according to the nature of risk, working and exposure conditions, and technical possibilities.
- a time-weighted OEL (VME, Valeur limite de Moyenne d'Exposition), to protect workers from middle-term or long-term effects. It is measured or estimated on a 8-hour shift, and can be exceeded during short periods, provided the VLE is not exceeded.

In the general case, these values are indicative : only 8 are regulatory in France. There is at this time no biological index of occupational exposure, with exceptions for lead and its compounds

NATO ASI Series, Vol. G 21
Health Related Effects of Phyllosilicates
Edited by J. Bignon

(in application of EEC Directive 82/605) and for carboxyhemoglobine (exposure to carbon monoxide).

2. SAMPLING AND ANALYSIS

Assessing occupational individual exposure to air contaminants in a representative way is a challenging task. Rapid fluctuations in contaminant concentration or large variations over small distances are commonplace : site, moment and duration of sampling are decisive. Some measurements on a given day or period may give an insufficient view of the real variability of individual exposure conditions.
The sampling train often introduces its own limitations, sometimes critical as in aerosol fractions assessments, and the analytical steps add further difficulties or uncertainties, eg insufficient identification or separation of chemical species, or interferences. In this complex context, sampling strategy is responsible for representativeness at the lowest possible cost. This is the reason why the French Ministry of Labour is about to publish a document on atmosphere control at the workplace, to explicitate sampling strategies and encourage their adoption.

Some sampling and analysis methods have been published by the French standardization body (AFNOR). In the other cases, it is considered that published methods (e.g. HSE or NIOSH methods) actually allow a sufficient means to solve a majority of problems.

3. NON- COMPLIANCE

Official texts allow work inspectorate to consider that health and safety conditions are not acceptable when OELs are "obviously" exceeded. If no corrective measure is applied in order to obtain an acceptable situation, inspectors are entitled to propose the administration to impose a control of the quality of workplace atmospheres.

4. OTHER VALUES

regulatory values
These values have been established by the French Council of State, some in application of EEC Directives. The texts generally define places to be controlled, methods to be applied, and frequency of

sampling. Concerned substances are : asbestos, benzene, hydrogen phosphide, hydrogen cyanide, methyl bromide, dusts, vinyl chloride and lead.

values recommanded by the national health insurance body
These values are adopted by technical bodies at a national level.

5. SPECIAL CASES

1) Carcinogens.
Several lists of OELs from different countries include carcinogens, with or without limit values The French list of carcinogens include chemicals, agents or processes proven as carcinogenic to humans, as well as those for which animal data are considered demonstrative. Exposure should always be maintained at the lowest reasonably possible level. This principle is not always sufficient or operational, so limit values (presently in 15 cases) have been determined taking technical possibilities into account.

2) Mixtures.
If the effects of the constituants are known as independant, each one is considered separately. On the contrary, particularly in the case of exposure to solvent vapours, the "additivity rule" of exposure indices (i.e. summing the ratios concentration/OEL) may be applied (the sum should be <1). Interpretation of the results should be made with maximum cautiousness, preferably by a trained industrial hygienist.
In the important case of mixtures of various forms of crystalline silica, the VME will be calculated as $10/(2 + q + 2c + 2t)$ where q, c, and t are respectively the percentages of quartz, cristobalite, and trydimite in the respirable (alveolar) fraction of the aerosol.

3) Dusts.
Regulatory OELs have been defined for dusts not otherwise classified : 10 mg/m^3 for total dust, and 5 mg/m^3 for respirable (alveolar) dust.

4) Ill-defined mixtures or agents.
For specific mixtures of complex or variable composition such as thermal degradation fumes, oil mists, wood dusts, vapours of mixtures of hydrocarbons (e.g. white spirit, kerosene, gasoline, turpentine, etc...), a reflection has yet to be made.
The situation concerning items of the kind "metal and its compounds" has to be cleared.

CONCLUSIONS

Professor J.C. McDonald (UK)

We have come to the end of what I sense has been one of the most stimulating and enjoyable scientific meetings for some time. The reasons are clear enough: it has not been too large, it has been widely multidisciplinary, many of us are old friends and Paris is a wonderful place to meet. Few of us had known what phyllosilicates were and when we found out it seemed unlikely that they would prove very exciting - but how wrong we were! We know now that the geology of these minerals is incredibly complicated, that their production and use are infinitely varied, that the mechanisms of fibrogenesis and carcinogenesis are extremely complex and that the health effects have been poorly defined and inadequatly studied.

Overall, the evidence presented is reassuring; such adverse effects of exposure as have been described are more probably caused by contaminating chrystalline silica or mineral fibres than by the phyllosilicates perse. This implies that preventive measures should rationally depend on the application of appropriate control limits for the toxic contaminants. That is obviously more easily said than done a) because of the enormous range of natural variation in level of occurance and b) because few studies have so far made any attempt to quantify the intensity of exposure. Despite these uncertainties official bodies are forced to make decisions and take action. Until the scientific world can provide better data our only advice can be that mineral dusts are potentially dangerous and all possible measures should be taken to control them.

I speak for you all in thanking Professor Jean Bignon and members of the Local Organizing Committee for organizing a first class conference; we look forward to the next.

NATO ASI Series G

Vol. 1: **Numerical Taxonomy.** Edited by J. Felsenstein. 644 pages. 1983. (out of print)

Vol. 2: **Immunotoxicology.** Edited by P.W. Mullen. 161 pages. 1984.

Vol. 3: **In Vitro Effects of Mineral Dusts.** Edited by E.G. Beck and J. Bignon. 548 pages. 1985.

Vol. 4: **Environmental Impact Assessment, Technology Assessment, and Risk Analysis.** Edited by V.T. Covello, J.L. Mumpower, P.J.M. Stallen, and V.R.R. Uppuluri. 1068 pages. 1985.

Vol. 5: **Genetic Differentiation and Dispersal in Plants.** Edited by P. Jacquard, G. Heim, and J. Antonovics. 452 pages. 1985.

Vol. 6: **Chemistry of Multiphase Atmospheric Systems.** Edited by W. Jaeschke. 773 pages. 1986.

Vol. 7: **The Role of Freshwater Outflow in Coastal Marine Ecosystems.** Edited by S. Skreslet. 453 pages. 1986.

Vol. 8: **Stratospheric Ozone Reduction, Solar Ultraviolet Radiation and Plant Life.** Edited by R.C. Worrest and M.M. Caldwell. 374 pages. 1986.

Vol. 9: **Strategies and Advanced Techniques for Marine Pollution Studies: Mediterranean Sea.** Edited by C.S. Giam and H.J.-M. Dou. 475 pages. 1986.

Vol. 10: **Urban Runoff Pollution.** Edited by H.C. Torno, J. Marsalek, and M. Desbordes. 893 pages. 1986.

Vol. 11: **Pest Control: Operations and Systems Analysis in Fruit Fly Management.** Edited by M. Mangel, J.R. Carey, and R.E. Plant. 465 pages. 1986.

Vol. 12: **Mediterranean Marine Avifauna: Population Studies and Conservation.** Edited by MEDMARAVIS and X. Monbailliu. 535 pages. 1986.

Vol. 13: **Taxonomy of Porifera from the N.E. Atlantic and Mediterranean Sea.** Edited by J. Vacelet and N. Boury-Esnault. 332 pages. 1987.

Vol. 14: **Developments in Numerical Ecology.** Edited by P. Legendre and L. Legendre. 585 pages. 1987.

Vol. 15: **Plant Response to Stress. Functional Analysis in Mediterranean Ecosystems.** Edited by J.D. Tenhunen, F.M. Catarino, O.L. Lange, and W.C. Oechel. 668 pages. 1987.

Vol. 16: **Effects of Atmospheric Pollutants on Forests, Wetlands and Agricultural Ecosystems.** Edited by T.C. Hutchinson and K.M. Meema. 652 pages. 1987.

Vol. 17: **Intelligence and Evolutionary Biology.** Edited by H.J. Jerison and I. Jerison. 481 pages. 1988.

Vol. 18: **Safety Assurance for Environmental Introductions of Genetically-Engineered Organisms.** Edited by J. Fiksel and V.T. Covello. 282 pages. 1988.

Vol. 19: **Environmental Stress in Plants. Biochemical and Physiological Mechanisms.** Edited by J.H. Cherry. 369 pages. 1989.

Vol. 20: **Behavioural Mechanisms of Food Selection.** Edited by R.N. Hughes. 886 pages. 1990.

Vol. 21: **Health Related Effects of Phyllosilicates.** Edited by J. Bignon. 462 pages. 1990.

MIX
Papier aus verantwortungsvollen Quellen
Paper from responsible sources
FSC® C105338
FSC
www.fsc.org

If you have any concerns about our products,
you can contact us on
ProductSafety@springernature.com

In case Publisher is established outside the EU,
the EU authorized representative is:
Springer Nature Customer Service Center GmbH
Europaplatz 3, 69115 Heidelberg, Germany

Printed by Libri Plureos GmbH
in Hamburg, Germany